AF760610

Epicyclic Gearing

This book provides a broad introduction to the optimization techniques used in the design and manufacturing of epicyclic gearing.

A wide variety of optimization techniques are covered, with a strong focus on practical application. The formulation of the underlying mathematical models and the algorithms for solving them are explained, which are then applied to solve real-world problems in epicyclic gearing. Figures and charts are provided to convey the intuition behind the various approaches. Each chapter includes a detailed case study based on a real-world application of epicyclic gearing. The case studies highlight the realities and challenges in the design optimization of epicyclic gearing systems, illustrate the application of optimization techniques in a real-world context, compare different optimization techniques in terms of performance, ease of use, etc., and identify the areas of future work.

Suitable for gear professionals and researchers alike, the book will be of interest to those in the fields of mechanical engineering, statistics, computer science, aerospace, automotive engineering, and operations research.

Epicyclic Gearing

Optimization Techniques

Chandrasekaran G

CRC Press
Taylor & Francis Group
Boca Raton London New York

CRC Press is an imprint of the
Taylor & Francis Group, an **informa** business

Designed cover image: Shutterstock

First edition published 2025
by CRC Press
2385 NW Executive Center Drive, Suite 320, Boca Raton FL 33431

and by CRC Press
4 Park Square, Milton Park, Abingdon, Oxon, OX14 4RN

CRC Press is an imprint of Taylor & Francis Group, LLC

© 2025 Chandrasekaran G

Reasonable efforts have been made to publish reliable data and information, but the author and publisher cannot assume responsibility for the validity of all materials or the consequences of their use. The authors and publishers have attempted to trace the copyright holders of all material reproduced in this publication and apologize to copyright holders if permission to publish in this form has not been obtained. If any copyright material has not been acknowledged please write and let us know so we may rectify in any future reprint.

Except as permitted under U.S. Copyright Law, no part of this book may be reprinted, reproduced, transmitted, or utilized in any form by any electronic, mechanical, or other means, now known or hereafter invented, including photocopying, microfilming, and recording, or in any information storage or retrieval system, without written permission from the publishers.

For permission to photocopy or use material electronically from this work, access www.copyright.com or contact the Copyright Clearance Center, Inc. (CCC), 222 Rosewood Drive, Danvers, MA 01923, 978-750-8400. For works that are not available on CCC please contact mpkbookspermissions@tandf.co.uk

Trademark notice: Product or corporate names may be trademarks or registered trademarks and are used only for identification and explanation without intent to infringe.

ISBN: 978-0-367-70066-9 (hbk)
ISBN: 978-0-367-70070-6 (pbk)
ISBN: 978-1-003-14446-5 (ebk)

DOI: 10.1201/9781003144465

Typeset in Times
by SPi Technologies India Pvt Ltd (Straive)

Contents

Preface

Optimization techniques have found use in the design and manufacture of epicyclic gearing in recent years, supported by the rapid advancement in computing technology and the availability of more and more powerful computing devices. There is an ever-increasing demand for engineers to raise efficiency and lower production costs in view of global competition. This has prompted engineers to look for rigorous techniques for optimization and decision-making in order to design and produce gearing products economically and efficiently.

This book provides a broad introduction to the optimization techniques used in the design and manufacturing of epicyclic gearing. A wide variety of optimization techniques are covered, with a strong focus on practical application. The formulation of the underlying mathematical models and the algorithms for solving them are explained, which are then applied to solve real-world problems in epicyclic gearing. Figures and charts are provided to convey the intuition behind the various approaches.

Each chapter includes a detailed case study based on a real-world application of epicyclic gearing. The case studies highlight the realities and challenges in the design optimization of epicyclic gearing systems, illustrate the application of optimization techniques in a real-world context, compare different optimization techniques in terms of performance, ease of use, etc., and identify the areas of future work.

The problems discussed in the case studies involve the simultaneous optimization of up to 14 objective functions and eight design variables, with a multitude of constraints, using various multi-objective optimization techniques. A diverse range of optimization algorithms are covered, spanning popular stochastic, population-based, path-based, and evolutionary algorithms. For each algorithm, a step-by-step approach is followed in developing the software procedure, along with a complete pseudo-code. The algorithms are implemented in order to handle the constrained, multi-objective optimization of real-valued and multivariate functions.

Chapter 1 provides an introduction to epicyclic gearing, its various types, modern-day applications, and the need for design optimization.

Chapter 2 introduces the concept of computational optimization and the different types of optimization algorithms. It covers the procedure for mathematical modeling of an epicyclic gearbox, its formulation as a multi-objective optimization problem, the software used, etc.

Chapter 3 covers the design optimization of a power take-off gearbox using an evolutionary algorithm – non-dominated sorting genetic (NSGA-II) algorithm.

Chapter 4 covers the design optimization of a wind turbine generator gearbox using a multi-objective particle swarm optimization (MOPSO) algorithm.

Chapter 5 covers the design optimization of a helicopter gearbox using a multi-objective simulated annealing (MOSA) algorithm.

Chapter 6 covers the design optimization of a two-speed electric vehicle transmission gearbox using the multi-objective ant colony optimization (MOACO) algorithm.

Chapter 7 covers the design optimization of a Wolfrom gearbox for a robotic actuator involving two stages, using three different multi-objective optimization algorithms, namely, particle swarm optimization, simulated annealing, and ant colony optimization, and compares the relative efficacy of these algorithms.

Chapter 8 covers various heuristic optimization and decision-making techniques involved in the sustainable design and manufacture of epicyclic gearing, complete with real-world case studies.

This book utilizes the **OPTIMO** software package in order to run the case studies and optimization problems. The software, written in the Python programming language, provides efficient implementations of the algorithms covered in this book, along with complete problem sets for each case study. The software is open source, published under the MIT license, and can be accessed from https://github.com/epicyclic-book/optimo/. **Chapter 2** provides detailed instructions on how to use the software and the included problem sets. Gokul Krishnan – the author of the software – has also co-authored a few chapters, covering some of the computational aspects.

Although some books are available that treat the subject of epicyclic gearing, there appears to be a need for a book that provides a practical treatment of the subject of optimization techniques and their application in the design and manufacture of epicyclic gearing, aimed at a broader audience, ranging from industry professionals to scientists and students. This book has been written to address this need.

This book would be especially useful for different disciplines, which would include mechanical engineering, statistics, computer science, artificial intelligence, aerospace, automotive engineering, medical, and operations research. Stochastic and population-based methods like genetic algorithm, particle swarm optimization, ant colony optimization, and simulated annealing applied for the design and manufacturing optimization of epicyclic gearing are discussed.

Acknowledgments

I would like to extend my heartfelt gratitude to Dr. R. Saravanan and Dr. K. Sivakumar, whose input and support has given me the confidence required to tackle a project like this. In addition, I'd like to thank Dr. N.R. Srinivasan for his support during the preparation of this book. I also thank the team at Taylor & Francis for their patience and understanding. Finally, I'd like to thank my son, Gokul Krishnan, who, upon my request, has provided the software used in this book and also co-authored a few chapters.

Author

Dr. Chandrasekaran G obtained his UG degree in Mechanical Engineering from The Institution of Engineers (India) in 1986, a PG degree in Engineering Design from Anna University, Chennai, India, in 1989 and PhD in Mechanical Engineering from Anna University, Chennai, India, in 2020. He has vast academic experience, having worked in prominent engineering colleges in India and taught subjects related to mechanical engineering and automobile engineering to undergraduate and postgraduate students. His research interests include gear design, optimization techniques, sustainable design and manufacturing, renewable sources of energy, and automotive engineering. He is a co-author of four research papers and one book chapter published in specialized international journals.

Contributing Author Biography

Gokul Krishnan is a software consultant, open-source advocate, and the creator of RVGL (https://rvgl.org). He has a degree in computer science and engineering from The Institution of Engineers (India). His areas of interest include computer graphics, game design, and artificial intelligence.

List of Symbols

Normal module (mm)	m_n
Number of teeth of planet gear	z_1
Number of teeth of sun gear	z_2
Number of teeth of ring gear	z_3
Normal pressure angle (°)	α_n
Transverse pressure angle (°)	α_t
Working transverse pressure angle (°)	α_{wt}
Addendum circle pressure angle (°)	α_{at}
Helix angle (°)	β
Individual shift in the basic profile (mm)	x_n
Base diameter (mm)	d_b
Reference pitch diameter (mm)	d
Working pitch diameter (mm)	d_w
Root diameter (mm)	d_f
Addendum diameter (mm)	d_a
Addendum (mm)	h_a
Dedendum (mm)	h_f
Whole depth (mm)	h
Circular pitch (mm)	p_c
Base pitch (mm)	p_b
Radius of curvature (mm)	ρ
Reduced radius of curvature (mm)	ρ_{redc}
Face width (mm)	b
Center distance (mm)	a
Modified center distance (mm)	a'
Top land thickness (mm)	S_a
Number of planets	n_p
Efficiency	η
Coefficient of friction	μ
Mean coefficient of friction	μ_m
Gear ratio	i
Base contact ratio	ε_α
Addendum contact ratio	ε
Power transmitted (kW)	P
Torque transmitted (Nm)	T
Speed (rpm)	N
Tangential force (N)	F_t
Normal force (N)	F_n
Tangential velocity (m/s)	v_t
Angular velocity (rad/s)	ω

Sliding velocity (m/s)	v_s
Rolling velocity (m/s)	v_r
Entraining velocity (m/s)	v_e
Slide roll ratio	SRR
Specific sliding	γ
Sum velocity at operating pitch circles (m/s)	$\gamma_{\varepsilon c}$
Lubricant correction factor	X_L
Hertz contact stress (N/mm^2)	σ_H
Bending strength (N/mm^2)	σ_F
Semi-width of Hertzian contact band (mm)	b_H
Young's modulus (N/mm^2)	E
Poisson's ratio	ν
Material density (kg/m^3)	ρ / ρ_m
Flash temperature (°C)	θ_{fl}
Heat conductivity of the planet/sun gear material (W/m.K)	λ_m
Specific heat of planet/sun gear material per unit of mass (J/(kg.K))	c_m
Lubricant pressure viscosity coefficient (m^2/N)	α'
Dynamic oil viscosity at oil temperature (mPa.s)	η_{oil} / η_o
Thermal contact coefficient	B_m
Minimum oil film thickness (µm)	h_{min}
Relative lubricant film thickness (Lambda ratio)	λ
Arithmetic mean roughness (µm)	R_a
AGMA	American Gear Manufacturers Association
JGMA	Japan Gear Manufacturers Association

1 Introduction to Epicyclic Gearing

Chandrasekaran G

1.1 EPICYCLIC GEARING

Epicyclic gearing systems have a relative motion of axes, wherein the axis of at least one of the gears also moves relative to the fixed frame. This type of gearing is termed epicyclic gearing because of the fact that one gear wheel rolls outside another gear wheel and traces an epicyclic path. When the fixed wheel is annular and another gear wheel rolls inside it, then the path traced will be a hypocyloid. Still, it has become common to call all gears, in which one of the axes rotates about a fixed axis as epicyclic gear. Large speed reductions are possible with epicyclic gearing, and because of the fixed annular wheel, a more compact unit could be obtained. Major applications of epicyclic gears are in power transmission, robotic systems, etc.

Epicyclic gear trains are known to provide high power density over conventional gear trains through their innovative planet gear load-sharing configuration. In a typical epicyclic gear train, a carrier connects the centers of the two gears and rotates to carry one gear, called the planet gear, around the other, called the sun gear. As the planet and sun gears mesh, their pitch circles roll without slipping.

All the planets are mounted to a single rotating member, called a cage, spider arm, or carrier. As the planet carrier turns, typically, it delivers low-speed, high-torque output (Table 1.1).

In some systems, every member rotates, but in many systems, at least one component is fixed.

Any epicyclic gear set has three main components:

- The sun gear
- The planet gears and the planet gears' carrier
- The ring gear

Each of these three components can be the input, the output, or can be held stationary. Choosing which piece plays which role decides the gear ratio for the gear set. One or more of the planet gears of the gear train are now carried on an arm that itself rotates about the axis of the sun gear. Due to the analogy with our solar system, the epicyclical gear train's central gear is called the *sun* gear. The gears that rotate around it are called *planet* gears. A *ring* gear, or annulus, a gear of internal teeth, meshes with the planet gears. The element that supports the

DOI: 10.1201/9781003144465-1

TABLE 1.1
Epicyclic Gearing Properties

Type of Epicyclic Gear Train	Sun Gear	Ring Gear	Carrier	Speed	Torque	Output Direction	Range of Ratios Normally Used
Planetary	Input	Fixed member	Output	Maximum reduction	Multiplication	Same as input	3:1–12:1
	Output		Input	Maximum increase	Reduction	Same as input	
Star	Input	Output	Fixed member	Reduction	Multiplication	Reverse of input	2:1–11:1
	Output	Input		Increase	Reduction	Reverse of input	
Solar	Fixed member	Input	Output	Minimum reduction	Multiplication	Same as input	1.2:1–1.7:1
		Output	Input	Minimum increase	Reduction	Same as input	

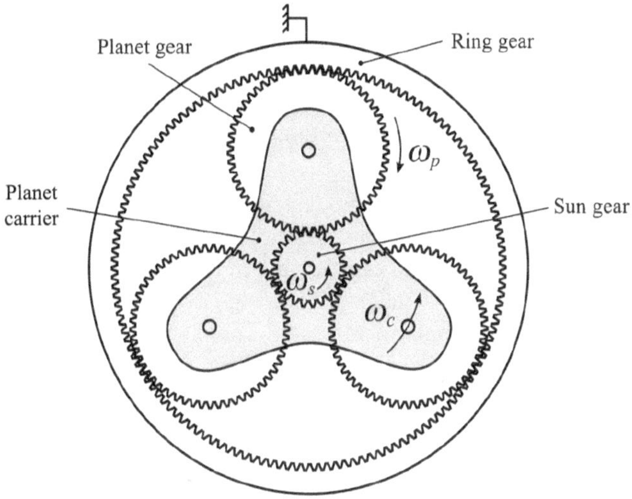

FIGURE 1.1 Elements of epicyclic gearing (Leaman et al., 2021).

movable axis of the planet gears and that rotates around the main axis of the epicyclical gear train is called a support, arm, or planet carrier. Figure 1.1 (Leaman et al., 2021) shows these elements.

The ring gear has a housing with integrated internal teeth, and in the majority of cases, the housing is fixed. It is a type of gear that looks like a ring and has annular cut teeth at its inner surface and is placed in the outermost position in an epicyclic gearbox. The inner teeth of the ring gear are in constant mesh at the outer point with the set of planetary gears.

The sun gear has annular cut teeth and is placed in the middle of the epicyclic gearbox. The sun gear is in constant mesh at the inner point with the planetary gears and is connected with the input shaft of the epicyclic gearbox. One or more sun gears can be used to achieve different output levels.

The planet gears are small gears used in between the ring and sun gear, and the teeth of the planet gears are in constant mesh with the sun and the ring gear at both the inner and outer engagement points, respectively. The shaft axes of the planet gears are attached to the planet carrier, which, normally, is carrying the output shaft of the epicyclic gearbox. The planet gears not only rotate about their axes but also can revolve between the ring and the sun gear. The number of planets can also vary. As the number of planetary gears increases, the distribution of the load increases, and therefore the torque that can be transmitted. Increasing the number of tooth engagements also reduces the rolling power loss. Since only part of the total output has to be transmitted as rolling power loss, an epicyclic gear is extremely efficient. The planet carrier is attached to the axis of the planet gears and is responsible for the final transmission of the output to the output shaft. The planet gears rotate on the planet shaft that is fixed to the carrier, and the revolution of the planetary gears in turn causes rotation of the carrier.

Epicyclic gearing finds wide application in areas where high output levels and fast speeds must be transmitted with favorable mass inertia ratio adaptation. High transmission ratios can also easily be achieved with epicyclic gearboxes. Epicyclic gearing provides a perfect balance of efficiency, torque, and flexibility. Because of their positive properties and compact design, the gearboxes find many potential uses in industrial applications.

Compared to conventional gear trains, epicyclic gear trains offer numerous benefits, and over the past several decades, their use has proliferated significantly in a variety of branches of mechanical engineering. In comparison to standard parallel axis gear trains, epicyclic gear trains provide high power density. Epicyclic gear trains are characterized by multiple kinematic combinations, purely torsion reactions, and coaxial shafting. Disadvantages of epicyclic gearing include constant lubrication requirements, high bearing loads, inaccessibility and design complexity.

The high efficiency of epicyclic gear trains ensures that a high proportion of the energy being input is transmitted through the gearbox, rather than being wasted on mechanical losses inside the gearbox. The load in an epicyclic gear train is shared among multiple planets; therefore, torque capability is greatly increased. The more the planets in the system, the greater the load ability and higher the torque density. The epicyclic gear train also provides stability due to an even distribution of mass and increased rotational stiffness. Torque applied radially onto the gears of an epicyclic gear train is transferred radially by the gear, without lateral pressure on the gear teeth.

The method of motion of an epicyclic gearing is different from the conventional parallel gears. In conventional parallel axis, the driving force is transferred through a small number of contact points between two mating gears. In this case, since all the loading is concentrated on a few contacting surfaces, the gears wear quickly and sometimes may even crack. But the epicyclic speed reducer has multiple gear contacting surfaces with a larger area that can distribute the loading evenly around the central axis. Since multiple gear surfaces share the load during normal running, and also any sudden instantaneous impact load evenly, epicyclic gearing are more resistant to damage from higher torque. Because of the above advantages, the housing and bearing parts are also less likely to be damaged and crack from high loading. Because of the inherent advantages of lightweight, compactness, differential drive and above all high-speed ratios possible with limited number of elements, epicyclic gear trains are often used in transmission systems. In an epicyclic gearing system, the input and output axes coincide. Usually two or more planet gears are used in order to balance the load evenly in epicyclic gearing systems. Epicyclic gear system is compact in size, but complex in design and need a high-quality manufacturing process. The interference of the internal gear, the load division between planet gears, the balance and vibration of the rotating arm, and the hazard of jamming, etc. are inherent problems to be solved.

1.2 TYPES OF EPICYCLIC GEARING SYSTEMS

The three main types of simple epicyclic gearing are the *planetary*, the *star*, and the *solar*. The diagram in Figure 1.2 shows the three arrangements.

These three types differ only in arrangement. In the epicyclic gearing system, the transmission ratio and the direction of rotation would be changed according to which member is fixed. The properties of the three types of epicyclic gearing are given in Table 1.1.

1.2.1 Planetary Type of Epicyclic Gearing

In this type, if the internal gear (ring gear or annulus) is fixed, then the input is the sun gear and the output is spider arm or carrier D. The transmission ratio is calculated as in Table 1.2.

In a typical application, for example, an aircraft turbine engine planetary gearing, the drive power connects to the sun gear. The sun gear then drives the planetary gears assembled with the external gear ring to operate. The whole set of planetary gear systems revolves on its own axis, and along the external gear ring, where the output shaft connected to the planetary carrier achieves the goal of speed reduction. A higher reduction ratio can be achieved by doubling the multiple-staged gears and planetary gears.

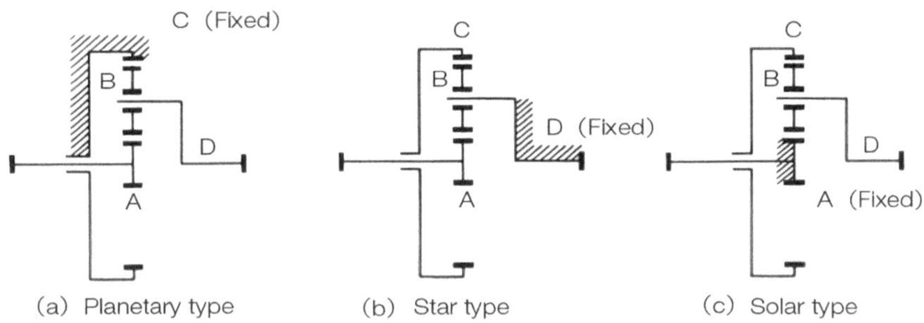

FIGURE 1.2 Types of epicyclic gearing.

TABLE 1.2
Calculation of Transmission Ratio

Operation	Spider Arm, D	Sun Gear, Z_A	Planet Gear, Z_B	Ring Gear, Z_C
1. Spider arm D is locked and sun wheel S is given +1 revolution.	0	+1	$-\frac{Z_A}{Z_B}$	$-\frac{Z_A}{Z_B}$
2. Multiply by x (System is fixed as a whole while rotating)	0	x	$-\frac{Z_A}{Z_B}x$	$-\frac{Z_A}{Z_B}x$
3. Add y to all columns	y	$y + x$	$y-\frac{Z_A}{Z_B}x$	$y-\frac{Z_A}{Z_B}x$

In the case of a wind turbine planetary gearbox, the drive power connects to the planetary carrier. The planetary carrier and planet gears then drive the sun gear to operate. The output shaft connected to the sun gear achieves the goal of speed increase.

1.2.2 Star Type of Epicyclic Gearing

In the case of the star type of epicyclic gearing, the carrier *D* is fixed. The planet's gears rotate only on fixed axes. The sun gear is on the input axis, and the internal gear is the output. Since the carrier is fixed, effectively there is no planetary action, namely, planets revolving around the sun gear, and consequently, the gear train action is similar to an ordinary gear train. In a typical example, in the case of a water power station, epicyclic gearing with a star type of arrangement (Immonen, Lahdelma, and Juuso, 2012), the input is received by the ring gear and the output is by the sun gear, achieving a speed increase.

Similarly, an epicyclic gearing system in a star arrangement is used for the speed reduction from the low pressure (LP) turbine blade to the fan blade in Pratt and Whitney Geared Turbofan™ PurePower® engine arrangement (Mohd Tobi and Ismail, 2016). The LP turbine shaft attached to the sun gear acts as a driver, and the ring gear is attached to the fan blade. The planet carrier holds five planet gears and does not rotate relative to the engine nacelle. Speed reduction of about 3:1 (9000 rpm from the turbine to 3200 rpm on the fan) is achieved and results in optimum working conditions on the LP turbine blade and the fan blade (J. Shapiro, 2008).

1.2.3 Solar Type of Epicyclic Gearing

In the case of a solar-type epicyclic gearing, the sun gear is fixed. The planet's gears rotate around the sun's gear. The ring gear is the input, and the carrier D is the output. For speed increase, carrier D is the input, and the ring gear is the output. The transmission ratio is calculated in Table 1.2. This arrangement finds applications where a low reduction ratio is required, as in the case of an electric vehicle gearbox.[1]

1.3 A Brief History of Epicyclic Gearing

Epicyclic gearing was used in the Antikythera Mechanism, used by ancient Greeks, which displayed celestial time for the sun, the moon, and five planets visible to the naked eye – Mercury, Venus, Mars, Jupiter, and Saturn. The mechanism, dating back to 200 BC, had several dials and clock faces, each of which served a different function for measuring the movements of the sun, moon, stars, and planets.

In the second-century AD treatise, Almagest, Ptolemy used epicyclic gear trains to predict the motions of the planets. Richard of Wallingford, an English abbot of St Albans monastery, a skilled mathematician and astronomer, is credited for reinventing epicyclic gearing for an astronomical clock in the fourteenth century.

In 1588, Italian military engineer Agostino Ramelli invented the book wheel, a vertically revolving bookstand containing epicyclic gearing with two levels of planetary gears to maintain the proper orientation of the books and to hold a number of large books. With this machine, a reader can see and turn through a large number of books without moving from one spot and could bring each into his view in turn.

James White, between 1786 and 1801, invented several epicyclic mechanisms like perpetual wedges, pulley systems, windmill regulators, differential combinations of wheels, hypocycloidal gear trains for linear to rotary conversion, bevel epicyclic torque meters, and rotary churns (G. White, 1989).

1.4 MODERN APPLICATIONS OF EPICYCLIC GEARING

Epicyclic gearing has gained much attention recently and has led mechanical engineers to feel confident that the epicyclic gear units will continue to be applied in a wide range of applications to meet numerous challenges that would not be solved using conventional gearboxes. Modern applications of epicyclic gearing span industries such as robotics, aerospace, defense, transportation, semiconductors, medical, food and beverages, agricultural machinery, energy, packaging, construction, manufacturing, machine tools, propulsion, automotive, and automation. Some applications of epicyclic gearing include wind turbines, automobiles running under automatic transmission, hybrid cars, car mirrors, tractors, helicopter transmission, space shuttle rudder/speed brake (RSB) actuators, bicycles, conveyor belts, toys, clocks, gas turbine engine transmission, hydroelectric power stations, lifting equipment, cranes, winches, hoists, 3D printing machines, press brakes, robots, laser cutting machines, automated guided vehicles, rotary tables, height adjustment devices, mobile satellite receivers, gear head motors, and so on. The number of devices/equipment/implements that can be fitted with epicyclic gearbox systems is inexhaustible, and many engineers and scientists have been undertaking numerous research projects in an attempt to come up with more complex systems that can make work even easier.

1.5 NEED FOR OPTIMIZATION

In the design of epicyclic gearing, both analysis and optimization are employed before any prototype work is started. The design of epicyclic gearboxes involves multiple planet branches, which also reduces efficiency. The main aim of optimization is to reduce the weight and power loss of the system and produce gearbox with enhanced service life of the individual components.

In the design of an epicyclic gearing for any application, because of strong inter-dependencies among all components, the variation of one parameter can result in different system performances. Optimization methods are aimed at improvements in load-carrying capacity, efficiency, noise, and weight for the given transmission ratios, component lifetimes and safeties, or space restrictions.

Because of the complex shape and geometry of gears, the design of epicyclic gear trains is a highly complicated task. The design of the epicyclic gear train involves consideration of many design variables, mathematical formulations, constraints, and many other influencing factors. Optimization techniques that use intelligent optimization algorithms are used to solve the design problem to provide an optimum solution. Optimization techniques are rigorous methods of decision-making that help the ever-increasing demand on engineers to lower production costs to withstand global competition and to design and produce products and systems economically and efficiently. Optimization techniques are being used in a wide spectrum of industries, including aerospace, automotive, chemical, electrical, construction, and manufacturing industries.

A multitude of factors have to be considered during epicyclic gearbox design and manufacturing, and optimization techniques are used to address the core concerns such as product performances, product qualities, manufacturing costs, safety aspects, environmental impact, recycling issues, and the satisfaction of product consumers for a successful gearbox design and manufacturing. Optimization techniques can concurrently consider a broad range of factors pertaining to gearbox manufacturing and streamline decision-making for gearbox design and manufacturing.

1.6 OPTIMIZATION TECHNIQUES FOR SUSTAINABLE DESIGN AND MANUFACTURE

Rapid industrialization has led to the degradation of the earth's natural environment, and nearly half of earth's natural resources have been destroyed in just thirty years. The manufacturing sector is a major contributor to global environmental pollution, as it is responsible for over 20% of accumulated greenhouse gas emissions. Gear manufacturers must innovate gear design, production, and distribution to minimize waste and conserve resources because of ever-escalating environmental concerns and resource depletion by adopting transformative approaches like sustainable manufacturing and the circular economy. They are expected to apply regenerative and sustainable practices to foster a restorative relationship between industries and the environment. There must be a shift from the traditional manufacturing model to product-as-a-service (PaaS) and integrated product-service (PS) models, promoting the circular economy paradigm that revives used materials, shunning disposal, reducing waste, and extending product lifecycle. The application of optimization techniques for the design and manufacture of gears will help in mitigating the problem of global environmental damage. Optimization techniques can be used for sustainable product design by redesigning products to enhance their environmental performance, reduce disposal costs, and promote reusability and recyclability.

NOTE

1 Refer to Chapter 6.

REFERENCES

Immonen, J., Lahdelma, S., Juuso, E., "Condition monitoring of an epicyclic gearbox at a water power station," *Conference: 53rd International Conference of the Scandinavian Simulation Society, SIMS 2012,* October 2012, Reykjavik, Iceland.

Leaman, F., Vicuña, C.M. & Clausen, E. "A Review of Gear Fault Diagnosis of Planetary Gearboxes Using Acoustic Emissions." *Acoust Aust.* 49, 265–272 (2021). https://doi.org/10.1007/s40857-021-00217-6. Used under the terms and conditions of the Creative Commons Attribution (CC BY) license (https://creativecommons.org/licenses/by/4.0/)

Mohd Tobi, A. L., Ismail, A. E., "Development in Geared Turbofan Aeroengine," *IOP Conference Series: Materials Science and Engineering*, 131 (2016) 012019. https://doi.org/10.1088/1757-899X/131/1/012019. Used under the terms and conditions of the Creative Commons Attribution 3.0 license (CC BY) (https://creativecommons.org/licenses/by/3.0/)

Shapiro, J. "Green Technology: Jets Gear up to fly Greener." (2008). Available: https://machinedesign.com/news/green-technology-jets-gear-fly-greener

White, G., "Early Epicyclic Reduction Gears", *Mech. Mach. Theory*. 24, 2, 127–142, 1989.

2 Introduction to Design Optimization of Epicyclic Gearing

Chandrasekaran G and Gokul Krishnan

2.1 OPTIMIZATION

The optimization of epicyclic gear systems involves a multi-disciplinary approach that integrates engineering design, materials science, advanced manufacturing techniques, and computational optimization. The goal is to strike a balance between performance, cost, durability, and efficiency to meet specific application requirements.

Engineering optimization is a decision-making process to identify a few appropriate solutions and then decide which one best meets the need of the client. The appropriate solutions are feasible solutions that satisfy all the constraints in the optimization problem. In an optimization problem, a function, known as the objective function, is to be maximized or minimized, for example, to minimize the cost of a process or maximize the efficiency of a system. The objective function acts as a performance index to obtain the best result under given circumstances.

2.2 DESIGN OPTIMIZATION

Design is the process of determining design parameters that lead to the *best performance* of a mechanical structure, device, or system. As a design tool, optimization assists engineers to automatically identify the optimal design from an infinite set of design options. Design optimization is increasingly applied in various fields since it provides engineers/scientist/researchers an affordable and flexible means to identify optimal designs before physical deployment. An optimal design is one that is most cost-effective, reliable, robust, and safe one achievable with available resources under prevailing operating conditions. A design optimization problem is a problem in which certain parameters (design variables) need to be determined to achieve the best measurable performance (objective function) under given constraints. The design optimization process provides a systematic, logical design procedure based on scientific data. In the design optimization process, we seek to refine the design to improve the performance and to automate the optimization procedure. Design optimization helps in speeding up the process of design, and a solution procedure written once can then be reapplied to other problems.

DOI: 10.1201/9781003144465-2

2.2.1 Procedure for Solving a Gear Design Optimization Problem

The design of epicyclic gearing systems can be a complex process, and it is a challenge for gear design engineers to design efficient and cost-effective gearing systems without compromising their integrity. During the problem formulation, many possibilities and factors must be considered and several assumptions must be made to develop realistic models. The developed models must be subjected to mathematical analysis by the available methods and also must be verified by experiments. Many disciplines form the core of the optimization process, and the essential metrics might include cost-effectiveness, efficiency, safety, reliability, and accuracy, keeping in mind the environmental concerns also. The gear designer has to formulate the problem, quantifying the merits of potential design in the form of a specification that details the parameters, objectives, and constraints that are to be achieved.

The following procedure needs to be followed to solve the gear engineering optimization problem:

- *Formulation of the optimization problem*: this involves simplifying the gear train problem and identifying the major factors that determine the performance or outcome of the gear train, such as costs, weight, efficiency, and reliable service life, known as objectives. Next is to find the primary parameters that determine the above major factors, called the design variables. Then the relations between design variables and the identified major factors (objective functions) have to be modeled. Finally, the constraints imposed on the design variables are identified, and their relationship is modeled in the form of constraint functions.
- *Selecting the most suitable optimization technique or algorithm to solve the formulated optimization problem*: this requires an in-depth know-how of various optimization techniques.
- Determining search control parameters like the initial points, step size, and stopping criteria of the numerical optimization.
- Analyzing, interpreting, and validating the calculated results from the optimization program in order to prove the result mathematically and to verify the result using checkpoints.
- Integration with gear design and analysis.

The common aspects in gear design optimization problems are:

- A problem has multiple solutions and the optimal solution is to be identified.
- There exists more than one objective to accomplish and a measure of how well these objectives are accomplished (measurable performance).
- Constraints of different forms are imposed.
- Several key influencing variables are present and the change of their values will influence (either improve or worsen) the *measurable performance (objectives)* and the degree of violation of the constraints.

2.3 MULTI-OBJECTIVE OPTIMIZATION PROBLEMS (MOPS)

In many real-world applications, MOPs exist where a decision maker (DM) has several conflicting objectives to be solved, preventing simultaneous optimization of each objective, and wants to determine an optimal trade-off among them. Decision-making is the process of selecting a possible course of action from all the available alternatives, and often there are multiplicities of criteria for judging the alternatives. The main difference between the single-objective optimization problem (SOP) and MOP is that the SOP usually provides a single optimum solution corresponding to the objective function, whereas the MOP requires the simultaneous optimization of more than one objective function, thus yielding a set of optimal solutions, rather than a single global optimum, each of which are *non-dominated* with respect to one another, based on their corresponding objective values. The set of solutions together form the *Pareto-optimal* solution set.[1]

2.4 MATHEMATICAL FORMULATION OF A MULTI-OBJECTIVE PROBLEM

To apply the optimization techniques for gear engineering problems, it is necessary to clearly identify the objectives of the system to be optimized. The next step is to select the system variables that will be used to characterize or identify candidates. The last step is to delineate the boundaries and constraints within which the system operates. This set of composite activities constitutes the process of formulating the gear engineering optimization problem. Good problem formulation is essential to the success of an optimization study and is learned through practice. Finally, a suitable optimization technique has to be chosen that is based on the knowledge of the strengths, weaknesses, and peculiarities of the particular optimization technique.

For computational design optimization, the objective function and constraints must be expressed as a function of design variables (or design vector x):

Objective function: $f(x)$
Constraints: $g(x)$
Optimization statement: minimize $f(x)$

Subject to,

$$g(x) \geq 0$$
$$h(x) = 0$$

where $f(x)$ is the objective function to be minimized; $g(x)$ is the inequality constraints; $h(x)$ is the equality constraints; x is the design variables.

2.4.1 Objective Functions, Constraints, and Design Variables

The objective function is a measure of a quantity that is to be optimized, and it is in the form of a formula expressed in terms of design variables, whose values can be chosen to attain the desired maximum or minimum, subject to a set of certain constraints.

2.4.1.1 Objective Functions

An objective function represents the desires of the DMs (designers) to minimize or maximize a quantity by exploring a search space under the imposed constraints. The design objective acts as a criterion for the best design (or goodness of a design) to compare whether or not a given design is better than others. The objectives are the quantitative criterion on the basis of which candidate solutions will be ranked to determine the best solution. In design optimization, an objective function for a design objective is represented by a mathematical function in terms of design variables. There are two types of objective functions: (i) maximization objective functions and (ii) minimization objective functions.

2.4.1.2 Constraints

Since there is a limitation in the resources that can be used in the product development, the design engineer's desires (for example, improving the efficiency) cannot be optimized infinitely. The constraints (design requirements) that must be satisfied are expressed in a mathematical form in terms of design variables. Constraints may be inequality constraints or equality constraints.

2.4.1.3 Design Variables

Since each objective function is a function of several parameters, it is necessary to determine the variables for the definition of the mathematical model. A design variable is under the control of a DM (designer), and hence it is also called a decision variable or control variable. The choice of design variables could have an impact on the solution of the optimization problem. Essentially, a design is determined by a set of design variables, *controlled* by the designers, which are parameters that are chosen to describe the design of a system, and different combinations of design variables represent different designs. There are many design variables to be determined as part of the solution for the problem; for example, the gear geometrical parameters, like module, pressure angle, addendum modifications, face width, and outside diameter.

2.5 MULTI-OBJECTIVE OPTIMIZATION OF GEARBOX

Gear design optimization involves multiple and conflicting objectives, having design variables in the forms of continuous, real-valued, and discrete (integer) variables, making it a very complex design problem. Epicyclic gear drives are used in various industries, such as, for example, power-transmitting machine elements. A compound group of gears is employed to achieve a specific gear ratio between the driver and driven shafts. By using conventional methods, designing an epicyclic gear drive as a reduction unit requires a very large number of calculations based on recommended gear standards, trial and error methods, etc. This time-consuming process very often ends up with inadequate design outcomes.

There is a need for light weight construction of gearbox designs with high load capacity and power density with high reliability of the gears, taking resources into consideration. All these conditions result in special design criteria with consideration of maximum load, in addition to dynamic loads under different load situations. Hence, a procedure to automate the optimization of the design of individual pairs, considering several objective functions and constraints, is required. Mathematical optimization models are difficult to implement for gear design due to the discrete nature of the variables involved and the complexity of the objective function. To achieve better solutions, more robust and reliable optimization techniques must be used. To optimize the epicyclic gear design, various optimization methods have been employed in the past, in order to achieve the best possible set of solutions.

Optimization of epicyclic gearing design is a complicated task, and conventional optimization techniques would have real-life difficulties in achieving a global optimum. Since every design challenge and quality factor is different, there is a need for diversified techniques to effectively optimize epicyclic gear design.

2.6 OPTIMIZATION TECHNIQUES

The optimization techniques, which are used for solving optimization problems, involve algorithms that tend to be iterative, and thus thousands or even millions of evaluations of objectives are needed. Traditional optimization methods are usually local-search algorithms, and they are not suitable for highly constrained problems as they cannot prove optimality.

For the optimization of epicyclic gearing design and manufacture, stochastic optimization techniques like heuristic and meta-heuristic algorithms are useful since randomization provides a good way to move away from local search to global search. Nature-inspired heuristic and meta-heuristic algorithms have two major properties: (i) selection of the best solutions by convergence to the true optimum and (ii) randomization or homogeneous diversity, so that very good exploration of search space is carried out. The selection of the best ensures convergence of solutions to the optimality, while the randomization avoids the solutions being trapped at local optima and, at the same time, increases the diversity of the solutions.

Modern optimization techniques can be applied to gear design problems with several variables and constraints. But there are also challenges associated with using optimization for epicyclic gearing design since we are generally limited in our computational resources and time, and the required algorithms have to be selected based on how they explore the design space. Also, the optimization algorithms are limited by the designer's ability to properly specify the problem. Epicyclic gear design optimization problems often involve combinatorial[2] complexity, whereby finding the optimal solution within a reasonable time frame may be impractical for large instances, and hence precise mathematical modeling skills are required for formulating real-world problems accurately within the optimization framework.

For epicyclic gearing design and manufacture, meta-heuristic optimization techniques like GA, PSO, ACO, and SA prove to be very effective. GAs are population-based algorithms, so also are PSO and ACO, which are also referred to as agent-based algorithms as they use a set of strings. SA is a trajectory-based optimization technique, as the solutions trace a trajectory in the design space or search space to reach the global optimum.

2.7 OPTIMIZATION ALGORITHMS

Optimization algorithms are powerful tools for solving complex problems by incrementally improving the design until it can no longer be improved or until the budgeted time/cost has been reached. Optimization algorithms have become indispensable tools in modern gear design and manufacturing systems. The three broad classifications of optimization algorithms are exact algorithms, approximate algorithms, and heuristic/meta-heuristic algorithms.

2.7.1 Exact Algorithms

Exact algorithms are computational methods that guarantee a mathematically precise solution, an optimal solution, or the best possible solution to a given problem. These algorithms can be time-consuming and computationally expensive for complex problems because they are designed to find the optimal solution by evaluating all possible solutions. Examples of exact algorithms include Branch and Bound and dynamic programming, which divide a problem into subproblems and solve it by combining the solutions to its subproblems.

2.7.2 Approximate Algorithms

Whenever finding the exact solution is too computationally intensive or infeasible within a reasonable amount of time, approximate algorithms are used, which provide a solution that is close to the optimal solution, but not necessarily exact, by trading off some level of precision for efficiency. Approximation methods find a suboptimal solution, providing an approximation guarantee on the quality of the solution found. Examples of approximate algorithms include greedy algorithms, sequential algorithms, local-search algorithms, linear programming, and relaxation-based algorithms.

2.7.3 Heuristic/Meta-Heuristic Algorithms

A heuristic is a technique to find promising solutions by using information about the problem being solved. Heuristic algorithms are used to find good enough solutions to complex problems when the objective is not necessarily finding the optimal solution. Meta-heuristic algorithms use some general strategies to explore the search space and escape from local optima, but unlike heuristic algorithms, which are specific to a problem, meta-heuristic algorithms are generic and can deal with different problems and can be adapted to different situations.

2.7.3.1 Heuristic Algorithms

Heuristic algorithms can solve a problem faster compared to traditional algorithms (or classical algorithms), which are too slow. Heuristic algorithms determine near-optimal solutions to an optimization problem at a faster rate, but this is achieved by trading optimality, completeness, accuracy, or precision.

The word "heuristic" originates from the Greek "heuriskein" meaning "to discover". Heuristic is the process of gaining knowledge or obtaining some desired result by intelligent guesswork by the "trial-by-error" method, or by using a "rule-of-thumb" rather than by following some pre-established formula. Heuristic methods use shortcuts for solving a problem in a given limited time frame to produce almost perfect solutions. It is a versatile technique for making rapid decisions to solve complex real-world problems. Heuristics are used to get workable, practical, and quick solutions, which are sufficient in a given limited timeframe, with an acceptable accuracy range.

Heuristics are flexible techniques that involve intelligent or educated guesswork for making informed decisions. Heuristic algorithms are often used when there is no known algorithm that can provide an exact solution, or when it is too computationally expensive to use an exact algorithm. The capabilities of existing heuristics are continually being expanded with the development of hybrid algorithms that combine selected features from various types of heuristic algorithms to solve complex problems.

2.7.3.2 Meta-Heuristic Algorithms

A meta-heuristic is a higher-level technique or heuristic that seeks, generates, or selects a heuristic that may provide a sufficiently good solution to an optimization problem. When a heuristic algorithm can be thought of as a problem-based algorithm, the meta-heuristic algorithm can be thought of as a problem-independent algorithm. Meta-heuristics, originating from "meta" (meaning beyond) and "heuristics" (meaning to find or to discover), are high-level strategies or guiding principles used to develop and apply heuristic methods. Meta-heuristics are instrumental in effectively navigating large search spaces to tackle dynamic environments with variable constraints and objectives and in facilitating intelligent decision-making processes. Meta-heuristics are flexible, heuristic-driven methodologies that harness the power of collective intelligence, evolutionary principles, and stochastic search mechanisms to tackle increasingly intricate optimization challenges.

2.8 META-HEURISTIC OPTIMIZATION ALGORITHMS FOR GEAR DESIGN

Since every design challenge and quality factors are different, there is a need for diversified techniques to effectively optimize gear design. The selection of the right optimization algorithm for epicyclic gear design and manufacture depends heavily on what type of problem needs solving, as well as the type of constraints or limitations present. To ensure better decision-making and improved results, a thorough understanding of all available options is important. Whichever algorithm one may choose, careful examination of both theoretical foundations and practical implementation details will help determine if it's suitable for attaining the particular objectives.

In recent decades, various types of meta-heuristic combinatorial optimization algorithms have emerged as powerful and popular methods that have been developed by researchers. These methods have been expanded by mimicking some well-known processes, primarily in biology, physics, chemistry, math, society, and nature. These algorithms are based on what we learn from the world around us and from nature and enable us to build new technology and new artificial intelligence techniques.

The meta-heuristic optimization algorithms inspired by nature can be divided into four main categories, including evolutionary algorithms (EAs), swarm intelligence (SI) algorithms, physics-chemistry-based (PCB) algorithms, and human-based (HB) algorithms. These search algorithms attempt to find the global minimum/maximum of a function in a large search space. These techniques have been shown to work well with complex gear engineering optimization problems. There are plenty of opportunities and challenges for researchers to improve and advance in the design and optimization of gearing products and systems by the use of EAs and other nature-inspired meta-heuristic algorithms.

2.8.1 Evolutionary Algorithms

EAs are the most commonly used population-based meta-heuristic algorithms. They are flexible to solve global optimization problems because they are able to perform a global search and also local exploitation of the population of solutions. There are many benefits associated with EAs, and one such advantage is its flexibility because most EA concepts are adaptable to even complex problems. With EAs, better optimization is possible, as the population of solutions prevents the algorithm from getting locked in a particular solution. Other techniques would take far too long to exhaustively process.

EAs are gaining increasing importance as computational intelligence methods, which are very useful for global optimization problems. EAs that come under the stochastic group have been developed by a combination of rules and randomness that mimics several natural phenomena. Population-based optimization algorithms can deal easily with multi-objective problems because of their ability to effectively handle this class of problems. These advantages, together with the steady improvement of computer performance, enable their increased use in research and industry for solving a wide variety of engineering problems.

2.8.2 Evolutionary Algorithms for Multi-Objective Optimization

Multi-objective evolutionary algorithms (MOEAs) are well-suited for solving MOPs. They can process a set of solutions in parallel, because of their population-based framework, thereby obtaining an approximation of the Pareto front in a single run. MOEAs simulate the evolution of individual structures via the processes of selection and reproduction to form computational models of evolutionary processes.

Gradient-based and simplex-based methods, which arc conventional optimization techniques, were not designed with multiple solutions in mind and are difficult to extend to the true multi-objective problems. On the other hand, EAs have been recognized to be well suited for solving MOPs since early in their development.

EAs mimic the optimization process found in nature as they optimize biological species in order to maximize the survival of the fittest. There are plenty of opportunities and challenges for researchers to improve and advance in the design and manufacturing optimization of epicyclic gearing by the use of EAs and nature-inspired meta-heuristics. Recently, EAs, such as GAs and PSO, have been applied successfully to gear design (Deb and Jain, 2003; Chandrasekaran et al., 2019; Younes et al., 2022; Sedak and Rosić, 2021), and the use of EAs for optimization of gear design appears to be a promising approach.

2.8.2.1 Genetic Algorithms

GAs are used to find minima of multivariable functions through a discretization of the domain, using bit strings. GAs define a set of methods and starts from an initial population of points and improves it through the iterative application of three transformations: selection, crossover, and mutation, which resemble nature's evolution of biological systems. A genetic representation of the solution domain and also a fitness function are defined, and by iterating the fitness function for each generation, the algorithm produces a population of candidate solutions and then selects the optimal solution.

2.8.2.2 Swarm Intelligence Algorithms

SI-based algorithms involve several characteristics that make them particularly suitable for solving multi-objective optimization of epicyclic gearing design and manufacture. This type of algorithm is a population-based stochastic algorithm that is inherently decentralized; that is, the members of the swarm can be in charge of different objectives, and also different levels and types of interactions can be defined in order to share individual search experiences with the rest of the swarm. SI refers to the collective actions of autonomous systems, and both natural and artificial systems can be represented by using it. PSO is one such powerful meta-heuristic optimization algorithm which can be used for the design and manufacturing optimization of epicyclic gearing. PSO is inspired by swarm behavior observed in nature, such as fish and bird schooling, and the PSO algorithm simulates the social system of unpredictable choreography of a school of fish or a bird flock. Another promising and efficient SI-based algorithm for the design and manufacturing optimization of epicyclic gearing is the ACO algorithm, which has unique advantages compared to the other related algorithms.

2.8.2.2.1 *Particle Swarm Optimization*

The PSO algorithm uses inputs inspired by the behavior of social organisms, or systems, such as birds, ants, lions, whales, grey wolves, and bees. It is based on the collective behavior of decentralized and self-organized systems, which typically comprise a population of individuals (agents) interacting locally and with their environments. The cooperation found in these agents, between or among themselves, creates a collective intelligence that delivers great results when applied to solving optimization problems.

PSO is a population-based technique in which each candidate solution is referred to as a particle, and multiple particles form the swarm. Each particle in the swarm looks for the best solution in the search area, which is the set of possible solutions. By changing the position of each particle using the information of its current position, velocity, the distance between the current position and personal optimum, and the current position and swarm optimum, the swarm of particles moves in the search space until getting the global optimum, which will solve the optimization problem.

2.8.2.2.2 *Ant Colony Optimization*

The ACO algorithm is a kind of discrete optimization algorithm which is modeled on the collective behavior of ant colonies which are capable of solving complex problems and tasks in order to survive in an ever-changing hostile environment. Ant colonies are able to find the shortest path from their nest to a food source, and if the food source is depleted, the ant colony adapts itself in a way that it will explore the environment and find new food sources.

In the process of searching for food, when ants find food, they leave a trail of pheromones behind them. Other ants, having smelt the scent of pheromone, follow the same path to food, gather it, and ultimately return to the anthill. All the ants choose the trail that has a stronger pheromone scent, because of pheromone evaporations and also for the reason that the strength of the scent is directly proportional to the number of ants that traversed that path. As more ants choose a given path, the bigger the probability that the other ants will follow the same path. ACO is modeled as a search algorithm that uses artificial ants, inspired by real ants, to search for minimum cost paths in a graph.

2.8.3 PCB Algorithms

Physics- and chemistry-based algorithms aim at the simulation of natural laws of physics (including those of electric charge, gravity, and leverage) and chemistry. Black hole algorithms, for example, absorb the worst individual elements or solutions in a population to derive better median solutions. Chemistry-based algorithms look to the molecular level of interaction in nature. Chemical reaction optimization, for example, is a population-based meta-heuristic algorithm based on the principles of the chemical reaction process of transforming molecules through a sequence of reactions into products.

2.8.3.1 Simulated Annealing

SA, a PCB algorithm, can also be effectively used to solve problems in the design and manufacture of epicyclic gearing. SA is a stochastic optimization technique that takes inspiration from the physical process of annealing in metallurgy. This method models the physical process of heating a metal and then slowly lowering the temperature to decrease defects in metals, thus minimizing the system energy.

The annealing process in metallurgy can make particles arrange themselves in the position with minimum energy configurations by increasing grain size as the temperature is slowly decreased. In the process of annealing, the metal is heated to a particularly high temperature, above its recrystallization temperature, held there for a period of several hours at that temperature, and allowed to cool slowly inside the furnace. The annealing process is used to soften materials or to obtain other desirable physical and chemical properties such as good machinability, suitable electrical properties, dimensional stability, increased toughness, improved elongation, and homogeneity of metals.

Annealing alters the microstructure of the metal by promoting the formation of a uniform, fine-grained structure, which helps to remove internal stresses. Annealing results in a softer, more ductile metal with improved machinability and reduced internal stresses. It is often used to prepare the metal for further processing or to restore ductility after cold working. In order to reduce energy requirements, steel is annealed before cold forming.

In SA, this process is mimicked to run the algorithm iteratively to generate a new solution in the neighborhood of the current solution in each iteration. Then the fitness value of the current and neighbor solutions is compared, and if the fitness value of the new solution is better, then the position of the current solution is updated.

2.8.4 Human-Based Algorithms

HB algorithms are social phenomenon-inspired algorithms and are more robust because human society is more complex than other groups. These algorithms are based on the civilization development phenomena such as leadership and influence, teaching and learning, alliance formation, social labeling of individuals, and social emotion. HB algorithms are inspired by human behavior since humans are the most ingenious species in the world, having high intelligence, abundant cognitive capabilities and processing power, and thought supremacy. Individuals in human society have their own aims, goals, objectives, etc., toward which they try to progress. The HB algorithms try to capture this rapid movement of humans toward their goal and the factors associated with this environment, using them as different resources of inspiration.

2.8.4.1 TLBO Algorithm

Another algorithm that can be used for the design and manufacture optimization of epicyclic gearing is the teaching–learning-based optimization (TLBO) algorithm, which is an HB algorithm that is derived from the relations between an educator, students, and their peers in a learning environment. In a classroom, students learn from the teacher and gain knowledge and then interact among themselves in sharing, modifying, and improving the gained knowledge with each other. The TLBO algorithm is inspired by the classroom environment by simulating these processes to optimize a given objective function.

2.9 OPTIMO SOFTWARE PACKAGE

This book utilizes the OPTIMO software package (Gokul Krishnan, 2024) in order to run the case studies and optimization problems. The software, written in the Python programming language, provides efficient implementations for the multi-objective optimization algorithms covered in this book, along with complete problem sets for each case study.

OPTIMO takes three well-known single-objective optimization algorithms, that is, techniques that optimize a single fitness function, and extends them in order to handle constrained MOPs and the generation of a Pareto-optimal solution set. Each of these techniques belongs to a unique category in computational optimization.

- *PSO* is a stochastic, population-based algorithm.
- *SA* involves the simulation of physical phenomena.
- *ACO* is a discrete, graph-based algorithm.

The manner in which each algorithm is adapted for multi-objective optimization is explained along with step-by-step implementation details in Chapters 4–6. In addition to the aforementioned algorithms, an implementation of the popular NSGA-II MOEA (K. Deb et al., 2002) is also included.

The salient features of this software package are the following:

- Ability to handle a large number of objectives, constraints, and design variables.
- No need to provide separate weights and/or niching parameters.
- Filters out redundant solutions to obtain spaced out Pareto front.
- Generate spreadsheets and 2D/3D plots for visualization of results.
- Cross-platform – runs on any platform that supports the Python runtime.
- Complete problem sets are supplied for each case study.

The software is open source, published under the Massachusetts Institute of Technology (MIT) license, and can be accessed from https://github.com/epicyclic-book/optimo/. The software repository can either be cloned with Git[3] or downloaded as a prepackaged source archive.

2.9.1 Requirements

The software has the following requirements:

- Platform that supports Python – Windows, macOS, and Linux.
- Python 3.6 or above.
- Additional Python packages: *matplotlib*.

Installation steps vary depending on the operating system and/or package manager. For example, these commands are used to install Python and matplotlib on an Ubuntu system (Listing 2.1):

LISTING 2.1 PYTHON INSTALLATION COMMANDS ON UBUNTU LINUX

```
sudo apt install python3
sudo apt install python3-matplotlib
```

2.9.2 Usage

In order to run the included problem sets, navigate to the root of the repository and execute *run.py* with the required options (Listing 2.2).

LISTING 2.2 COMMAND TO EXECUTE THE ALGORITHM RUN

```
# macOS / Linux
python3 run.py -a <algorithm> -p <problem>
# Windows
python.exe run.py -a <algorithm> -p <problem>
```

where

`<algorithm>` is one of – pso, sa, aco, nsga2.
`<problem>` is one of – ch3, ch4, ch5, ch6a, ch6b, ch7a, ch7b.

Optionally, specify the output path and a list of formats for plot generation. The path must exist on the device. If more than one format is specified, separate plots are saved in each of the specified formats. The supported formats are – *png, pdf,* and *svg*. For example, the command in Listing 2.3 saves the results into a results subfolder and generates both *png* and *pdf* plots.

LISTING 2.3 EXTENDED OPTIONS FOR ALGORITHM EXECUTION

```
# macOS / Linux
python3 run.py -a nsga2 -p ch3 -o results -f png pdf
# Windows
python.exe run.py -a nsga2 -p ch3 -o results -f png pdf
```

At the end of the simulation run, a *results.csv* file is generated, along with plots that are defined in the problem specification. The *results.csv* file can be imported into spreadsheet software such as Microsoft Excel or LibreOffice Calc.

In order to batch execute all of the included problems, run *batch.py*. All the results are saved into a *results* subfolder (Listing 2.4).

LISTING 2.4 COMMAND TO EXECUTE THE BATCH PROCESSOR

```
# macOS / Linux
python3 batch.py
# Windows
python.exe batch.py
```

2.9.3 Defining New Problems

To define a new problem, consult the example provided in *sample.py*. The basic steps are given below.

2.9.3.1 Import the Desired Algorithm

Import one or more of the required algorithm classes using the lines in Listing 2.5.

LISTING 2.5 ALGORITHM IMPORT

```
from optimo.algorithm.pso import PSO
from optimo.algorithm.sa import SA
from optimo.algorithm.aco import ACO
from optimo.algorithm.nsga2 import NSGA2
```

2.9.3.2 Provide the Config Information

Problem configuration is provided in the form of key-value pairs in a Python dictionary (Listing 2.6).

LISTING 2.6 CONFIG DEFINITION

```
config = {
  "num_vars": 2,
  "num_obj": 2,
  "num_constr": 2,
  "bounds_min": [0, 0],
  "bounds_max": [5, 3]
}
```

The following values are required to be defined:

- *num_vars*: Number of variables.
- *num_obj*: Number of objectives.
- *num_constr*: Number of constraints.
- *bounds_min*: Minimum bounds – list of values, one for each variable.
- *bounds_max*: Maximum bounds – list of values, one for each variable.

In addition, the following optional values can be defined:

- *ref_data*: Reference data for debugging and testing – list of values, one for each variable. This is used to run a test evaluation of the problem function and obtain a reference solution. If unspecified, data is assumed based on the bounds provided.

To specify control parameters for the algorithm:

- *num_iter*: Total number of iterations (the default is 100).
- *num_pop*: Population size – total number of particles/individuals/candidates (the default is 200).
- *num_archive*: Archive size – total number of non-dominated solutions (the default is 200).

To provide information for plot generation:

- *obj_names*: Objective names – list of strings, one for each objective. If unspecified, objective names are assumed to be *f1, f2, f3, ...*
- *obj_types*: Objective types – list of strings, one for each objective. The type is either `"min"` (minimizing) or `"max"` (maximizing). If unspecified, all objectives are assumed to be minimized.
- *constr_names*: Constraint names – list of strings, one for each constraint. If unspecified, constraint names are assumed to be *c1, c2, c3*, etc.
- *var_names*: Variable names – list of strings, one for each variable. If unspecified, constraint names are assumed to be *x1, x2, x3*, etc.
- *var_types*: Variable types – list of strings, one for each variable. The type is either `"int"` (integer) or `"float"` (real valued). If unspecified, all variables are assumed to be real valued.

- *plot_data*: Data for plot generation – list of 2- or 3-valued tuples, one for each plot. Each tuple specifies the names of objectives to be plotted along the X and Y axes (for the 2D plot) or the X, Y, and Z axes (for the 3D plot). See Listing 2.7 for an example.

LISTING 2.7 PLOT SPECIFICATION

```
config = {
  ...
  "plot_data": (
    # plot 1: 2D plot of f1 and f2
    ("f1", "f2"),

    # plot 2: 3D plot of f1, f3 and f4
    ("f1", "f3", "f4")
  )
}
```

2.9.3.3 Provide the Problem Definition

The problem definition is provided as a function with the following signature (Listing 2.8).

LISTING 2.8 PROBLEM DEFINITION

```
def Problem(v, obj, con):
  x1, x2 = v[0], v[1]

  # Objective 1
  f1 = ... ...
  obj.append(f1)

  # Objective 2
  f2 = ... ...
  obj.append(f2)

  # Constraint 1
  c1 = ... ...
  con.append(c1)

  # Constraint 2
  c2 = ... ...
  con.append(c2)
```

where `v` is the list of variables (input data); `obj` is an empty list to which objective values are to be outputted; `con` is an empty list to which constraint values are to be outputted.

Because the software performs minimization of objectives, objectives that are to be maximized must be provided in negated form (Listing 2.9).

LISTING 2.9 MINIMIZATION VS. MAXIMIZATION OF OBJECTIVES

```
obj.append( f1)   # This objective will be minimized
obj.append(-f2)   # This objective will be maximized
```

Likewise, constraint equations must be reformulated into ≥ 0 forms (Listing 2.10).

LISTING 2.10 REFORMULATION OF CONSTRAINTS

```
# Constraint equation: x₁ + x₂ ≥ 25
c1 = x1 + x2 - 25
con.append(c1)
```

Discrete variables (integers) must be rounded off with the appropriate rounding scheme before usage. It is recommended to round up discrete variables with the `math.ceil()` function (Listing 2.11).

LISTING 2.11 HANDLING DISCRETE VARIABLES

```
x1, x2 = v[0], v[1]  # x₁ and x₂ are real valued.
x3 = math.ceil(v[2]) # x₃ is an integer.
```

2.9.3.4 Instantiate the Algorithm Object

Create the algorithm object, providing the previously defined configuration and problem definition (Listing 2.12).

LISTING 2.12 ALGORITHM INSTANTIATION

```
alg = PSO(Problem, config)
```

2.9.3.5 Run the Solver

Begin the algorithm execution using `run()` (Listing 2.13).

LISTING 2.13 ALGORITHM EXECUTION

```
alg.run()
```

2.9.3.6 Save Output

Save out the results using `save()`. This will produce a *results.csv* file and a sequence of one more plot. Optionally, before saving, specify the path for storing the results using `set_path()`. The path must exist on the device.

A format list for plot generation can be supplied using `set_formats()`. This method takes a list of strings, one for each format. The supported formats are `"png"`, `"pdf"`, and `"svg"`. If more than one format is specified, separate plots are saved in each of the specified formats. For example, the code in Listing 2.14 saves the results into a *results* subfolder and generates both *png* and *pdf* plots.

LISTING 2.14 SAVING THE RESULTS

```
alg.set_path("results") # set output path
alg.set_formats(["png", "pdf"]) # set output formats
alg.save()
```

2.9.4 Computational Complexity

OPTIMO provides multi-objective optimization through a selection of stochastic optimization algorithms. At the core of each optimization procedure is a Pareto archival system, which involves the maintenance of an external archive of Pareto-optimal solutions. The overall runtime complexity is governed by this archival system.

Archiving refers to the extraction of non-dominated solutions from the population set into a separately maintained external archive. The goal of archiving is twofold:

i To obtain a set of non-dominated solutions which form the Pareto-optimal solution set.
ii To promote diversity of solutions in the Pareto solution set.

Archiving is performed at the end of each iteration of the algorithm. The following is a quick rundown of the steps involved in OPTIMO's archival process along with an analysis of the runtime complexity at each step. We assume a population size of N and the total number of objectives M.

2.9.4.1 Preprocessing

Initially, the archive starts out as an empty set. The solutions in the archive are merged with the present population to produce a merged population. This merged population is preprocessed in order to filter out obviously undesirable solutions. At first, infeasible solutions, if any, are filtered out. This takes $O(1)$ per-solution, which amounts to $O(N)$ going through the entire merged solution set, whose size is assumed to be of the same order as N.

Because the solution's last recorded best position is considered for archiving, rather than the instantaneous position, there is a high chance that solutions have not *moved* between iterations. This raises the possibility of the same solution being accepted multiple times into the archive because a solution is non-dominated with respect to itself, that is,

Consider two objective vectors P_1 and P_2.

Assuming $P_1 \equiv P_2$, then, P_1 is non-dominated w.r.t. P_2, and P_2 is non-dominated w.r.t. P_1.

Therefore, OPTIMO uses a hashed set-based lookup procedure in order to eliminate duplicate solutions prior to dominance computation. This process is $O(M)$ for hash computation and $O(1)$ for hash lookup per solution, leading to $O(MN)$ overall complexity.

2.9.4.2 Dominance Computation

The extraction of non-dominated solutions involves computing the Pareto-dominance state of each solution vis-à-vis every other solution in the population set. For a population size N and a total number of objectives M, this is a $O(MN^2)$ process for the extraction of a single non-dominated front. Because we limit to extracting the first non-dominated front alone, the overall complexity of this step remains $O(MN^2)$.

2.9.4.3 Crowding Distance Computation

When the number of non-dominated solutions exceeds the desired archive size, some of the new solutions must be rejected, while others replace existing solutions in the archive. In order to encourage population diversity and prevent entanglement within local optima, acceptance of solutions is either based on segmentation of the objective space, which requires the specification of a niching radius, or the assignment of a crowding factor to each solution. OPTIMO makes use of the *crowding distance* factor originally proposed for NSGA-II (K. Deb et al., 2002). Crowding distance computation requires the population to be sorted M times on each objective. Therefore, the process takes $O(MN \log N)$.

2.9.4.4 Crowded Filtering

In order to filter solutions based on crowding distance, two approaches have been considered. Let the size of the solution set be S_C, and the maximum archive size be S. Then, assuming $S_C > S$, the number of surplus solutions that need to be removed is,

$$S_R = S_C - S$$

The first approach considered is to reverse sort the solution set based on the crowding distance factor and keep the first S solutions only. This ensures that the solutions with the largest crowding distances, that is, solutions from the least crowded regions, are preserved and the rest are unconditionally rejected. The sorting process typically has a complexity of $O(N \log N)$, although it could approach $O(N^2)$ in the worst case.

However, OPTIMO uses a slightly different approach that introduces randomness in the crowded filtering criterion. We use a roulette-wheel selection process with the inverse crowding distances taken as weights, and the solutions selected thus are removed from the archive. This prioritizes solutions from denser regions for removal. The process requires S_R selections, where each selection takes $O(N)$. In the worst case, where $S_R \simeq N$, the overall complexity could approach $O(N^2)$.

While the *probabilistic crowded filtering* approach is theoretically less efficient when compared to the deterministic approach, our empirical tests reveal that the probabilistic approach is better at preserving good quality solutions with very similar execution times in practice.

In the end, the overall computational complexity of the Pareto archival system, and therefore the overall complexity of OPTIMO algorithms, arrives at $O(MN^2)$. This is of the same order of complexity as some of the popular multi-objective GAs such as NSGA-II. In addition, these algorithms are relatively simple when compared to MOGAs. Depending on the use case and the population size, this could result in a smaller computational footprint.

NOTES

1 The idea of pareto-optimality is explored in detail in Chapter 3, Section 3.4.
2 Combinatorial optimization techniques focus on finding effective ways to combine elements from distinct sets.
3 Open source version control software. https://git-scm.com/.

REFERENCES

Chandrasekaran, G., et al., "Multiobjective Optimisation of Bevel Gear Pair design using NSGA-II," *Materials Today: Proceedings*, 16, Part 2, 2019, 351–360, https://doi.org/10.1016/j.matpr.2019.05.101

Deb, K. and Jain, S., "Multi-Speed Gearbox Design Using Multi-Objective Evolutionary Algorithms," *Journal of Mechanical Design*, 125, 3, September 2003, 609–619 (11 pages), https://doi.org/10.1115/1.1596242

Deb, K., et al., "A Fast and Elitist Multiobjective Genetic Algorithm: NSGA-II," *IEEE Transactions On Evolutionary Computation*, 6, 2, 2002, 182–197, https://doi.org/10.1109/4235.996017

Krishnan, G., OPTIMO software – Algorithms for Multi-Objective optimization. 2024. https://github.com/epicyclic-book/optimo/. Used under the terms and conditions of the MIT license.

Sedak, M. and Rosić, B., "Multi-Objective Optimization of Planetary Gearbox with Adaptive Hybrid Particle Swarm Differential Evolution Algorithm." *Applied Science*, 11, 2021, 1107, https://doi.org/10.3390/app11031107

Younes, B., et al., "Multi-Objective Optimization of Gear Unit Design to Improve Efficiency and Transmission Error," *Mechanism and Machine Theory*, 167, January 2022, 104499, https://doi.org/10.1016/j.mechmachtheory.2021.104499

3 Design Optimization of Power Take-Off Gearbox Using Multi-Objective Genetic Algorithm

NSGA-II

Chandrasekaran G

3.1 INTRODUCTION

Epicyclic gear trains are widely used in power transmission, aeronautical and industrial applications since they have positive features including compact structure, self-centering capability and high power density. Epicyclic gears are an integral part of many kinds of complex machinery. There are different gear configurations, and it is vital to arrange them and assemble them in specific ways to achieve particular results. In epicyclic gear trains, several design considerations, such as planet spacing, number of planets, choices for tooth numbers, and the flexibility of the ring gear, need to be addressed to attain long life.

The challenges in the design optimization of epicyclic gearing are:

- To find the Pareto fronts (PFs) using appropriate optimization techniques.
- To find the global optimum value which gives the best solution for the problem.
- Reduce the weight of gears and hence cost savings.
- Reduce volume and hence less space requirement.
- High efficiency of transmission and hence better performance.
- Reliable operation during service life.

The application of multi-objective optimization methods for choosing an optimal solution from the PF is carried out in the following sections. Evolutionary multi-objective optimization algorithms combine two major disciplines of evolutionary computation and the theoretical frameworks of multi-criteria decision-making. A Pareto-optimal solution is an optimal solution that is not dominated by any other

DOI: 10.1201/9781003144465-3

solution in the search space, and the entire set of such optimal trade-off solutions is called a Pareto-optimal set. The decision-making (trade-off) process is required to obtain the optimal solution, and evolutionary methods maintain a set of solutions as a population during the course of the search and thus result in a set of Pareto optimal solutions in a single run.

3.2 OPTIMIZATION WITH GENETIC ALGORITHMS

To solve a problem, we are usually looking for some solution which will be the best among other solutions. The search space with all the feasible solutions contains the required set of solutions with minimum or maximum fitness value. There are many methods to find suitable solutions, such as hill climbing, tabu search, simulated annealing, and genetic algorithms. Genetic algorithms belong to the family of evolutionary algorithms and are much more unique and intelligent than other traditional and random search algorithms. Genetic algorithms explore the solution space faster by improving the initial solutions to get local and global optimal solutions for the problem.

3.2.1 Genetic Algorithms

Genetic algorithms are a class of evolutionary algorithms, also called evolutionary computation methods, which include evolution strategies, evolutionary programming, differential evolution, and genetic programming. A genetic algorithm, a simulated algorithm, and a method of random search borrowed from the laws of evolution in biology were first proposed by Professor John H. Holland (1975) and later developed by De Jong K. A. (1975). Genetic algorithms are used to find minima of multivariable functions through a discretization of the domain using bit strings.

A genetic algorithm is a particular class of evolutionary algorithms which uses adaptive processes of natural systems to find true or approximate solutions to global search/optimization problems. Genetic algorithms use efficient and effective techniques for optimization that retain the robustness of natural systems, inspired by evolutionary biology such as inheritance, mutation, selection, and crossover. A genetic algorithm is an artificial intelligence technique in which a population of abstract representations (called chromosomes) of candidate solutions (called individuals) to an optimization problem evolves toward better solutions.

The basic idea of genetic algorithms is learning, through iterations, what regions of the domain contain better solutions and focusing the search on those regions. Whereas other optimization methods take a solution (point) and improve it by means of iterative transformations (e.g., steepest gradient), genetic algorithms start from a population of randomly chosen points and transform them in a manner similar to Natural Genetics.

Genetic algorithms start from an initial population of points and improve it through the iterative application of three transformations: selection, crossover, and mutation, which resemble nature's evolution of biological systems. During the first stage, selection, the objective function is evaluated for each member of the actual population. Then a new set of points is extracted from the actual one by means of a stochastical sampling in which better elements have a higher probability of extraction. The second stage, crossover, acts like chromosome recombination, which combines the extracted solutions in order to search different areas of the parameter space. Since some regions of the space could not be reached by crossover, mutation adds a random variation of points inside the population. By successively applying these three transformations in each generation till a termination criterion is satisfied, a final set of solutions is achieved. The final set of solutions thus obtained contains the best solutions of previous generations because the traits of a stronger individual tend to be carried forward into the next generation.

Genetic algorithms, when applied to multi-objective optimization problems (MOPs), where there is more than one objective function to be minimized or maximized, help in finding the PF of non-dominated solutions. Since the genetic algorithm searches many peaks in parallel, the probability of getting trapped in local minima is much lower in solving MOPs. A multi-objective genetic algorithm uses a crowding distance comparison operator to create a homogeneous distribution of the non-dominated solutions on the PF to find a final set of solutions till a required termination criterion is met.

Figure 3.1 shows a flow chart of the genetic algorithm procedure.

3.2.1.1 Selection

Genetic algorithms use many different techniques to select the individuals to be copied over into the next generation. Roulette-wheel selection, elitist selection, fitness-proportionate selection, scaling selection, rank selection, generational selection, hierarchical selection, etc., are some of the selection methods. After choosing fit individuals through selection, they must be randomly altered in order to improve their fitness for the next generation. The process of selection is also termed as inheritance.

3.2.1.2 Crossover

In natural systems, crossing-over is the recombination of genetic material to form variability in the population. Crossover operator exchanges substrings between chromosomes in genetic algorithm. Initially, a population of *parent* strings is generated by the selection operator from which two individuals are chosen at random. After that, string segments are exchanged by choosing crossover points at one or more string locations. Lastly, two resultant offspring individuals are produced by exchanging and then combining parent string segments. Crossover creates new offspring by operating on selected genes from parent chromosomes. Generating offspring from two selected parents may be done by a single crossover point, by N-point crossover, or by uniform crossover. Generally, crossover probability should be high.

3.2.1.3 Mutation

Mutation, which involves reordering the list, prevents the population from stagnating at any local optimum. Mutation randomly changes the new offspring. One

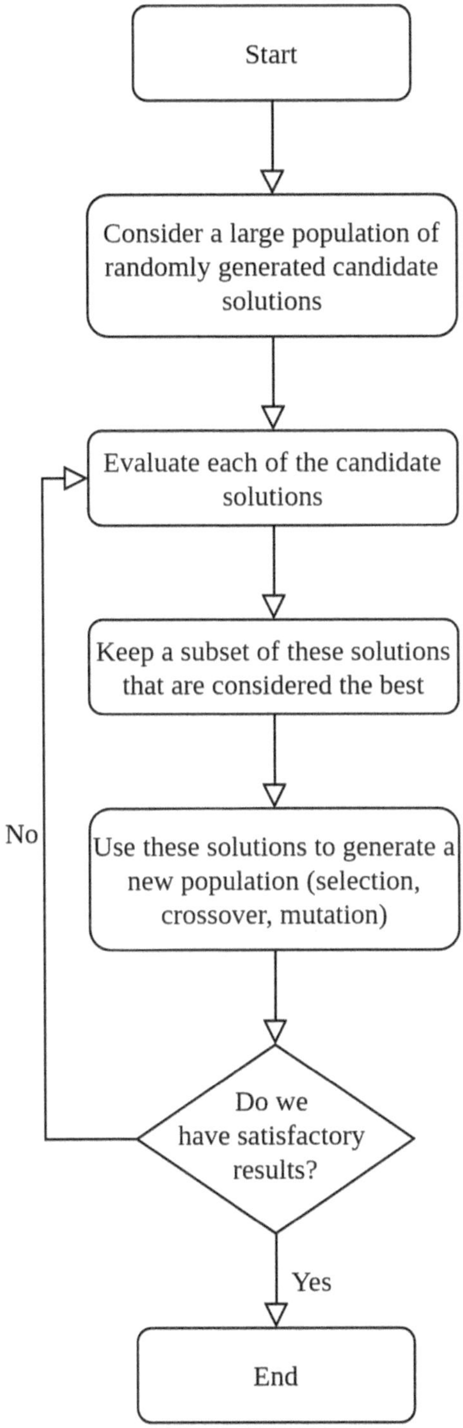

FIGURE 3.1 Genetic algorithm procedure.

or more gene values are altered from their initial state with the help of mutation, which results in entirely new gene values. Mutation may improve the quality of the solution. Mutation probability should be low to prevent GA from becoming a random search. Mutation helps to maintain diversity within the population and to prevent premature convergence.

3.2.1.4 Termination

The algorithm is terminated when the population has converged, providing a set of solutions to our problem. The termination condition is such that our solution is close to the optimal at the end of the run. The termination condition is usually to keep one of the following:

- After X iterations, when there is no improvement in the population.
- After reaching an absolute number of generations.
- When we obtain certain pre-determined best solutions for the objectives.

3.3 MULTI-OBJECTIVE DESIGN OPTIMIZATION OF EPICYCLIC GEARING

The volume or weight of a gearbox is an important parameter in many applications which require a small size. Also, efficiency is an important factor that decides the running cost of the gear drive during its lifetime. There are several objectives in the design optimization of epicyclic gearing, making it an MOP. A set of objective functions that are essential for the proper functioning of the system are identified, which are subjected to several constraints. The optimization procedure is carried out based on international standards. The problem definition and mathematical formulation are given with case study design examples.

3.4 PARETO OPTIMALITY

Pareto optimality is named after the Italian sociologist Vilfredo Pareto. A Pareto optimal set is a set of solutions that are non-dominated with respect to each other in the solution space. A Pareto optimal solution cannot be improved with respect to any objective since there is always some amount of sacrifice in one objective to achieve some amount of gain in the other.

A Pareto optimal solution is a candidate solution for achieving the best trade-off, and Pareto optimal solution sets are often preferred to single solutions because they can be practical when considering real-life problems. Pareto optimal sets can be of different sizes, but the size of the Pareto set (PS) enlarges with the increase in the number of objectives. An MOP may have many, even an infinite number of Pareto-optimal solutions. The PS is the set of all the Pareto-optimal solutions in the decision/objective space, and its image in the objective space is the PF. Without the decision maker's preference information, all the Pareto-optimal solutions are equally good. To make his final decision, the DM often wants to have a good approximation of the PS and/or the PF available.

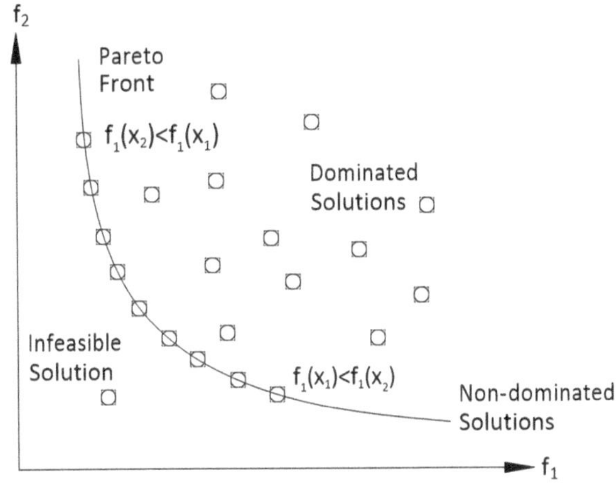

FIGURE 3.2 Two-dimensional Pareto space.

Figure 3.2 depicts a two-dimensional case, where there is a trade-off between minimizing both objective functions f_1 and f_2. The goal of a Pareto evolutionary algorithm is to find a set of solutions along the PF, which is optimal for a combination of criteria.

For solving multiple objective problems, there are two general approaches:

1. Combine the individual objectives into a composite single-objective function.
2. Determine a Pareto optimal solution set from multi-objective solution vectors.

The first method uses weighted sums or utility functions, but the selection of parameters is difficult because small perturbations in the weights can lead to very different solutions. The second method finds a set of solutions that are non-dominated with respect to each other. Multi-objective optimization is the task of obtaining the PF by automatically extracting non-dominated solutions from a set of multi-objective and multidimensional solutions.

3.5 NSGA-II

By using multi-objective genetic algorithm (MOGA) optimization techniques, problems with more than one objective function are concurrently optimized. Several Pareto optimal points can be simultaneously obtained with an even distribution of solutions from the Pareto optimal sets, which can reduce the production time and cost. There are many new multi-objective optimization techniques which are enhancements from single-objective optimization GA, and among all MOGA techniques, NSGA-II has been most widely applied for optimizing gear design parameters. The following sections review the application of the NSGA-II in optimizing gear design parameters.

3.5.1 NSGA-II Functionality

NSGA-II represents the state of the art of multi-objective evolutionary optimization. Proposed by *K*. Deb (2002) on the basis of NSGA, NSGA-II improves the non-dominated sorting algorithm and reduces the computational complexity. It sorts the combination of parent and children populations with an elitist strategy, introduces the crowded comparison operator to improve the diversity of solutions, and avoids the use of niched operators.

3.5.2 Concept of NSGA-II

NSGA-II is characterized by (i) a faster non-dominated sorting approach, (ii) an elitist strategy, and (iii) no niching parameter.[1] The elitist strategy is the inclusion of the best individuals of a generation in the population of the next generation. Diversity in the population is preserved by the use of crowded comparison criteria in the tournament selection and in the phase of population reduction.

3.5.3 NSGA-II Procedure

The procedure of the NSGA-II algorithm is given below (Virginia Yannibelli et al., 2020). The flow chart in Figure 3.3 depicts the entire process.

3.5.3.1 Population Initialization

The NSGA-II algorithm begins creating a random initial population with *s* solutions, where *s* is a given number. In each generation, the algorithm selects *s* solutions from the current population to compose the mating pool by applying the traditional binary tournament selection process. This process considers the non-domination level of the solutions, which is defined by applying non-dominated sorting to the current population.

The process also considers the crowding distance of the solutions, which represents the distance between a solution and its neighbors. Then, the process

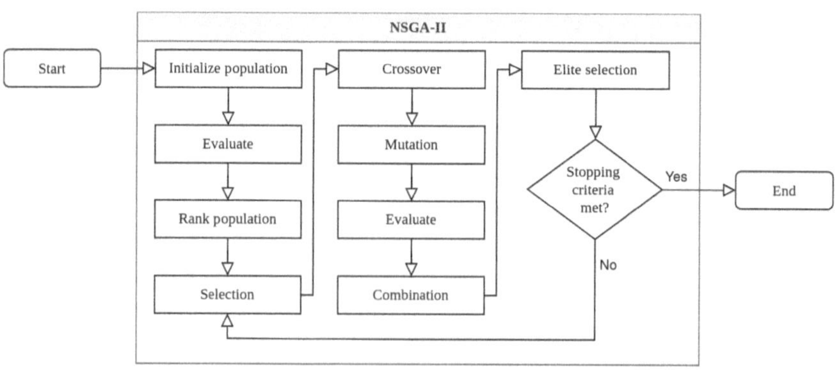

FIGURE 3.3 NSGA-II flow chart (Pedro P. C. Bhagubai et al., 2020).

considers the following criterion to determine the best solution for each of the s tournaments. When a tournament involves two solutions with different non-domination levels, the solution with the lower (better) level is preferred and selected for the mating pool. Otherwise, when the two solutions of a tournament have the same non-domination level, the solution with the higher crowding distance is preferred and selected in order to promote the selection of diverse non-dominated solutions for the mating pool.

3.5.3.2 Crossover and Mutation

After the mating pool is composed, the algorithm creates an offspring population from the mating pool by applying the known simulated binary crossover operator and then the known polynomial mutation operator (K. Deb et al., 2007).

The crossover operator is applied with a crossover probability of P_c and a crossover distribution index of D_c, and the mutation operator is applied with a mutation probability of P_m and a mutation distribution index of D_m.

Once the offspring population is created, the algorithm combines the current and offspring populations and then selects the best s solutions from this combined population to create a new population for the next generation. To select these s solutions, the algorithm applies the non-dominated sorting to the combined population.

3.5.3.3 Non-dominated Sorting

The solutions in the combined population are grouped according to their non-domination levels as $\{F_1, F_2, \ldots\}$. Then, each non-domination level is selected one at a time to create the new population, beginning from F_1, until the size of the new population is equal to s or higher than s. When the size of the new population is equal to s, additional selection operations are not needed, and the next generation starts from this population. When the size of the new population is higher than s, the last non-domination level selected, F_L, is not fully included in this population. In this respect, the solutions from level F_1 to level $F_L - 1$ are included in this population, and the r remaining solutions $\left[r = s - \left| F_1 \cup \ldots \cup \left(F_L - 1 \right) \right| \right]$ are selected from level F_L.

3.5.3.4 Crowding Distance Sorting

To select the r remaining solutions from the level F_L, the algorithm utilizes a selection process, which considers the crowding distances of the solutions in F_L. This process calculates the crowding distance of each solution in F_L. Then, the process sorts the solutions in F_L based on these distances in descending order and selects the r solutions with the highest distances. Thus, this process aims to promote the selection of diverse, non-dominated solutions, with the aim of preserving the diversity of the new population.

The algorithm ends its execution once a predefined termination criterion is achieved (i.e., a given number of evaluations for the generated solutions), providing the PS (i.e., the set of non-dominated solutions) of the population corresponding to the last generation.

3.6 CASE STUDY: DESIGN OPTIMIZATION OF A POWER TAKE-OFF GEARBOX OF A HEAVY TRUCK USING NSGA-II

The following sections explain a case study design problem for the design optimization of a Power Take-off gearbox of a heavy truck. The purpose of the study is to explore the possibility of improving the existing design and also to automate the design optimization problem so that the procedure can be applied to any other similar situation or for the design of a new gearbox.

3.6.1 The Case Study Approach

A case study is an in-depth, detailed, and multifaceted exploration of a complex real-world situation, undertaken in order to understand its complexities and gain insights into a particular instance or situation. In the present example, a case study on the design optimization of a Power Take-off gearbox used in heavy trucks is undertaken in order to evaluate the product and objectively examine it in its natural, real-life context and to explore the key characteristics, meanings, and implications of this case.

A case study is an empirical inquiry that is used to present a problem, allowing individuals to interpret it and provide a solution. It is a form of scientifically rigorous research, using one or more types of research methods to learn as much as possible about the present problem so that the information can be generalized to many others. Similarly, the case study provides designers/researchers the chance to collect information on why one strategy might be chosen over another. It permits researchers to develop hypotheses that can be explored through their insights and to develop additional ideas that might be explored in future studies. A case study is a method of minute and detailed study of a situation and should be carefully conceptualized and thoughtfully undertaken and reported to yield powerful insights into many important aspects of the given problem (Crowe et al., 2011).

3.6.2 Overview of Power Take-Off Gearbox

A power take-off (PTO) is a gearbox, which is usually installed on a truck and which can be used to provide power to an attachment or to a separate hydraulic or mechanical device. It is designed in such a way that it can be easily connected and disconnected to draw energy from the engine. The PTO is a method of transferring high power and torque from the engine via the transmission of trucks. The PTO gearbox under consideration is a two-stage gearbox, the first-stage reduction being a spiral bevel gear pair and the second-stage reduction being a planetary gear train. A schematic diagram for the arrangement of various gears of the PTO gearbox under consideration is shown in Figure 3.4.

3.6.3 PTO Gearbox Specifications

The design optimization is attempted for the second-stage planetary gear train with the following specifications (Table 3.1).

Figure 3.5 shows a schematic diagram of the gear elements in PTO planetary gearing.

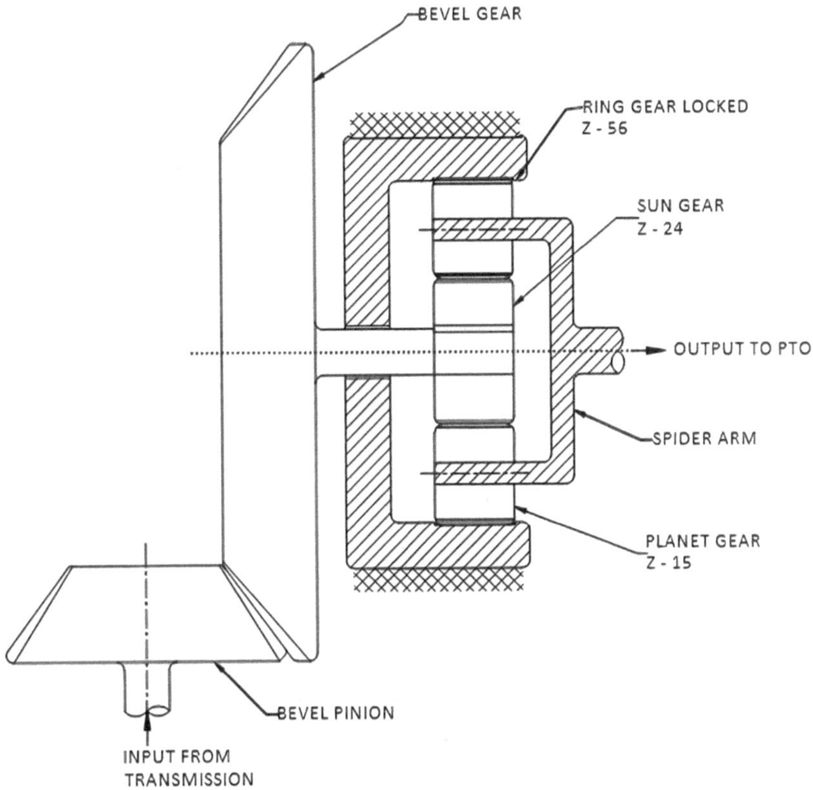

FIGURE 3.4 Elements of PTO gearbox.

TABLE 3.1
Power Take-Off Planetary Gearing Specifications

	Sun Gear	Planet Gear	Ring Gear
Normal module (mm)	4	4	4
Number of teeth	24	15	56
Normal pressure angle (°)	25	25	25
Helix angle (°)	0	0	0
Individual shift in the basic profile *X*	−0.042	+0.57	+0.1
Base diameter (mm)	87.006	54.378	203.013
Pitch diameter (mm)	96	60	224
Face width (mm)	60	52.9	62
No. of planets	–	5	–

3.6.4 PTO Gearbox Design Optimization Procedure Using NSGA-II

The purpose of this case study is to automate and present a GA-based design optimization for the planetary gearing stage of a PTO two-stage gearbox. The procedure reported herein provides an optimization of the basic type of planetary gear train. The presented approach is based on suitably applying the NSGA-II algorithm to the mathematical model. The advantage of using an optimization technique is that it enables the designer to consider a spectrum of possible designs. The method employed in the following procedure starts by listing the design parameters available, the equality constraints which must be satisfied, the inequality constraints which define the limits of acceptable designs, and the objective functions which are used to compare the relative merit of each possible design with the existing design.

The input power to the PTO gearbox from the truck transmission is given to the sun gear. The output shaft of the PTO gearbox is the spider arm shaft. Hence, this arrangement comes under the first type of epicyclic gearing, namely the planetary gearing.[2]

To find the speed of individual elements of the existing planetary gear train (Table 3.2), let

Number of teeth on sun wheel, $Z_2 = 24(Z_s)$
Number of teeth on planet gears, $Z_1 = 15(Z_p)$
Number of teeth on the annular ring gear, $Z_3 = 56(Z_a)$
Spider arm is L

The last row of Table 3.2 provides us with the equation for the speed of individual elements.

Input speed of bevel pinion = 500 rpm

Since the bevel reduction is 4.778, the speed of the bevel gear is,

TABLE 3.2
Speed Calculation for Planetary Elements

Operation	Spider Arm L	Sun Wheel Z_s	Planet Wheel Z_p	Annular Wheel Z_a
1. Arm L is locked and sun wheel S is given +1 revolution.	0	+1	$-\frac{Z_s}{Z_p} = -\frac{24}{15}$	$-\frac{Z_s}{Z_a} = -\frac{24}{56}$
2. Multiply by x	0	x	$-\frac{Z_s}{Z_p}x = -\frac{24}{15}x$	$-\frac{Z_s}{Z_a}x = -\frac{24}{56}x$
3. Add y to all columns	y	$y + x$	$y - \frac{Z_s}{Z_p}x = y - \frac{24}{15}x$	$y - \frac{Z_s}{Z_a}x = y - \frac{24}{56}x$

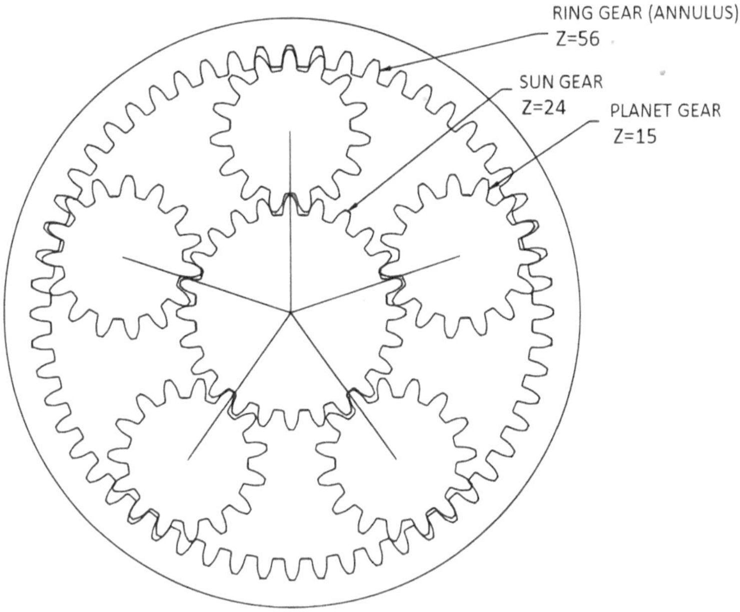

FIGURE 3.5 Schematic diagram of elements in PTO planetary gearing.

$$\frac{500}{4.778} = 104.651 \text{ rpm}$$

Since the sun gear is mounted on the same shaft of the bevel gear, input speed to the sun gear = 104.651 rpm.

Speed of planets,

$$y - \frac{24}{15}x = 31.3953 - \frac{24}{15} \times 73.2557 = -85.814 \text{ rpm}(\text{ACW})$$

Annulus is locked,

$$y - \frac{24}{56}x = 0 \tag{3.1}$$

$$y + x = 104.651$$

$$y = 104.651 - x \tag{3.2}$$

Substituting Equation (3.2) in Equation (3.1),

$$104.651 - x - \frac{24}{56}x = 0$$

$$x\left(1+\frac{24}{56}\right)=104.651$$

$$\therefore x=73.2557$$

Speed of the spider arm

$$y=31.3953 \text{ rpm}(\text{CW})$$

Power input = 50 HP = 37.285 kW

Power input is given by,

$$P=\frac{2\pi NT}{60}$$

Therefore, torque input is,

$$T=\frac{60P}{2\pi N}$$

Substituting the relevant values,

$$T=3402.2\,\text{Nm}$$

3.6.5 Calculation of Gear Parameters

The transverse pressure angle is given by,

$$\alpha_t=\tan^{-1}\left(\frac{\tan\alpha_n}{\cos\beta}\right)$$

The involute function of α_t is given by,

$$inv\alpha_t=\tan\alpha_t-\left(\alpha_t\frac{\pi}{180}\right)$$

The involute function of the working transverse pressure angle between the planet gear and the sun gear, in terms of profile shift coefficients, is given by,

$$inv\alpha_{wt_1}=2\tan\alpha_n\left(\frac{x_{n_1}+x_{n_2}}{z_1+z_2}\right)+inv\alpha_t$$

By substituting the values of the relevant parameters in the above equation, we can find out the value of $inv\alpha_{wt_1}$.

Hence, from the above-known value of $inv\alpha_{wt_1}$, we can find out the value of the working transverse pressure angle α_{wt_1} by the following method:

Let ξ_1 be defined as given below:

$$\xi_1=inv\alpha_{wt_1}=2\tan\alpha_n\left(\frac{x_{n_1}+x_{n_2}}{z_1+z_2}\right)+inv\alpha_t$$

When the involute function value ξ_1 of the working transverse pressure angle between the planet gear and the sun gear is known, the empirical equation to find out the value of the working transverse pressure angle α_{wt_1} between the planet gear and the sun gear is given by,

$$\alpha_{wt_1} = \frac{\xi_1^{1/3}}{0.693357 + 0.192484\xi_1^{2/3}} \frac{180}{\pi}$$

In a similar manner, the involute function of the working transverse pressure angle α_{wt_2} between the planet gear and the ring gear is given by,

$$inv\alpha_{wt_2} = 2\tan\alpha_n \left(\frac{x_{n_3} - x_{n_1}}{z_3 - z_1} \right) + inv\alpha_t$$

By substituting the values of the relevant parameters in the above equation, we can find out the value of $inv\alpha_{wt_2}$.

As before, let ξ_2 be defined as given below:

$$\xi_2 = inv\alpha_{wt_2} = 2\tan\alpha_n \left(\frac{x_{n_3} - x_{n_1}}{z_3 - z_1} \right) + inv\alpha_t$$

Again, by applying the empirical equation in terms of ξ_2, as given below,

$$\alpha_{wt_2} = \frac{\xi_2^{1/3}}{0.693357 + 0.192484\xi_2^{2/3}} \frac{180}{\pi}$$

We can find out the value of the working transverse pressure angle α_{wt_2} between the planet gear and the ring gear.

The reference circle diameters of the planet gear, the sun gear, and the ring gear are given by,

$$d_1 = \frac{z_1 m_n}{\cos\beta}$$

$$d_2 = \frac{z_2 m_n}{\cos\beta}$$

$$d_3 = \frac{z_3 m_n}{\cos\beta}$$

The base circle diameters of the planet gear, the sun gear, and the ring gear are given by,

$$d_{b_1} = d_1 \cos\alpha_t$$
$$d_{b_2} = d_2 \cos\alpha_t$$
$$d_{b_3} = d_3 \cos\alpha_t$$

The base circle radii of the planet gear, the sun gear, and the ring gear are given by,

$$r_{b_1} = \frac{d_{b_1}}{2}$$

$$r_{b_2} = \frac{d_{b_2}}{2}$$

$$r_{b_3} = \frac{d_{b_3}}{2}$$

The working pitch circle diameters of the planet gear, the sun gear, and the ring gear are given by,

$$d_{w_1} = \frac{d_{b_1}}{\cos \alpha_{wt_1}}$$

$$d_{w_2} = \frac{d_{b_2}}{\cos \alpha_{wt_1}}$$

$$d_{w_3} = \frac{d_{b_3}}{\cos \alpha_{wt_2}}$$

The working pitch circle radii of the planet gear, the sun gear, and the ring gear are given by,

$$r_{w_1} = \frac{d_{w_1}}{2}$$

$$r_{w_2} = \frac{d_{w_2}}{2}$$

$$r_{w_3} = \frac{d_{w_3}}{2}$$

The theoretical center distance between the planet gear and the sun gear is given by,

$$a = \frac{d_1 + d_2}{2}$$

The theoretical center distance between the planet gear and the ring gear is given as,

$$a = \frac{d_3 - d_1}{2}$$

The sun gear and the ring gear have the same axis of rotation.

The working center distance between the planet gear and the sun gear (or between the planet gear and the ring gear) is given by,

$$a' = \frac{d_{w_1} + d_{w_2}}{2}$$

The change in center distance is given by,

$$\Delta a = a' - a$$

The center distance modification coefficient for the planet gear and the sun gear pair is given by,

$$y_1 = \frac{a'}{m_n} - \frac{z_1 + z_2}{2\cos\beta}$$

The addendum of the planet gear teeth is given by,

$$h'_{a_1} = \left(1 + y_1 - x_{n_2}\right) m_n$$

The addendum circle diameter of the planet gear is given by,

$$d'_{a_1} = d_1 + 2h'_{a_1}$$

Hence, the addendum circle radius of the planet gear is,

$$r'_{a_1} = \frac{d'_{a_1}}{2}$$

The addendum of the sun gear is given by,

$$h'_{a_1} = \left(1 + y_1 - x_{n_1}\right) m_n$$

The addendum circle diameter of the sun gear is given by,

$$d'_{a_2} = d_2 + 2h'_{a_2}$$

Hence, the addendum circle radius of the sun gear is,

$$r'_{a_2} = \frac{d'_{a_2}}{2}$$

The working transverse pressure angle of the planet gear and sun gear pair, in terms of the center distance modification coefficient, is given by,

$$\alpha_{wt_1} = \cos^{-1}\left(\frac{\cos\alpha_t}{\dfrac{2y_1\cos\beta}{z_1 + z_2} + 1}\right)$$

The above equation can be used to check the correctness of the previous calculations for the value of α_{wt_1}.

The center distance modification coefficient for the planet gear and the ring gear pair is given by,

$$y_2 = \frac{(z_3 - z_1)}{2}\left(\frac{\cos \alpha_n}{\cos \alpha_{t_2}} - 1\right)$$

Also, the center distance modification coefficient for the planet gear and the ring gear pair in terms of modified center distance a' is given by,

$$y_2 = \frac{a'}{m_n} - \frac{z_3 - z_1}{z \cos \beta}$$

The whole depth of the involute tooth of the planet gear is given by,

$$h_1' = \left(2.25 + y_2 - \left(x_{n_3} - x_{n_1}\right)\right) m_n$$

The dedendum circle diameter of the planet gear is given by,

$$d_{f_1}' = d_{a_1}' - 2h_1'$$

The working transverse pressure angle of the planet gear and the ring gear pair, in terms of the center distance modification coefficient, is given by,

$$\alpha_{wt_2} = \cos^{-1}\left(\frac{\cos \alpha t}{\frac{2y_2 \cos \beta}{z_3 - z_2} + 1}\right)$$

The above equation can be used to check the correctness of the previous calculations for the value of α_{wt_2}.

The addendum circle diameter of the ring gear is given by,

$$d_{a_3}' = 2\left(a' + \frac{d_{f_1}'}{2} + \frac{m_n}{4}\right)$$

Hence, the addendum circle radius of the ring gear is given by,

$$r_{a_3}' = \frac{d_{a_3}'}{2}$$

3.6.6 Design Space

The existing gear design problem can be reduced to an eight-dimensional design problem where the eight free parameters are the number of teeth on the planet gear z_1, the number of teeth on the sun gear z_2, the number of teeth on the ring gear z_3, the module m_n, the three profile shift coefficients x_{n1}, x_{n2}, and x_{n3} of the planet gear, the sun gear, and the ring gear, respectively, and the number of planets n_p. The inequality constraints which bound this design space are the bending stress, Hertz stress, contact ratio, top land thickness, operating interference of gears, and involute and tip interference of internal gears. The gear pair ratios, conditions for proper assembly of planetary gearing, etc., form equality constraints.

3.6.7 Objective Functions and Constraints

For the design optimization of the PTO planetary gear train, 14 objective functions for minimization/maximization with an array of inequality and equality constraints have been considered, and the problem can be stated as follows:

To optimize the objective functions,

$$f_{obj} = \begin{cases} Min.center\ distance\ between\ planetary\ gear\ pair \\ Min.weight\ of\ the\ planetary\ gear\ train \\ Max.efficiency\ of\ planetary\ gear\ train \\ Min.slide\ roll\ ratio \\ Balancing\ the\ specific\ sliding \\ Balancing\ the\ Almen\ product \\ Balancing\ the\ flash\ temperature \\ Balancing\ the\ lambda\ ratio \end{cases}$$

The independent design variables of the planetary gear train include a module, the number of teeth of the sun gear, the planet gear, and the ring gear; the three profile shift coefficients x_{n1}, x_{n2}, and x_{n3} of the planet gear, the sun gear, and the ring gear, respectively; and the number of planets n_p. The vector of design variables is given below:

$$x(i) = \begin{cases} planet\ gear\ no.of teeth, Z_1 \\ sun\ gear\ no.of teeth, Z_2 \\ ring\ gear\ no.of teeth, Z_3 \\ planet\ gear\ profile\ shift\ coefficient, x_{n_1} \\ sun\ gear\ profile\ shift\ coefficient, x_{n_2} \\ ring\ gear\ profile\ shift\ coefficient, x_{n_3} \\ normal\ module, m_n \\ number\ of\ planets, n_p \end{cases}$$

where $i = 1, 2, \ldots,$ the number of variables.

A number of constraints, which act as sub functions to restrict the objective function, are introduced into the objective function to be optimized and are used to provide suitable design choices. The very often encountered failure types in a transmission gear are tooth fracture, surface fatigue, abrasion, and plastic deformation. Thus, the following constraints are considered for the design optimization of the planetary gear train:

$$g(j)=\begin{cases} \textit{Bending stress constraint} \\ \textit{Surface contact compressive stress constraint} \\ \textit{Contact ratio constraint} \\ \textit{Top land thickness constraint} \\ \textit{Operating interference of planet gear constraint} \\ \textit{Operating interference of sun gear constraint} \\ \textit{Operating interference of ring gear constraint} \\ \textit{Involute interference of internal gear constraint} \\ \textit{Tip interference of internal gear constraint} \\ \textit{Proper assembly of planetary gearing constraint} \end{cases}$$

where $j = 1, 2, \ldots,$ the number of constraints.

3.6.8 Establishing a Mathematical Model

The gear design problem presented here has the following requirements and constraints:

1. Requirements: Fixed reduction ratio and design torque.
2. Constraints: Minimum outer diameter, minimum face width, and maximum stress capacity.

Besides these requirements and constraints, there is an array of discrete and continuous variables, which have to be taken care of while developing the optimization model.

To establish a mathematical model for the planetary gear train design optimization, mathematical equations for the design variables, objective functions, and necessary constraints are to be developed.[3]

The various assumptions that were made in the development of the mathematical model formulation of epicyclic gearing are as follows:

1. Contact is always along the line of action during the duration of a mesh cycle.
2. The tooth is assumed to be a non-uniform cantilever beam in both bending and shear analysis.
3. The load is assumed to be uniformly distributed along the tooth face width.

3.6.9 Objective Functions for the Epicyclic Gearing

An optimization procedure uses the objective function of a mathematical program to select better solutions over poorer solutions. For designing a new planetary gearbox or for improving the characteristics of an existing gearbox design, there are many international gear standards available, for reference, to define the macro geometry gear parameters. For the performance evaluation of the epicyclic gearing, the important factors to be considered are strength, durability, reliability, and NVH (noise, vibration, and harshness) characteristics. The strength of the gear system is guaranteed by the initial macro geometry design, which mainly consists of the definition of the number of teeth, module, gear tooth width, pressure angle, helical angle, tooth profile modification coefficient, gear material, etc. The efficiency of the power transmission system is also an important factor to be considered, and it is also improved with the reduction of NVH characteristics. Additionally, the gear system should be compact in size and light in weight. The methodology described below could be expanded to include additional unique design factors and failure modes.

3.6.9.1 Minimization of Center Distance

Since the planet pinion is the smallest of all the three gears, and also since all the gears are made of the same material, the design is based on the planet pinion. The first objective function is for the minimization of center distance between the planet pinion and the ring gear. The size of the gear set is a design parameter which affects cost. The gear mesh center distance, a, is a measure of the size of the gear set. Obviously, a smaller gear set will use less material for the gears and for the surrounding housing. If the same life and reliability can be achieved in a small package, a more economical design is achieved.

The objective function for minimization of center distance is given by,

$$f_1 = \min a = \frac{m_n\left(z_2 + z_1\right)}{2}$$

3.6.9.2 Minimization of Weight

To reduce energy consumption and save material, the weight reduction design (WRD) of the gear drive system is of great significance. Also, lower exhaust emission norms require a light weight design of gear systems. The target of net-zero global greenhouse gas emissions by global nations, by mid-century (by 2050) or soon after, aims at the world's climate goals to eliminate all emissions from heavy-duty trucks, aviation, shipping, and heavy industry and to limit the rise in global temperatures to 1.5°C.

To develop a mathematical model for the second objective function of minimization of the weight of the planetary gear train, the second objective function may be written as,

$$Minimize\ f_2 = \min \text{total weight}$$

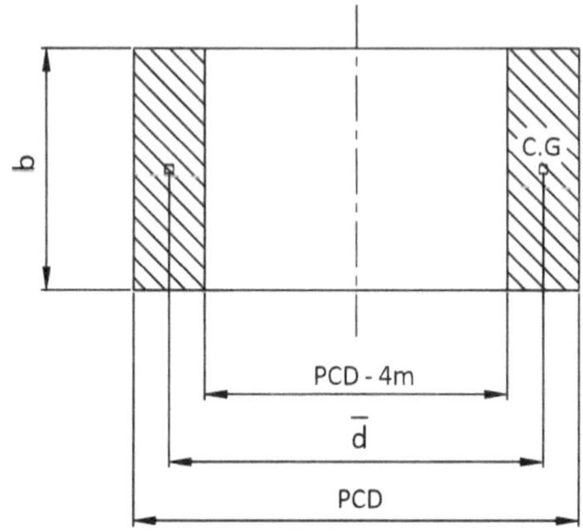

FIGURE 3.6 Gear parameters for weight calculation.

$$f_2 = W = W_1 + W_2 + W_3$$
$$= n_w \rho \pi \bar{d}_1 A_1 + \rho \pi \bar{d}_2 A_2 + \rho \pi \bar{d}_3 A_3$$

where W_1, W_2, and W_3 are the weight of the planet gear, the sun gear, and the ring gear, respectively, in kg; A is the swept area, mm²; $\bar{d}$ is the diameter at C.G. mm; ρ is the material density, kg/mm².

Referring to Figure 3.6, assuming a rim thickness of $1m_n$ for a gear:

For the sun gear and the planet gears,

$$\bar{d} = d_w - 2m_n$$

For the ring gear,

$$\bar{d} = d_w + 2m_n$$

The swept area is given by,

$$A = b \times 2m_n$$

3.6.9.3 Maximization of Efficiency

The efficiency of a planetary speed reducer is an important performance factor, and high-efficiency gearing cuts the cost of gear drives and their operation. Efficient speed reducers can achieve significant savings, both in speed reducer costs and energy costs. However, another reason to use them is to reduce health risks from greenhouse gases in our environment. From the point of view of conservation of energy as well as a reduction in the amount of heat rejected to the geared system, the most efficient gear designs are required. Maximization of efficiency is an important objective in order to optimize the losses in a gearbox and thereby decrease the fuel consumption and exhaust emissions of vehicles. Also, increasing the efficiency of gearboxes has great market potential. The efficiency of the gear drive depends on the tooth loads and velocity of the tooth arrangements. The product of tooth load and engagement velocity gives the measure of equivalent power being transmitted by the mesh.

The overall efficiency of gears is dependent on three separate and distinct types of losses, commonly known as windage and churning losses, bearing losses, and gear mesh losses. The efficiency equations used here are based only on the losses of the gear mesh.

Power losses due to gear meshing are due to the sliding friction losses and rolling friction losses of the two mating tooth flanks of the pinion and the gear wheel against each other along the path of contact. The instantaneous sliding frictional power loss is calculated by multiplying the sliding velocity[4] v_s, the coefficient of friction f, and the tooth normal force F_n. The instantaneous rolling frictional power loss is calculated as the product of the rolling velocity[4] v_r and the rolling frictional force F_R. However, power loss due to rolling is often negligible because the sliding power loss dominates the gear mesh losses.

The efficiency of the gear pair will be increased by reducing the frictional power loss, which consequently reduces the risk of temperature increase between the mating tooth flanks.

The objective function for efficiency increase is given by,

Maximize

$$f_3 = \eta = \left[1 - \left(\mu_{m_1} Hv_1 + \mu_{m_2} Hv_2\right)\right]100$$

where μ_{m_1} is the coefficient of friction between the sun gear and the planet gear pair engagement; Hv_1 is the power loss factor for the sun gear and the planet gear pair engagement; μ_{m_2} is the coefficient of friction between the planet gear and the ring gear pair engagement; Hv_2 is the power loss factor for the planet gear and the ring gear pair engagement.

Here, Hv_1 and Hv_2, the power loss factors, introduced by Ohlendorf (1958), depend only on gear geometry.

The power loss factor Hv_1 for the planet gear and sun gear pair engagement is given by,

$$Hv_1 = \frac{\pi}{z_1 \cos \beta_b} \cdot \frac{1 + i_1}{i_1}\left(1 - \varepsilon_{\alpha_1} + \varepsilon_1^{\,2} + \varepsilon_2^2\right)$$

where ε_{α_1}, the base contact ratio for the planet gear and sun gear pair engagement, is given by,

$$\varepsilon_{\alpha_1} = \varepsilon_1 + \varepsilon_2$$

The power loss factor Hv_2 for the planet gear and ring gear pair engagement is given by,

$$Hv_2 = \frac{\pi}{z_1 \cos \beta_b} \cdot \frac{1+i_2}{i_2}\left(1 - \varepsilon_{\alpha_2} + \varepsilon_1'^2 + \varepsilon_3^2\right)$$

where ε_{α_2}, the base contact ratio for the planet gear and ring gear pair engagement, is given by,

$$\varepsilon_{\alpha_2} = \varepsilon_1' + \varepsilon_3$$

The efficiency of the internal gear is expressed in terms of the coefficient of friction because of power loss due to frictional heat developed. The normal force and the principal torque either assist or oppose gear rotation. The lever arms of the driver are lesser on approach than on the recess.

The base pitch and the base transverse pitch of the planet gear are given by,

$$p_b = \frac{\pi d_{b_1}}{z_1}$$

$$p_{b_t} = \frac{p_b}{\cos \beta}$$

Referring to Figure 3.7, for a clockwise rotation of the input sun gear, the contact path with the mating planet gear is along the common tangent to the respective base circles. The contact takes place along the line MN, and point A denotes the start of gear pair contact, and point B denotes the end of gear pair contact. The

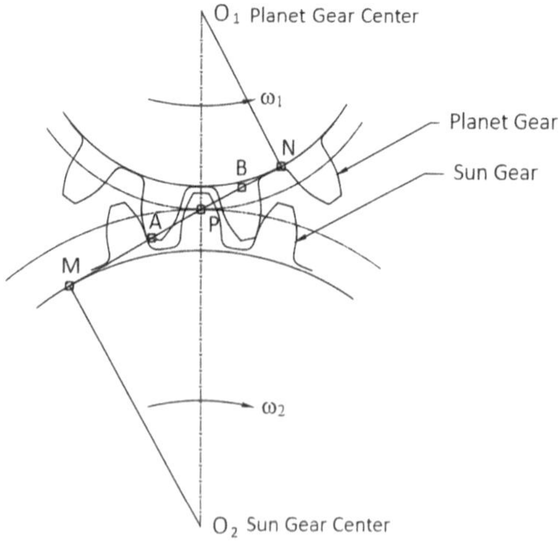

FIGURE 3.7 External gear mesh contact path.

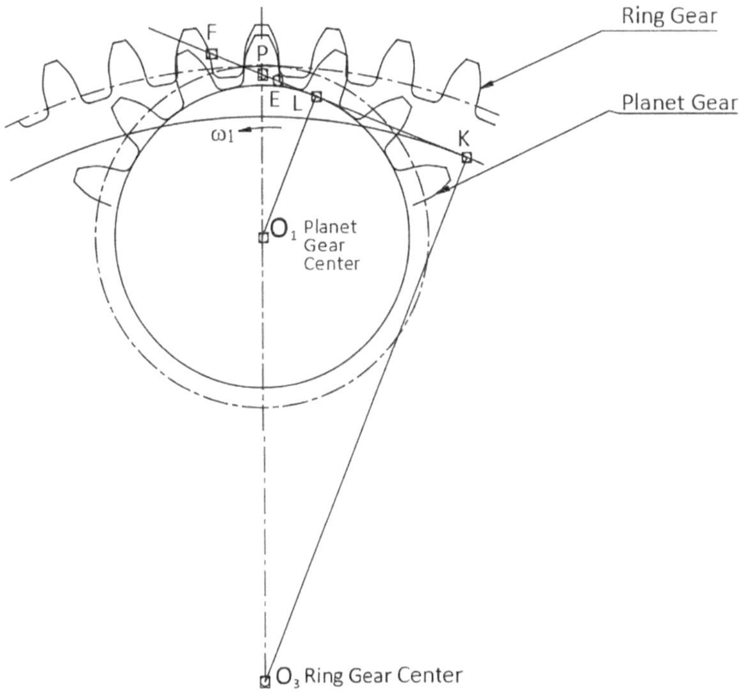

FIGURE 3.8 Internal gear mesh contact path.

pitch point P is where the pitch circles of the mating gear pair meet and also at the line connecting their centers of rotation.

The addendum contact ratio for planet gear ε_1 and sun gear ε_2 in the planet and the sun gear engagement is given by,

$$\varepsilon_1 = \frac{AP}{p_{b_t}}$$

$$\varepsilon_2 = \frac{BP}{p_{b_t}}$$

Referring to Figure 3.8, for the planet gear and the ring gear pair engagement, as the input planet gear rotates anticlockwise, the contact path is along the common tangent to the respective base circles. The contact takes place along the line KF, and point E denotes the start of gear pair contact, and point F denotes the end of gear pair contact. As before, the pitch point P is where the pitch circles of the mating gear pair meet and also at the line connecting their centers of rotation.

Similarly, the addendum contact ratio for planet gear ε_1' and ring gear ε_3 in the planet gear and the ring gear engagement is given by,

$$\varepsilon_1' = \frac{FP}{p_{b_t}}$$

$$\varepsilon_3 = \frac{EP}{p_{b_t}}$$

The base helix angle is given by,

$$\beta_b = \tan^{-1}\left(\frac{d_{b_1}\tan\beta}{\frac{m_n z_1}{\cos\beta}}\right)$$

Also,

$$\beta_b = \sin^{-1}\left(\sin\beta\cos\alpha_n\right)$$

The mean coefficient of friction is assumed to be constant for every contact point of the engagement region, even though it varies slightly along the path of contact. The approximate value of the mean coefficient of friction, proposed by Schlenk (1995), takes into account determining factors such as oil parameters, sum velocity, normal force, line load, the reduced radius of curvature, and surface roughness, and is given by,

$$\mu_m = 0.048\left(\frac{F_{t_b}/b}{\gamma_{\varepsilon c}\rho_{redc}}\right)^{0.2}\eta_{oil}^{-0.05}R_a^{+0.25}X_L$$

where F_{t_b} is the circumferential force at the base circle (N); b is the face width for spur gears (mm); $\gamma_{\varepsilon c}$ is the sum velocity at the operating pitch circle (m/s); ρ_{redc} is the reduced radius of curvature at pitch point (mm); η_{oil} is the dynamic oil viscosity at oil temperature (mPa.s); R_a is the arithmetic mean roughness (μm); X_L is the lubricant correction factor.

The load-dependent losses are influenced by both gear geometry and lubricant properties.

The coefficient of friction for the planet gear and the sun gear pair engagement is given by,

$$\mu_{m_1} = 0.048\left(\frac{F_{t_{b_1}}/b}{\gamma_{\varepsilon c_1}\rho_{redc_1}}\right)^{0.2}\eta_{oil_1}^{-0.05}R_{a_1}^{+0.25}X_{L_1}$$

The tangential force at the base circle for the planet gear and the sun gear engagement is,

$$F_{t_{b_1}} = \frac{1000T_2}{r_{b_2}n_p}$$

where T_2 is the torque for the sun gear and is calculated as,

$$T_2 = \frac{60000P}{2\pi N_2}$$

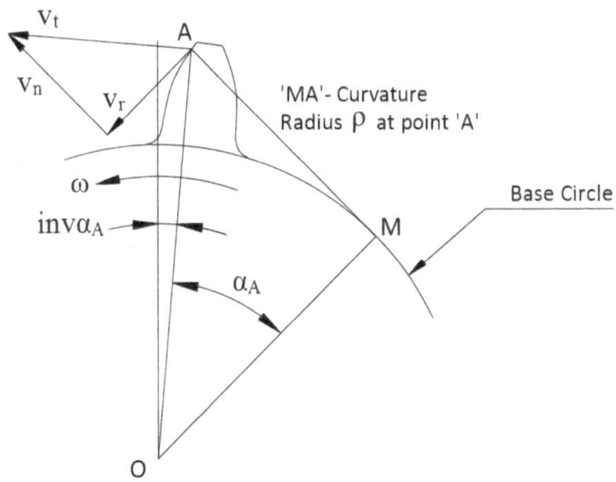

FIGURE 3.9 Direction of velocities at the contact point.

The sum velocity at operating pitch circles for the planet gear and sun gear pair engagement is,

$$\gamma_{\varepsilon c_1} = 2v_{t_2} \sin \alpha_{wt_1}$$

Figure 3.9 shows the direction of velocities at the contact point A, wherein the contact point A is the momentary contact point during the rotation of the mating gears. For the involute profile tooth, MA represents the curvature radius at point A. The angle subtended at the center α_A is the pressure angle at point A, and the remaining angle with the vertical axis of the gear forms the involute function of the angle α_A, that is, $inv\alpha_A$. The tangential velocity v_t is perpendicular to the line OA. The tangential velocity v_t is the resultant of two velocities perpendicular to each other. The velocity v_r is the rolling velocity[1] and v_n is the normal velocity.

Here, the tangential velocity at the pitch circle is calculated as,

$$v_{t_2} = r_{w_2} \omega_2$$

And angular velocity of the sun gear is,

$$\omega_2 = \frac{2\pi N_2}{60}$$

The curvature radius, or transverse profile radius, for the planet gear (in the sun gear, planet gear pair engagement) at pitch point is,

$$\rho_1 = \frac{r_{w_1} \sin \alpha_{wt_1}}{\cos \beta_b}$$

The curvature radius for the planet gear (in the ring gear, planet gear pair engagement) at the pitch point is,

$$\rho_1' = \frac{r_{w1} \sin \alpha_{wt_2}}{\cos \beta_b}$$

The curvature radius for the sun gear (in the sun gear, planet gear pair engagement) at the pitch point is,

$$\rho_2 = \frac{r_{w2} \sin \alpha_{wt_1}}{\cos \beta_b}$$

The curvature radius for the ring gear is (in the sun gear, planet gear pair engagement) at the pitch point is,

$$\rho_3 = \frac{r_{w3} \sin \alpha_{wt_2}}{\cos \beta_b}$$

Then the reduced radius of curvature (in the sun gear, planet gear pair engagement) at the pitch point is,

$$\rho_{redc_1} = \frac{\rho_1 \rho_2}{(\rho_1 + \rho_2)\cos \beta_b}$$

The lubricant correction factor X_{L_1}, which is dependent on the type of lubricant, for the planet gear sun gear pair engagement is given by,

$$X_{L_1} = \frac{1}{\left(\frac{F_{tb_1}}{b}\right)^d}$$

Here, it is assumed that lubricating oil is mineral oil for which the factor d = 0.0651 (Martinsa et al., 2006).

In a similar manner, the coefficient of friction for the planet gear and ring gear pair engagement is given by,

$$\mu_{m_2} = 0.048\left(\frac{F_{tb_2} / b}{\gamma_{\varepsilon c_2} \rho_{redc_2}}\right)^{0.2} \eta_{oil_2}^{-0.05} R_{a_2}^{+0.25} X_{L_2}$$

The tangential force at the base circle for the planet gear and the ring gear pair engagement is,

$$F_{tb_2} = \frac{1000 T_1}{r_{b_1}}$$

Here, the torque T_1 for the planet gear is calculated as,

$$T_1 = \frac{60000P}{2\pi N_1 n_p}$$

The sum velocity at operating pitch circles for the planet gear and ring gear pair engagement is,

$$\gamma_{\varepsilon c_2} = 2v_{t_1} \sin \alpha_{wt_2}$$

Here, the tangential velocity at the pitch circle is calculated as,

$$v_{t_1} = r_{w_1} \omega_1$$

And the angular velocity of the sun gear is,

$$\omega_1 = \frac{2\pi N_1}{60}$$

The reduced radius of curvature (in the ring gear, planet gear pair engagement) is,

$$\rho_{redc_2} = \frac{\rho_1' \rho_3}{\left(\rho_3 - \rho_1'\right) \cos \beta_b}$$

The lubricant correction factor X_{L_2} for the planet gear ring gear pair engagement is given by,

$$X_{L_2} = \frac{1}{\left(\frac{F_{tb_2}}{b}\right)^d}$$

For a typical gear lubricating oil, the following values are taken from standards:

$$\eta_{oil} = 50\,\text{mPa.s}$$
$$R_{a1} = R_{a2} = 0.8\,\text{microns}$$

3.6.9.4 Sliding and Rolling Velocities

The load-dependent losses in a gear pair mesh are due to sliding friction loss and rolling friction loss. Linear velocities of the contact point B (Figure 3.10) are denoted by v_{t_1} and v_{t_2} for the planet gear 1 and the sun gear 2, respectively, and are given by the following equations:

$$v_{t_1} = r_{B_1} \omega_1$$
$$v_{t_2} = r_{B_2} \omega_2$$

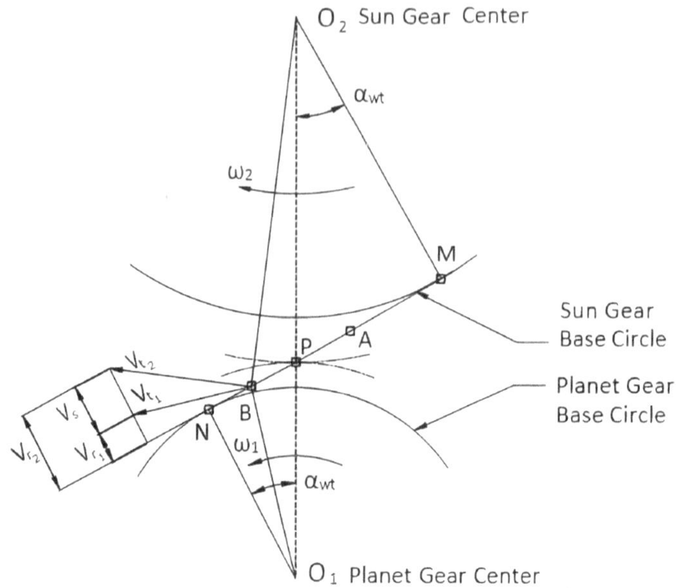

FIGURE 3.10 Sliding and rolling velocities.

Where r_{B_1} and r_{B_2} are the distances of contact point B from the gear centers O_1 and O_2, respectively. The direction of linear velocity is perpendicular to the line joining the gear center and the contact point, in the direction of rotation. The components of velocity normal to the line of action, and therefore tangential to the involute tooth profiles, are called rolling velocities and are denoted by v_{r2} and v_{r2}. Rolling velocity is the velocity of the contact point in the direction of sliding.

The rolling velocities of the gear pair are calculated by the following equations:

$$v_{r_1} = v_{t_1} \sin \alpha_{B_1} = \rho_1 \omega_1$$

$$v_{r_2} = v_{t_2} \sin \alpha_{B_2} = \rho_2 \omega_2$$

where

α_{B_1} is the pressure angle of planet gear at contact point B; α_{B_2} is the pressure angle of sun gear at contact point B.

Since these tangential components of velocity are unequal at all the contact points along the contact path, sliding action occurs everywhere except the pitch point, P. Sliding velocity is the relative tangential velocity of the tooth profiles and is defined as the difference between the rolling velocities of teeth in mesh.

Hence, the sliding velocity is calculated as:

$$v_{s_1} = v_{r_1} - v_{r_2}$$

where v_{s_1} is interpreted as the sliding velocity of the driving gear relative to the driven gear.

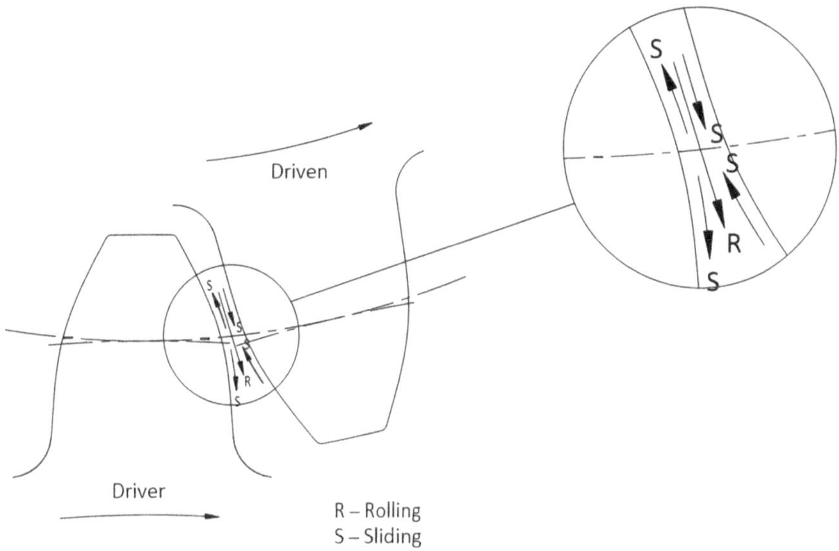

FIGURE 3.11 Direction of sliding and rolling velocities.

Similarly, v_{s_2} is the sliding velocity of the driven gear relative to the driving gear and is given by,

$$v_{s_2} = v_{r_2} - v_{r_1}$$

The above equations show that v_{s_1} and v_{s_2} have equal and opposite relationships, that is, $v_{s_1} = -v_{s_2}$.

Unlike sliding velocity, rolling velocity is beneficial because it entrains lubricant between contacting teeth, increases oil film thickness, and reduces the number and severity of asperity contacts.

Figure 3.11 shows the directions of the rolling and sliding velocities on the driving and driven gear teeth. The contact on the driving gear tooth starts near the root of the tooth, rolls up the tooth, and ends at the tooth tip. The sliding is away from the driving gear pitch line. Similarly, the contact on the driven gear tooth starts at the tooth tip, rolls down the tooth, and ends near the tooth root. The sliding is toward the driven gear pitch line.

As seen in Figure 3.11, the gear teeth addenda have negative sliding; that is, the direction of rolling velocity is opposite sliding velocity. Negative sliding is detrimental because it promotes Hertzian fatigue and allows oil to enter surface cracks and accelerate crack growth by the mechanism of hydraulic pressure propagation.

Cylindrical involute gearing is sometimes described as analogous to a cam-follower mechanism because of the sliding action of the mating tooth surfaces. As mentioned earlier, for two involute profile teeth of mating gears in mesh, at a given point on the line of action, the product of curvature radius and rotational speed, calculated for pinion and gear, respectively, are not equal. Therefore, the resultant rolling velocities are different. The pitch point, a location where the pitch circle intersects the line of action, is the only contact point for meshing tooth flanks where there is pure rolling motion momentarily.

Hence, while meshing, during the interaction of tooth profiles, the involute gear teeth are in a state of a pure roll at the pitch point, and at any other location, there is relative sliding. The tooth surfaces are sliding against each other in the approach and recess regions in the line of action. The direction of sliding reverses at the pitch point, and the amount of sliding increases as the contact point moves away from the pitch point.

During the mesh cycle of an involute gear pair, the involute tooth profiles have a starting point and an ending point in the contact region. The region of involute profile between these start and end points is termed the "active profile" and the start and end points are referred to as the "start of active profile" (SAP) and "end of active profile" (EAP). SAP and EAP are the points of contact furthest from the pitch point, and these are the points where the mating gear teeth are entering and exiting the mesh cycle. The active profile is formed in the plane of contact, which is the active portion of the line of action in a transverse section of a gear pair.

Sliding is of greatest concern at the SAP and EAP, since the sliding velocity increases as the distance of the point of contact from the pitch point increases in the line of action. Gear tooth sliding affects gear mesh efficiency, wear, scuffing, and noise. High sliding velocity is detrimental because it generates heat and increases flash temperature,[5] increases asperity distress, and increases the risk of scuffing. High sliding velocity also induces frictional power loss by shearing the oil film and shearing boundary films on contacting surfaces, thereby reducing the efficiency of transmission. The total friction multiplied by the sliding velocity provides the frictional power loss for each contacting tooth flank pair.

3.6.9.5 Entraining Velocity

The entraining velocity draws oil between contacting gear teeth and is responsible for the formation of any oil film by hydrodynamic action. Entraining velocity v_e is related to the geometry and angular speed of gears and is defined as the sum of the absolute rolling velocities.

The entraining velocity is given by,

$$v_{e_1} = v_{r_2} + v_{r_1}$$

Mean entraining velocity is given by

$$v_{e_{mean}} = \frac{v_{r_2} + v_{r_1}}{2}$$

The relative contact geometry and also sliding, and entraining velocities vary throughout the meshing cycle. In high-speed gear design, sliding velocity should be minimized by using rather small teeth while choosing teeth of sufficient size to resist the bending stresses caused by tooth loading. The required strength is obtained by using a relatively large face width.

3.6.9.6 Slide Roll Ratio

The ratio of the sliding velocity to the entraining velocity is known as the slide roll ratio (SRR) and is related to micro pitting, which can be thought of as a progressive fatigue wear process. Micro pitting differs from macro pitting, both caused by the mechanism of rolling contact fatigue, in that macro pitting occurs on the scale of the nominal Hertzian, or macro, contact, whereas micro pitting is characterized by pits of the order of overall contact size or occurs on the scale of the roughness. Micro pitting is mainly influenced by local contact pressures, sliding speeds, and lubrication conditions. Micro pitting damage causes a significant loss of material from the load-bearing surface of a gear tooth flank and is particularly troublesome in heavily loaded gears with hardened teeth. Increased SRR leads to increased wear (Everitt and Alfredsson, 2020). Increased SRR also leads to increased micro pitting damage by increasing the number of asperity stress cycles, which is proportional to the sliding distance (Rycerz and Kadiric, 2019).

The SRR is given by,

$$SRR = \frac{v_s}{v_{e_1}}$$

Hence, the objective function for SRR is given by,

$$\min f_4 = \min \text{SRR}$$

3.6.9.7 Specific Sliding

Specific sliding ratio (also known as specific sliding coefficient), which is a contact severity parameter, is defined as the ratio of the sliding velocity v_s to the rolling velocity v_r of the mating tooth flanks. Specific sliding is a geometrical parameter that affects scuffing and the gear teeth wear. The recommended value of specific sliding for gears is $|v_s| < 1.0$. However, for good resistance to macro pitting, micro pitting, and scuffing, a value of $|v_s| < 0.5$ is preferred.

Specific sliding is given by,

$$\gamma = \frac{v_s}{v_r}$$

where v_s is the sliding velocity of the mating tooth flanks; v_r is the rolling velocity of the mating tooth flanks.

Specific sliding for the sun gear and planet gear pair engagement is given by,

$$\gamma_1 = \frac{v_{s1}}{v_{r1}}$$

$$\gamma_2 = \frac{v_{s2}}{v_{r2}}$$

where γ_1 is the specific sliding for the planet gear; γ_2 is the specific sliding for the sun gear.

In terms of angular velocity and curvature radius, a specific sliding ratio is given by,

$$\gamma_1 = \frac{\omega_1 \rho_1 - \omega_2 \rho_2}{\omega_1 \rho_1}$$

$$\gamma_2 = \frac{\omega_2 \rho_2 - \omega_1 \rho_1}{\omega_2 \rho_2}$$

The value of the specific sliding ratio reaches its maximum at the SAP and EAP. Equalization of the specific sliding at SAP and EAP ensures achieving uniform wear of the tooth profile of the coupled gear pair. Uneven wear of tooth profiles of mating gears will give rise to several undesired phenomena, including vibrations and noise. Thus, equalization of the specific sliding at SAP and EAP also ensures maximum service life and wear resistance of the gear pair. Also, efficiency increases for balanced specific sliding on pinion and wheel.

For balancing or equalization of specific sliding, the following equations are used, which form the next set of objective functions:

$$\min\ f_5 = \left|\gamma_{1\max} - \gamma_{2\max}\right|$$
$$\min\ f_6 = \left|\gamma_{3\max} - \gamma'_{1\max}\right|$$

where $\gamma_{1\max}$ and $\gamma_{2\max}$ are the specific sliding ratios at SAP and EAP, respectively, for the planet gear and sun gear pair meshing.

To find out the values of $\gamma_{1\max}$ and $\gamma_{2\max}$, the following equations are used:

$$\gamma_{1\max} = \frac{\left(\tan\alpha_{at_2} - \tan\alpha_{wt_1}\right)}{\left(\left(i_1+1\right)\tan\alpha_{wt_1} - \tan\alpha_{at_2}\right)}\left(i_1+1\right)$$

$$\gamma_{2\max} = \frac{\left(\tan\alpha_{at_1} - \tan\alpha_{wt_1}\right)}{\left(\left(\frac{i_1+1}{i_1}\right)\tan\alpha_{wt_1} - \tan\alpha_{at_1}\right)}\left(\frac{i_1+1}{i_1}\right)$$

Here, α_{wt_1} is the working transverse pressure angle between the planet gear and the sun gear pair meshing.

Similarly, $\gamma_{1\max}{}'$ and $\gamma_{3\max}$ are the specific sliding ratios at SAP and EAP, respectively, for the planet gear and ring gear pair meshing.

To find out the values of $\gamma_{1\max}{}'$ and $\gamma_{3\max}$, the following equations are used:

$$\gamma_{1\max}{}' = \frac{\left(\tan\alpha_{at_1} - \tan\alpha_{wt_2}\right)}{\left(\left(1+i_2\right)\tan\alpha_{wt_2} - \tan\alpha_{at_1}\right)}\left(i_2+1\right)$$

$$\gamma_{3\max} = \frac{\left(\tan\alpha_{at_3} - \tan\alpha_{wt_2}\right)}{\left(\left(\frac{i_2+1}{i_2}\right)\tan\alpha_{wt_2} - \tan\alpha_{at_3}\right)}\left(\frac{i_2+1}{i_2}\right)$$

Here, α_{wt_2} is the working transverse pressure angle between the planet gear and the ring gear pair meshing.

Where i_1 and i_2 are the gear ratios between the planet gear and the sun gear pair and the planet gear and the ring gear pair, respectively, and are defined as below:

$$i_1 = \frac{z_1}{z_2}$$

$$i_2 = \frac{z_3}{z_1}$$

The addendum circle pressure angle of the planet gear is given by,

$$\alpha_{at_1} = \cos^{-1}\left(\frac{d_1\cos\alpha_t}{d_{a_1}{}'}\right)$$

The addendum circle pressure angle of the sun gear is given by,

$$\alpha_{at_2} = \cos^{-1}\left(\frac{d_2\cos\alpha_t}{d_{a_2}{}'}\right)$$

The addendum circle pressure angle of the ring gear is given by,

$$\alpha_{at_3} = \cos^{-1}\left(\frac{d_3\cos\alpha_t}{d_{a_3}{}'}\right)$$

The transverse pressure angle α_t is given by,

$$\alpha_t = \tan^{-1}\left(\frac{\tan\alpha_n}{\cos\beta}\right)$$

3.6.9.8 Profile Shifting

The profile shift is the quantity, in terms of the module, that is added to, or subtracted from, the gear teeth addendum to enhance the operational performance of the gear set or to meet fixed design criteria. Hence, it is also termed as an addendum modification. The sum of profile shift coefficients of the pinion and mating gear can be used as a design criterion to achieve performance characteristics. A sum of profile shift coefficients, which equals 0.0, ensures a large contact ratio for reduced noise and better dynamic loading. When $x_1 + x_2 = 0.25$, it provides a well-balanced design that ensures durability, efficiency, and a dynamically balanced design. Similarly, a sum of $x_1 + x_2 = 0.5$ ensures high load capacity and reduced sliding for better efficiency.

Profile shift, or addendum modification, is a technique essentially used to avoid undercut in pinions with less than 17 teeth. Profile shift is also widely used for gear design, which enables one to achieve nonstandard distances between shafts and to improve the loading capacity. Involute tooth profiles of spur and helical gears allow for center distance variation without affecting the nominal transmission ratio or conjugate action. The base circle diameter and reference circle diameter of a profile-shifted gear will not change in comparison to the standard gears. Profile shifting influences the tooth parameters such as tooth thickness, bottom clearance, and addendum and dedendum values. Profile shift, sometimes called tool shift, can be used in order to optimize the design according to the specific requirements of each problem because it influences design parameters like the contact ratio, the size of the gears, the expected life and gears' safety factors. Profile shift is used for many reasons, including:

- Balancing or equalization of specific sliding, which maximizes wear resistance and Hertzian fatigue resistance.
- Balancing or equalizing flash temperature, which maximizes scuffing resistance.
- Balancing or equalization of bending fatigue life of gears.
- Avoiding narrow top-land and pointed tooth to avoid case/core separation in carburized gears.

Out of the total sum of the profile shift needed for the required operating conditions of a gear pair, individual profile shift coefficients for the pinion and the gear should be chosen after careful consideration of several factors. There is a lower limit for the profile shift, below which value causes interference of mating tooth profiles leading to undercutting. On the other hand, the upper limit of the profile shift is controlled by the pointing of the tooth tip, which occurs when the tooth thickness at the tip circle, or in other words, the thickness of the top land, becomes equal to zero. Thinning of the top land of the gear pair is not permissible below a certain minimum value, which is about 0.2–0.3 times the module, as per the required operating conditions of a gear pair. Also, the choice of individual profile shift coefficients should not make the value of the transverse contact ratio fall below a predetermined minimum value, thus ensuring regularity of the transmission and reduction in noise and vibrations. Also, the individual profile shift coefficients influence the tooth shape and tooth thickness, which in turn reflect on the tooth root strength.

3.6.9.9 Almen Product

The Almen product, also called Almen multiplication or Almen factor, is a factor obtained by the multiplication of the contact stress and the sliding velocity at any contact point of the gear teeth profile. Minimizing its value is advantageous from the point of view of wear (Almen, J.O., 1950). Also, reducing the contact stress

(Hertzian stress) to a low level helps in reducing the risk of pitting. Equalization, or balancing the Alman product at SAP, and EAP balances specific sliding and flash temperature values at the same time (Botka, I., 1953).

For balancing or equalization of the Almen product, the following equations are used, which form the next set of objective functions:

Minimize $f_7 = \left|\sigma_{H_{a1}} v_{s1a} - \sigma_{H_{r1}} v_{s1r}\right|$ for the planet gear and the sun gear pair meshing.
Minimize $f_8 = \left|\sigma_{H_{a2}} v_{s2a} - \sigma_{H_{r2}} v_{s2r}\right|$ for the planet gear and the ring gear pair meshing.

For the planet gear and sun gear pair meshing, the contact stress at SAP is given by,

$$\sigma_{H_{a1}} = Z_{E_1}\left[F_{n_a}'\left(\frac{1}{\rho_{1_a}}+\frac{1}{\rho_{2_a}}\right)\right]^{0.5}$$

where Z_{E_1} is the elasticity factor, MPa; F_{n_a}' is the normal load per unit length at SAP, N/mm; ρ_{1_a} is the radius of curvature of the planet gear at SAP, mm; ρ_{2_a} is the radius of curvature of the sun gear at SAP, mm.

Similarly, for the planet gear and sun gear pair meshing, the contact stress at EAP is given by,

$$\sigma_{H_{r1}} = Z_{E_1}\left[F_{n_r}'\left(\frac{1}{\rho_{1_r}}+\frac{1}{\rho_{2_r}}\right)\right]^{0.5}$$

where Z_{E_1} is the elasticity factor, MPa; F_{n_r}' is the normal load per unit length at EAP, N/mm; ρ_{1_r} is the radius of curvature of the planet gear at EAP, mm; ρ_{2_r} is the radius of curvature of the sun gear at EAP, mm.

The elasticity factor, or reduced modulus of elasticity, Z_E, depends on Young's moduli E_1 and E_2 and Poisson's ratios ν_1 and ν_2 of the planet gear and sun gear material, respectively, according to the equation:

$$Z_{E_1} = \left[\frac{1}{\pi\left\{\dfrac{1-\nu_1^{2}}{E_1}+\dfrac{1-\nu_2^{2}}{E_2}\right\}}\right]^{0.5}$$

where E_1, E_2 are Young's moduli of the planet gear and sun gear material, MPa; ν_1, ν_2 are Poisson's ratios of the planet gear and sun gear material.

Referring to Figure 3.7, the normal load per unit length at SAP is,

$$F_{n_a}' = \frac{10^3 T_2}{\left(r_{b_2}{}^2 + AM^2\right)^{0.5}\cos\alpha_t b n_p}$$

The normal load per unit length at EAP is,

$$F_{n_r}' = \frac{10^3 T_2}{r_{a_2}' \cos\alpha_t b n_p}$$

For the planet gear and the sun gear engagement, the radius of curvature of the sun gear at SAP is,

$$\rho_{2_a} = \frac{MP - AP}{\cos\beta_b}$$

The radius of curvature of the sun gear at EAP is,

$$\rho_{2_r} = \frac{MP + PB}{\cos\beta_b}$$

The radius of curvature of the planet gear at SAP is,

$$\rho_{1_a} = \frac{NP + AP}{\cos\beta_b}$$

The radius of curvature of the planet gear at EAP is,

$$\rho_{1_r} = \frac{NP - PB}{\cos\beta_b}$$

Here,

$$MP = r_{w_2} \sin\alpha_{wt_1}$$

$$NP = r_{w_1} \sin\alpha_{wt_1}$$

$$AP = \left(r_{a_1}'^2 - r_{b_1}^2\right)^{0.5} - r_{w_1} \sin\alpha_{wt_1}$$

$$PB = \left(r_{a_2}'^2 - r_{b_2}^2\right)^{0.5} - r_{w_2} \sin\alpha_{wt_1}$$

$$AM = MP - AP$$

Similarly, for the planet gear and the ring gear engagement, the contact stress at SAP is given by,

$$\sigma_{H_{a2}} = Z_{E_2}\left[F_{n_a}''\left(\frac{1}{\rho_{1_a}'} - \frac{1}{\rho_{3_a}}\right)\right]^{0.5}$$

where Z_{E_2} is the elasticity factor, MPa; F_{n_a}'' is the normal load per unit length at SAP, N/mm; ρ_{1_a}' is the radius of curvature of the planet gear at SAP, mm; ρ_{3_a} is the radius of curvature of the ring gear at SAP, mm.

Here, the elasticity factor is,

$$Z_{E_2} = \left[\frac{1}{\pi \left\{ \frac{1-\nu_1^2}{E_1} + \frac{1-\nu_3^2}{E_3} \right\}} \right]^{0.5}$$

where E_3 is the Young's modulus of the ring gear material, MPa; ν_3 is Poisson's ratio of the ring gear material.

Referring to Figure 3.8, the normal load per unit length at SAP is,

$$F_{n_a}'' \frac{10^3 T_1}{\left(r_{b_1}^2 + LE^2\right)^{0.5} \cos\alpha_t b}$$

The contact stress at EAP is given by,

$$\sigma_{H_{r2}} = Z_{E_2} \left[F_{n_r}'' \left(\frac{1}{\rho_{1_r}'} - \frac{1}{\rho_{3_r}} \right) \right]^{0.5}$$

where F_{n_r}' is the normal load per unit length at EAP, N/mm.

The normal load per unit length at EAP is,

$$F_{n_r}'' = \frac{10^3 T_1}{r_{a_1}' \cos\alpha_t b}$$

The radius of curvature of the planet gear at SAP is,

$$\rho_{1_a}' = \frac{LE}{\cos\beta_b}$$

The radius of curvature of the planet gear at EAP is,

$$\rho_{1_r}' = \frac{LF}{\cos\beta_b}$$

The radius of curvature of the ring gear at SAP is,

$$\rho_{3_a} = \frac{KE}{\cos\beta_b}$$

The radius of curvature of the ring gear at EAP is,

$$\rho_{3_r} = \frac{KF}{\cos\beta_b}$$

Here,

$$LF = \left(r_{a_1}'^2 - r_{b_1}^2\right)^{0.5}$$

$$KE = \left(r_{a_3}'^2 - r_{b_3}^2\right)^{0.5}$$

$$EF = LF - LE$$

$$EP = KP - KE$$

$$PF = LF - LP$$

$$KP = \left(r_{w_3}^2 - r_{b_3}^2\right)^{0.5}$$

$$LP = \left(r_{w_1}^2 - r_{b_1}^2\right)^{0.5}$$

$$LE = KE - KL$$

$$KL = a' \sin \alpha_{wt_2}$$

$$KF = KP + PF$$

The sliding velocity for the planet gear and the sun gear pair meshing at SAP is,

$$v_{s_{1a}} = AP\left(\omega_1 + \omega_2\right)$$

The sliding velocity for the planet gear and the sun gear pair meshing at EAP is,

$$v_{s_{1r}} = PB\left(\omega_1 + \omega_2\right)$$

The sliding velocity for the planet gear and the ring gear pair meshing at SAP is,

$$v_{s_{2a}} = EP\left(\omega_2 - \omega_3\right)$$

The sliding velocity for the planet gear and the sun gear pair meshing at EAP is,

$$v_{s_{2r}} = PF\left(\omega_2 - \omega_3\right)$$

The angular velocity of the planet gear and the sun gear are given by,

$$\omega_1 = \frac{2\pi N_1}{60}$$

$$\omega_2 = \frac{2\pi N_2}{60}$$

Since the ring gear is stationary,

$$\omega_3 = 0$$

For the present case, the following material properties are taken from the standards:

Poisson's ratio and Young's modulus for the gear pair,

$$\nu_1 = \nu_3 = 0.3$$
$$E_1 = E_3 = 210GPa$$

3.6.9.10 Flash Temperature

During pinion and gearwheel meshing, the lubricant oil film may not adequately separate the contacting surfaces, when highly loaded conditions and high sliding velocities are involved. Hence, instantaneous high temperature is generated at contact points of the tooth surface. This instantaneous high temperature may cause localized damage to the surface of the gear tooth flanks instantaneously, which is called "scuffing". Scuffing is an adhesive action where the transfer of metal from one contact surface to the other occurs due to welding and subsequent tearing, resulting in severe adhesive wear. At the contact points, the surfaces weld together briefly because of this momentary high temperature and are torn apart again as the gears continue to revolve, which leads to partial destruction of the surfaces. Scuffing occurs locally when contact zone temperatures rise sharply ("flash temperatures"), and it depends on the load, peripheral speed, and oil sump temperature.

Scuffing is not a fatigue phenomenon and can occur instantaneously even during the early stages of the operation of the gearbox. The risk of scuffing damage is dependent on the material of the gear, the lubricant being used, the viscosity of the lubricant, the surface roughness of the tooth flanks, the sliding velocity of the mating gear teeth under load, and the geometry of the gear teeth. Any changes in any of these above factors can alter the scuffing risk.

In 1937, Professor Harmon Blok published his theory (Blok, 1937), about the relationship between contact temperature and scuffing, and his theory is based on what is termed "flash temperature". He stated that, due to friction at the tooth contact point, scuffing will occur when the local temperature at the point of contact rapidly rises (flashes). This highest momentary local maximum temperature instantly raises the lubricant temperature above the lubricant's ability to maintain its viscosity. This local temperature occurring in the contact area of the gear teeth ("flash temperature") limits permissible sliding velocity for a gear pair. To reduce the risk of scuffing in high-speed gears, the flash temperature has to be reduced. The rated maximum power transmission capability of a gear pair, against scuffing, is based on the maximum flash temperature.

This so-called "flash temperature" is different from the actual flash point temperature of a lubricant oil. The flash point temperature of a lubricant oil is the lowest temperature to which a lubricant must be heated before its vapor, in the air near its surface, "flash", that is, briefly ignites but does not continue to burn.

The flash temperature for a gear pair in engagement can be calculated for any local point on the involute flank along the path of contact, using the following equation (AGMA 925–A03):

$$\theta_{fl} = 31.62 K \mu_m \frac{X_r F_n}{\left(b_H\right)^{0.5}} \frac{\left|v_{r_1} - v_{r_2}\right|}{B_{m1}\left(v_{r_1}\right)^{0.5} + B_{m2}\left(v_{r_2}\right)^{0.5}}$$

where θ_{fl} is the flash temperature, °C; K is 0.80, a numerical factor valid for a semi-elliptic (Hertzian) distribution of frictional heat over the instantaneous width, 2 b_H, of the rectangular contact band; μ_m is the coefficient of friction; X_r is the load-sharing factor; F_n is the normal unit load, N/mm; b_H is the semi-width of the Hertzian contact band, mm; v_{r_1} is the rolling tangential velocity of the pinion, m/s;v_{r_2} is the rolling tangential velocity of the wheel, m/s; B_{ml} is the thermal contact coefficient of the pinion, N/(mm $s^{0.5}$K); B_{m2} is the thermal contact coefficient of the wheel, N/(mm $s^{0.5}$K).

The above equation is the modified form of Professor Harmon Blok's equation, by considering the influence of surface roughness and the loading actually applied on the gear tooth flank. A widely favored evaluation method for a new gearbox installation is to take the maximum flash temperature as the critical factor for the assessment of scuffing risk (Blok, 1963).

The semi-width of the Hertzian contact band is given by,

$$b_H = \left[\frac{4 F_n \rho_{redc}}{\pi E'}\right]^{0.5}$$

where F_n is the normal force at the contact point per unit width, N/mm; ρ_{redc} is the reduced radius of curvature at the contact point, mm; E' is the equivalent elastic modulus of gear pair materials, N/mm^2.

For balancing or equalization of flash temperature, the following equations are used, which form the next set of objective functions:

Minimize

$f_9 = \theta_{fla_1}$ for the planet gear and the sun gear pair meshing at SAP.
$f_{10} = \theta_{flr_1}$ for the planet gear and the sun gear pair meshing at EAP.
$f_{11} = \theta_{fla_2}$ for the planet gear and the ring gear pair meshing at SAP.
$f_{12} = \theta_{flr_2}$ for the planet gear and the ring gear pair meshing at EAP.

The flash temperature for the planet gear and the sun gear pair engagement at SAP is,

$$\theta_{fla_1} = 31.62 K \mu_{m_1} \frac{X_{r_1} F_{n_a}'}{\left(b_{H_{a1}}\right)^{0.5}} \frac{\left|v_{r_1 a}' - v_{r_2 a}'\right|}{B_{m1}\left(v_{r_1 a}'\right)^{0.5} + B_{m2}\left(v_{r_2 a}'\right)^{0.5}}$$

Here, the rolling tangential velocity of the planet gear at SAP is,

$$v_{r1a}' = \rho_{1a}\omega_1$$

The rolling tangential velocity of the sun gear at SAP is,

$$v_{r2a}' = \rho_{2a}\omega_2$$

The minimum Hertzian semi-contact width in one mesh cycle at SAP,

$$b_{Ha1} = \left[\frac{4F_{na}'\rho_{redc1a}}{\pi E_1'}\right]^{0.5}$$

The reduced radius of curvature at SAP (in the sun gear, planet gear pair) is,

$$\rho_{redc1a} = \frac{\rho_{1a}\rho_{2a}}{\left(\rho_{1a} + \rho_{2a}\right)\cos\beta_b}$$

Here, E_1' is the equivalent or effective modulus of elasticity, which depends on the modulus of elasticity and Poisson ratio of the sun gear material and planet gear material.

$$E_1' = \frac{1}{2}\left[\frac{1-v_1^2}{E_1} + \frac{1-v_2^2}{E_2}\right]$$

Here, E_1 and E_2 are in N/mm^2.

The thermal contact coefficient B_{m1} for the planet gear is given by the following equation:

$$B_{m1} = \frac{\left(\lambda_m\rho_m c_m\right)^{0.5}}{10^3}$$

where λ_m is the heat conductivity of the planet gear material, W/(m.K); ρ_m is the planet gear material density, kg/m^3; c_m is the specific heat of planet gear material per unit of mass, J/(kg.K).

For the present case, the following material properties are taken from the standards:

$\lambda_m = 40$ W/(m.K)
$\rho_m = 7850$ kg/m^3
$c_m = 465$ J/(kg.K)

The thermal contact coefficient B_{m2} for the sun gear is given by the following equation:

$$B_{m2} = \frac{\left(\lambda_m\rho_m c_m\right)^{0.5}}{10^3}$$

where λ_m is the heat conductivity of the sun gear material, W/(m.K); ρ_m is the sun gear material density, kg/m^3; c_m is the specific heat of sun gear material per unit of mass, J/(kg.K).

The flash temperature for the planet gear and the sun gear pair engagement at EAP is,

$$\theta_{flr_1} = 31.62K\mu_{m_1}\frac{X_{r_1}F_{n_r}'}{\left(b_{H_{r1}}\right)^{0.5}}\frac{\left|v_{r_1r}' - v_{r_2r}'\right|}{B_{m1}\left(v_{r_1r}'\right)^{0.5} + B_{m2}\left(v_{r_2r}'\right)^{0.5}}$$

The minimum Hertzian semi-contact width in one mesh cycle at EAP,

$$b_{H_{r1}} = \left[\frac{4F_{n_r}'\rho_{redc_{1r}}}{\pi E_1'}\right]^{0.5}$$

The reduced radius of curvature at EAP (in the sun gear, planet gear pair) is,

$$\rho_{redc_{1r}} = \frac{\rho_{1_r}\rho_{2_r}}{\left(\rho_{1_r} + \rho_{2_r}\right)\cos\beta_b}$$

The flash temperature for the planet gear and the ring gear pair engagement at SAP is,

$$\theta_{fla_2} = 31.62K\mu_{m_2}\frac{X_{r_2}F_{n_a}''}{\left(b_{H_{a2}}\right)^{0.5}}\frac{\left|v_{r_1a}'' - v_{r_3a}''\right|}{B_{m1}\left(v_{r_1a}''\right)^{0.5} + B_{m2}\left(v_{r_3a}''\right)^{0.5}}$$

The minimum Hertzian semi-contact width in one mesh cycle at SAP,

$$b_{H_{a2}} = \left[\frac{4F_{n_a}''\rho_{redc_{2a}}}{\pi E_2'}\right]^{0.5}$$

The flash temperature for the planet gear and the ring gear pair engagement at EAP is,

$$\theta_{flr_2} = 31.62K\mu_{m_2}\frac{X_{r_2}F_{n_r}''}{\left(b_{H_{r2}}\right)^{0.5}}\frac{\left|v_{r_1r}'' - v_{r_3r}''\right|}{B_{m1}\left(v_{r_1r}''\right)^{0.5} + B_{m2}\left(v_{r_3r}''\right)^{0.5}}$$

The minimum Hertzian semi-contact width in one mesh cycle at EAP,

$$b_{H_{r2}} = \left[\frac{4F_{n_r}''\rho_{redc_{2r}}}{\pi E_2'}\right]^{0.5}$$

Here, E_2' is the equivalent or effective modulus of elasticity, which depends on the modulus of elasticity and Poisson ratio of the sun gear material and planet gear material.

$$E_2' = \frac{1}{2}\left[\frac{1-\nu_1^2}{E_1} + \frac{1-\nu_3^2}{E_3}\right]$$

Here, E_1 and E_3 are in N/mm^2.

The reduced radius of curvature at SAP (in the planet gear, ring gear pair meshing) is,

$$\rho_{redc_{2a}} = \frac{\rho_{1_a}{}' \rho_{3_a}}{\left(\rho_{3_a} - \rho_{1_a}{}'\right)\cos\beta_b}$$

The reduced radius of curvature at EAP (in the planet gear, ring gear pair meshing) is,

$$\rho_{redc_{2r}} = \frac{\rho_{1_r}{}' \rho_{3_r}}{\left(\rho_{3_r} - \rho_{1_r}{}'\right)\cos\beta_b}$$

Rolling tangential velocity of the planet gear at SAP (in the planet gear, ring gear pair meshing) is,

$$v_{r_1a}'' = \rho_{1_a}{}' \omega_2$$

The rolling tangential velocity of the ring gear at SAP (in the planet gear, ring gear pair meshing) is,

$$v_{r_3a}'' = 0$$

The rolling tangential velocity of the planet gear at EAP (in the planet gear, ring gear pair meshing) is,

$$v_{r_1r}'' = \rho_{1_r}{}' \omega_2$$

The rolling tangential velocity of the ring gear at EAP (in the planet gear, ring gear pair meshing) is,

$$v_{r_3r}'' = 0$$

3.6.9.11 Minimum Oil Film Thickness

Lubricant oil film thickness refers to the distance between two contacting tooth flanks of mating gear teeth that are separated by a layer of lubricant. It is a critical parameter in gear lubrication which is determined by a few factors such as lubricant viscosity, operating conditions (temperature of contacting surfaces, speed, and load), gear geometry, and surface roughness. Optimal film thickness is also equally important for gear performance because too thin a film can lead to metal-to-metal contact and increased friction and wear, while an excessively thick film can result in energy losses and reduced efficiency. The lubricant film thickness in gears directly impacts gear performance, efficiency, and longevity.

The lubricant film acts as a protective barrier between mating tooth flanks, ensuring smooth gear operation and preventing surface damage. An appropriate lubricant oil film thickness helps dissipate heat generated during high-load and high-speed operations, facilitates load distribution across gear teeth, and optimizes power transmission efficiency by reducing energy losses due to churning and viscous drag.

The lubrication environment and mating tooth surface topography strongly influence the friction, wear, scoring and scuffing, micro pitting, and surface fatigue of gears. In order to ensure the proper sliding conditions and to ensure good efficiency, it is essential to maintain the required film thickness of the lubricant between the mating gear contact surfaces. The conditions of elasto-hydrodynamic lubrication[6] (EHL) also must be favorable to maintain proper film thickness of the lubricant. Wear on mating gear tooth flanks occurs as a result of high sliding speeds, high contact pressure, and low oil film thicknesses.

The formation of a lubricant film between the mating profile teeth is a complex phenomenon which depends on many factors, including gear pair geometry, gear pair materials, the surface quality of working flanks, loads applied on the teeth, kinematics of gear pair (including rolling and sliding speeds), clearance and alignment between the gear pair at the stage of erection and installation, running-in lubrication procedure, the lubricants applied at the stage of erection and installation, the type and conditions of operational lubricant used like grease and oil, lubricant's viscosity, piezo-viscosity, temperatures, and lubricant application method.

Normally, the lubricant film will be thicker if the gear tooth profile curvature radius is large, the applied loads are light, the rolling speeds are high, there is a greater quantity of appropriate lubricant in the contact area, the lubricant viscosity is high, the temperatures at the contact zone are low, and the surface quality of the mating tooth profile is good.

3.6.9.12 Lubricant Oil Film Thickness

The calculation of the lubricant film's thickness between two mating teeth depends on the hydrodynamics of viscous fluids in thin films, lubricant behavior under high pressure, and contact surface distortion. The generation of the lubricant film between contacting teeth surfaces is a function of the elastic properties of the mating gears, the geometry and accuracy of contacting teeth, the gears kinematics, the mode of dynamic load, the lubricant behavior, and also on the time during which teeth contact takes place. The lubricant squeeze action plays a major role in keeping a lubricant film between teeth.

Taking in to account the unsteady speed and load conditions, the minimum oil film thickness for a gear pair engagement is given by,

$$\frac{h_{\min}}{R_i} = 1.714\left(\frac{F_n}{bE''R_i}\right)^{-0.128}\left(\frac{\eta_o\left(v_1 \pm v_2\right)\cos\alpha_{wt}}{2\times 10^3 E''R_i}\right)^{0.694}\left(\alpha' E''\right)^{0.568}$$

where $h_{\min}$ is the minimum oil film thickness, m; F_n is the normal load at pitch point, N; b is the face width, m; E'' is the equivalent modulus of elasticity, N/m^2;R_i is the radius of curvature of the interface between the mating gears, m; η_o is the viscosity at working temperature, mPa.s; v_1, v_2 is the tangential velocity of mating gears at the pitch circle, m/s; α' is the lubricant pressure viscosity coefficient, m^2/N.

Here, the equivalent modulus of elasticity is given by,

$$E'' = \left[\frac{2}{\frac{1-v_1^2}{E_1} + \frac{1-v_2^2}{E_2}}\right]$$

Here, E_1 and E_2 are the modulus of elasticity of mating gears in N/m^2.

R_i is given by,

$$R_i = \frac{r_{w1} r_{w2} \sin\alpha_{wt_1}}{10^3\left(r_{w_1} \pm r_{w_2}\right)}$$

Here, the "+" sign holds good for external gear pair meshing, and the "–" sign holds good for internal gear pair meshing.

The minimum oil film thickness equation is modified in the following form:

$$h_{\min} = 1.714 R_i \left(A\right)^{-0.128}\left(B\right)^{0.694}\left(C\right)^{0.568}$$

where

$$A = \frac{10^3 F_n}{bE''R_i}$$

where F_n is the normal load at pitch point, N; b is the face width, mm; E'' is the equivalent modulus of elasticity, N/m^2; R_i is the radius of curvature of the interface between the mating gears, m.

$$B = \frac{\eta_o\left(v_1 \pm v_2\right)\cos\alpha_{wt_1}}{2\times 10^3 E''R_i}$$

where η_o is the viscosity at working temperature; v_1, v_2 is the tangential velocity of mating gears at the pitch circle, m/s; E'' is the equivalent modulus of elasticity, N/m^2; R_i is the radius of curvature of the interface between the mating gears, *m*.

Here, "+" sign holds good for external gear pair meshing, and "–" sign holds good for internal gear pair meshing.

$$C = \alpha' E''$$

where α' is the lubricant pressure viscosity coefficient, m^2/N.

The minimum oil film thickness for the planet gear and the sun gear pair engagement is given by,

$$h_{\min_1} = 1.714 R_{i_1} \left(A'\right)^{-0.128} \left(B'\right)^{0.694} \left(C'\right)^{0.568}$$

where

$$A' = \frac{10^3 F_{n_1}}{b E_1'' R_{i_1}}$$

Here, F_{n_1} is the normal load at pitch point, *N*; *B* is the face width, mm.

$$F_{n_1} = \left(\frac{10^3 T_2}{r_{w_2} n_p} \right)$$

The equivalent modulus of elasticity in this case is given by,

$$E_1'' = \left[\frac{2}{\frac{1-\nu_1^2}{E_1} + \frac{1-\nu_2^2}{E_2}} \right]$$

Here, E_1 and E_2 are in N/m^2

R_{i_1} is the radius of curvature of the interface between the planet gear and the sun gear and is given by,

$$R_{i_1} = \frac{r_{w_1} r_{w_2} \sin \alpha_{wt_1}}{10^3 \left(r_{w_1} + r_{w_2}\right)}$$

$$B' = \frac{\eta_o \left(\nu_1 + \nu_2\right) \cos \alpha_{wt_1}}{2 \times 10^3 E_1'' R_i}$$

where η_o is the viscosity at working temperature, mPa.s.

For the planet gear,

$$v_1 = \frac{r_{w1}\omega_1}{10^3} = \frac{v_{t1}}{10^3}$$

For the sun gear,

$$v_2 = \frac{r_{w2}\omega_2}{10^3} = \frac{v_{t2}}{10^3}$$

Here, v_1 and v_2 are in m/s.

$$C' = \alpha_1' E_1''$$

For the present case, the following values are taken from the standards:

$$\eta_o = 90\text{mPa.s}$$
$$\alpha_1' = 2\times10^{-8}\text{m}^2/\text{N}$$

Similarly, the minimum oil film thickness for the planet gear and ring gear pair engagement is given by,

$$h_{\min_2} = 1.714 R_{i_2}\left(A''\right)^{-0.128}\left(B''\right)^{0.694}\left(C''\right)^{0.568}$$

where

$$A'' = \frac{10^3 F_{n_2}}{b E_2'' R_{i_2}}$$

$$F_{n_2} = \left(\frac{10^3 T_1}{r_{w_1}}\right)$$

The equivalent modulus of elasticity in this case is given by,

$$E_2'' = \left[\frac{2}{\frac{1-v_1^2}{E_1}+\frac{1-v_3^2}{E_3}}\right]$$

$$R_{i_2} = \frac{r_{w1} r_{w3}\sin\alpha_{wt_2}}{10^3\left(r_{w_3}-r_{w_1}\right)}$$

$$B'' = \frac{\eta_o\left(v_1 - v_3\right)\cos\alpha_{wt_2}}{2\times10^3 E_2'' R_i}$$

Since the ring gear is stationary,

$$v_3 = 0$$
$$C'' = \alpha_2^{'} E_2^{''}$$

Since both the sun gear–planet gear pair and the planet gear–ring gear pair operate in the same lubricant oil environment, the lubricant pressure viscosity coefficient will be the same for both pairs.

$$\alpha_1^{'} = \alpha_2^{'}$$

3.6.9.13 Lambda Ratio

The ratio of the minimum lubricant film thickness and the surface roughness of the tooth flanks is known as relative lubricant film thickness λ. This ratio is popularly known as the lambda ratio. It is a conventional measure of the severity of lubrication between two mating tooth surfaces. Improvement in the lambda ratio value ensures better performance of the gear pair in engagement, and its value must be maximized.

Hence, the following equations form the next set of objectives:

To maximize lambda ratio, that is,

$$Maximize\ f_{13} = \lambda_1 = \frac{h_{\min_1}}{R_{a1}^{'}}$$
$$Maximize\ f_{14} = \lambda_2 = \frac{h_{\min_2}}{R_{a2}^{'}}$$

where

$$R_{a1}^{'} = R_{a1} + R_{a2}$$
$$R_{a2}^{'} = R_{a1} + R_{a3}$$

where R_{a1}, R_{a2}, and R_{a3} are the arithmetic mean roughness of the planet, sun, and ring gears, respectively (μm).

For the present case,

$$R_{a1}^{'} = R_{a2}^{'} = 0.1$$

3.6.10 Lubrication Regimes

For an efficient operation, a gear set requires appropriate lubrication at the Hertzian contact zone of two mating tooth flank surfaces. Proper lubrication helps to increase the longevity and reliability of gear drives. There are four lubrication regimes for gears depending on the speed of operation and the type of loading – boundary lubrication, mixed lubrication, EHL and hydrodynamic lubrication.

3.6.10.1 Boundary Lubrication Regime

Boundary lubrication regimes occur when the asperities of two lubricated gear tooth surfaces in relative motion may come into physical contact and the potential for abrasion and/or adhesion occurs. In boundary lubrication, surface interactions are very significant, and there are many cases of metal-to-metal contact between gear teeth sliding surfaces, and the surfaces are only separated by an oil film formed by the chemical reaction of additives contained in lubricants.

During initial start-up or shutdown of the gear set, or under heavily loaded conditions, this boundary lubrication situation is inevitable, and the metal surfaces in a lubricated gear system may actually come into severe contact with each other. However, boundary lubrication can be very dangerous when the gear set is operating because of the risk of quick and permanent deterioration of the contact surfaces, and as much as 70% of wear occurs during the start-up and shutdown phases of gear tooth contact. In a boundary lubrication regime, the asperities of the lubricated surfaces of the gear tooth flanks may come into physical contact, leading to potential abrasion and/or adhesion, because friction may be at its highest level. The boundary lubrication regime occurs at start-up, shutdown, low entrainment speed, or high load conditions, where friction may be at its highest level, and it should be avoided where possible.

A main method of reducing the boundary lubrication regime is to provide the correct lubricant viscosity. Another method of reducing this phenomenon of the boundary lubrication regime is with the use of a fully formulated lubricant that includes anti-wear or extreme pressure additives. These additives react with the gear tooth surface asperities that have come into contact and instantly form an altered ductile (pliable) film on the metal surface. This new film then acts sacrificially as the contacting tooth surfaces slide or roll over each other, and the chemical film formed by the additive is worn away instead of the gear tooth metal surface.

3.6.10.2 Mixed Lubrication Regime

A mixed lubrication regime is the gray area or an intermediary condition between boundary and hydrodynamic/EHL regimes, where in some metal-to-metal asperity loading at the gear profile contact zone is still occurring combined with loading on the lubricant. In this regime, both the hydrodynamics of the lubricant oil flow and the surface roughness of contacting gear tooth flanks actively play a role. As sliding speed increases from the initial start-up condition, a wedge of lubricant film is created between the gear tooth surfaces in motion, thereby dramatically reducing the boundary lubrication. The coefficient of friction drops dramatically as the potential for asperity contact is reduced and oil film thickness is increased, leading to the condition known as mixed lubrication. As the oil film thickness increases further, the system now moves into full film lubrication, namely either elasto-hydrodynamic or hydrodynamic lubrication regime.

In a mixed lubrication regime, the gear tooth load is partially carried by the pressure generated in the fluid and partially by the asperities. In mixed lubrication regimes, the external force was compensated by the hydrodynamic pressure of the lubrication film and the solid contact pressure of the asperities that were in contact (Maximilian Prölß et al., 2018).

3.6.10.3 Elasto-hydrodynamic Lubrication (EHL) Regime

EHL condition, which is a remarkable characteristic of lubricant and metal, occurs because of the rolling motion that exists between the mating gear teeth. The contact zone formed between the mating gear flanks has a low degree of geometrical conformity and thus has a small contact area (almost a point or line contact), which creates high contact pressures. As the oil enters the contact zone between a pair of mating tooth flanks (by rolling action), the oil's pressure rises sharply, and this high pressure in turn significantly increases the oil's viscosity and load-holding ability. Even though the resulting concentrated load will slightly deform (flatten) the metal of the gear tooth flanks in the contact zone, the metal elastically returns to its normal form as the rotation continues. In this lubrication regime of operation, the lubricant totally separates the mating contact surfaces. The EHD regime forms a thinner film compared to a hydrodynamic regime with an elastic distortion of the surfaces and with an ultimate lubrication goal of minimizing energy loss by guaranteeing efficient surface separation.

This lubrication regime depends on the metallurgy and heat treatment of the metal, and any incorrect or abnormal operating temperatures will interfere with the formation of the EHL film since an oil's viscosity is directly affected by temperature. The EHL film helps in avoiding asperity contact if operating conditions such as speeds, loads, and temperatures are not exceeded.

The formation and maintenance of an effective EHL film between the gear teeth has a significant impact on the performance and reliability of gears by ensuring smooth operation, reducing friction and wear, and enhancing the overall efficiency of the gear system.

3.6.10.4 Hydrodynamic Lubrication Regime

The hydrodynamic lubrication regime occurs when the lubricant completely separates the contact zone surfaces at the lubricant film thickness near or more than 1 μm. Comparing to other lubrication regimes, friction losses under hydrodynamic lubrication are very small. However, there is no surface deformation as in the case of EHL because of the very low hydrodynamic pressure generated in the lubricant film due to the relative motion and inner film friction. The hydrodynamic lubrication regime condition occurs after a pair of gears has begun to rotate, and the speeds and loads are such that a wedge of oil has been formed between the gear tooth flank contact zone surfaces. This wedge of oil creates enough working clearance between the gear tooth contact bearing surface so that there is little risk of asperity contact, and this is a desirable condition to avoid friction and wear. In order for hydrodynamic lubrication to be successfully and completely applied, there must be a high degree of geometric conformity between the mating gear tooth surfaces, like that of an internal gearing.

3.6.11 Stribeck Curve

The Stribeck curve (Stribeck, 1902a, 1902b, 1902c) is a plot showing the frictional characteristics of a liquid lubricant over the boundary, mixed, EHL, and

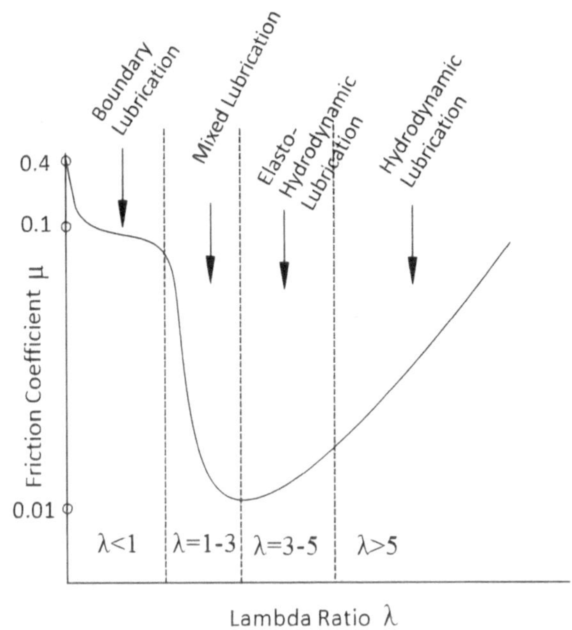

FIGURE 3.12 Stribeck curve.

hydrodynamic lubrication regimes. The ordinate of the plot is the coefficient of friction, μ, and the abscissa is the ratio of the lubricant film thickness to the surface roughness, or the λ ratio, as shown in Figure 3.12. The Stribeck curve concept shows the friction behavior in fluid-lubricated contacts as a nonlinear function of lubricant viscosity, entrainment velocity, surface roughness, and contact load.

3.6.12 Gear Lubricant Oil Selection

Selecting proper gearbox lubricant oil is very important to the long term efficient operation of gear drive. Gear lubricant oil is made up of two critical components: base oil and additives. Base oil is the main component of lubricants, and the type and quantity of base oils used determine the lubricating properties of the lubricant. The base oils can be either mineral oil or synthetic oil or even, in some cases, a combination of the two. The choice of the type of base oil used should be determined by the operating conditions, gear type, and other factors.

Additives are added to lubricating oils to provide properties that are not present in the base oil in order to enhance some of the natural properties of the lubricating oil. Additives are added to impart desirable properties and suppress undesirable ones. The additives package used in the base oil forms the backbone of the lubricant's performance and affects various key performance properties under operating conditions. The base oils and additives used determine the scuffing load capacity of a lubricant and the consequent lubricant film thickness.

Another important property of gear lubrication oil is viscosity, which is a measure of internal friction, or the resistance to relative motion between the molecules under shear stress. The correct viscosity is the most important consideration in choosing a proper lubricant for a gear pair application, and the viscosity for a gear lubricant is primarily chosen to provide a desired oil film thickness between interacting gear tooth surfaces at a given speed and load.

The viscosity of a gear lubricating oil changes with changes in pressure and temperature. The viscosity index (VI) is a measure of the rate of change of viscosity with the change of temperature. A high VI lubricating oil has the ability to maintain a small viscosity differential over the operating range of the gearbox, providing a more consistent lubricating film to the gears and a more predictable wear performance.

For proper viscosity selection of a gear lubricant oil, ISO Viscosity Grade, AGMA Lubrication Numbering, or SAE Grade (Gear Oils) methods are available. It is important to understand the methods for viscosity selection and the factors that affect the particular requirement. For selecting proper gearbox lubricant oil, selection charts are available for all types of gears (for enclosed and open gearing), using which the required ISO viscosity grade can be read from the chart. The selection of proper ISO viscosity grade for any particular application depends on the type of gear set, gear geometry, highest likely operating temperatures, and the pitch-line velocity of the slowest gear in the unit.

3.6.13 Environmentally Acceptable Lubricants (EALs)

The reduction of crude oil reserves, increasing oil prices, and environmental considerations have attracted attention to new and less environmentally harmful lubricants for gearbox applications. To reduce impact on the environment and contribute to improved environmental protection and sustainable development, industries started using safer, high performance lubricants called EALs since the early 1970s. Environmentally responsible companies, looking to fulfill regulatory requirements while promoting environmental protection, sustainability, and *greener* applications, use EALs for gearbox applications in sensitive ecological environments. EALs are lubricants that are biodegradable, non-bio accumulative, and minimally toxic.

Ester oils, having high biodegradability, low toxicity, and good thermal and oxidative stability, are the most common EALs used as industrial lubricants. Synthetic esters respond favorably to performance-enhancing additives such as antioxidants, corrosion inhibitors, anti-wear agents, and load-carrying additives and are used in lubricants for both high- and low-temperature applications. Industrial EALs can be substituted for mineral oils, especially for the gears working in the EHL regime. Ester molecules provide unique properties for finished lubricants across a range of viscosities because of their solvency level, low toxicity, and excellent biodegradability, and ester-based lubricants provide lower friction coefficients in full film and EHL regimes. Industrial EALs are a lesser burden on air, water, ground, and human health than conventional mineral oil-based gear lubricant oils.

For the cylindrical and bevel gears operating at low speeds and high loads, the lubricant usually fails to prevent the asperity contacts, so the lubrication is performed under the mixed EHL regime. Also, since most gears operate in a mixed EHL regime, in which both hydrodynamic lubricant films and rough surface asperity contacts coexist, it is imperative to identify the best choice of lubricant for the given application. Appropriate performance properties of lubricant oil must be evaluated after selecting the right viscosity, base oil, and type of lubricant. For failure prevention, design optimization, gear performance and efficiency improvement, a good understanding of mixed EHL characteristics is very essential.

Having completed formulating the equations for the above-mentioned 14 objective functions, since center distance and weight should be minimized and efficiency should be maximized, the following notation is denoted in this model since the software performs minimization of objectives:

$$f_1 = a$$
$$f_2 = W$$
$$f_3 = -\eta, \text{ etc.}$$

3.6.14 Design Variables and Constraints

The independent design parameters of a planetary gear drive include the module m_n, the number of teeth of the planet gear z_1, the number of teeth of the sun gear z_2, the number of teeth of the ring gear z_3, the three profile shift coefficients x_{n1}, x_{n2}, and x_{n3}, and the number of planets n_p.

Therefore, the design variables are,

$$X = \left[z_1, z_2, z_3, x_{n1}, x_{n2}, x_{n3}, m_n, n_p \right]^T = \left[x_1, x_2, x_3, x_4, x_5, x_6, x_7, x_8 \right]^T$$

A number of constraints, which are introduced into the objective functions to be optimized, are used to provide suitable design choices.

3.6.14.1 Types of Gear Failures

Essentially, gear failures can be of two types: material fatigue-related failures and non-fatigue failure modes. The non-fatigue failures are mainly due to tribological problems in the lubricated contact, such as, for example, scuffing. Scuffing wear is an adhesive-type wear, which occurs at normal temperatures where tooth pressures are high, oil films are in the boundary conditions, and high contact temperatures do not occur due to slow speeds. This type of wear can be reduced by increased oil viscosities or by reduced load.

Gear teeth may fail in bending fatigue at their root; they may fail by pitting fatigue on their surface; or they may fail by scoring. Bending fatigue can be found by the Lewis model of bending failure, and pitting is produced by the Hertz stress. Scoring failure is influenced by sliding velocity, surface finish, oil film characteristics and temperature, profile modification, and anti-weld agents. Generally, gears are case hardened to provide wear resistance and adequate fatigue strength. Also, the gears should have a tough core to prevent brittle failure under high impact loads.

3.6.14.2 Bending Strength

For the gear tooth failure by bending fatigue at its root, *T.* Wilfred Lewis in 1893 developed the basic model for bending failure of a gear tooth. Modern gear design procedures take into account compensating factors that influence gear tooth bending stresses that are now added, which make the Lewis model of bending failure reasonably accurate.

Maximum allowable bending stress is given by the bending strength of spur gears and helical gears (ISO 6336:2006-based JGMA 401-01 (1974) standard),

$$\sigma_F = F_t \frac{y_F y_\varepsilon y_\beta}{9.81 m_n b} \left(\frac{K_V K_O}{K_L K_{FX}} \right) S_F \quad \text{kgf / mm}^2$$

where F_t is the tangential force at the working pitch circle, kgf; y_F is the tooth profile factor; y_ε is the load sharing factor; y_β is the helix angle factor; K_V is the dynamic load factor; K_O is the overload factor; K_L is the life factor; K_{FX} is the size factor of root stress; S_F is the safety factor for bending failure.

For the sun gear–planet gear engagement,

$$F_t = F_{n_2}$$

The tooth profile factor y_F is found by the equation (Figure 3.13),

$$y_F = \frac{6\left(\dfrac{h_A}{m_n}\right)\cos\alpha_{wt}}{\left(\dfrac{S_A{}^2}{m_n{}^2}\right)\cos\alpha_n}$$

The load sharing factor y_ε is given by,

$$y_\varepsilon = \frac{1}{\varepsilon'_{\alpha_1}}$$

The helix angle factor y_β is given by,

$$y_\beta = 1 - \frac{\beta}{120}$$

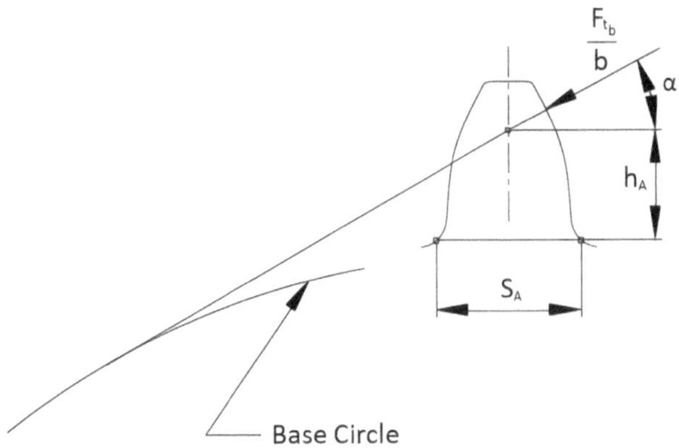

FIGURE 3.13 Parameters for calculating tooth profile factor Y_F.

The dynamic load factor, K_V, is based on the precision of the gear and the tangential speed of gears at the working pitch circle. The tangential speed of gears at the working pitch circle is,

$$v_{t_2} = \frac{\pi d_{w_2} N_2}{60000}$$

The dynamic load factor K_V is taken from the standards.

For tangential speed less than 1.0 m/s,

$$K_V = 1.0$$

The overload factor Ko is defined as,

$$K_O = \frac{\text{Actual tangential force}}{\text{Nominal tangential force } F_t}$$

Since the PTO gearbox is expected to be subjected to heavy input load, K_O is taken as 1.75 from the standards.

The life factor, K_L, depends on the number of cyclic repetitions or the total loaded meshing cycles during its lifetime. Its value is taken from the standards. For the number of cyclic repetitions of $>10^7$ cycles, K_L is taken as,

$$K_L = 1.0$$

The size factor of root stress K_{FX} is generally taken as 1.0.

S_F is the safety factor for bending failure. Since S_F is too complicated to be determined precisely, it is usually taken as at least 1.2.

Similarly, for the planet gear and the ring gear pair engagement,

$$F_t = F_{n_1}$$

The load sharing factor y_ε is given by,

$$y_\varepsilon = \frac{1}{\varepsilon_{\alpha_1}},$$

The helix angle factor y_β is given by,

$$y_\beta = 1 - \frac{\beta}{120}$$

The dynamic load factor, K_V, is based on the precision of the gear and the tangential speed of gears at the working pitch circle. The tangential speed of gears at the working pitch circle is,

$$v_{t_1} = \frac{\pi d_{w_1} N_1}{60000}$$

The dynamic load factor, K_V, is taken from the standards.
For tangential speed less than 1.0 m/s,

$$K_V = 1.0$$

The overload factor Ko is defined as,

$$\mathrm{K_O} = \frac{\text{Actual tangential force}}{\text{Nominal tangential force } F_t}$$

Since the PTO gearbox is expected to be subjected to heavy input load, K_O is taken as 1.75 from the standards.

The life factor, K_L, depends on the number of cyclic repetitions or the total loaded meshing cycles during its lifetime. Its value is taken from the standards. For the number of cyclic repetitions of $>10^7$ cycles, the life factor K_L is taken as,

$$K_L = 1.0$$

The size factor of root stress K_{FX} is generally taken as 1.0.

S_F is the safety factor for bending failure. Since S_F is too complicated to be determined precisely, it is usually taken as at least 1.2.

3.6.14.3 Contact Compressive (Crushing) Strength

In the epicyclic gear train of the planetary gear system of the second stage of reduction, the ring gear is stationary. The contact pressure (Hertz stress) is a parameter to predict the severity of both surface pitting and gear tooth tip scoring. Pitting is due to the fatigue process in the teeth surface layer. Scoring is caused by instability in oil film thickness at high speed and high load with inadequate cooling. This mode of failure is affected by the relatively large sliding velocity present at the gear tip and by the temperature of the teeth. Because of the sliding velocity present, some elasto-hydrodynamic effects are present in the contact, which alters the pressure distribution from the simple Hertzian model of the equation. Scoring can be prevented by proper oil cooling. Dynamic loading, which is largest at the initial point of contact, increases the tip pressure above that of the simple Hertzian model.

Different kinds of teeth flanks wear like pitting, squeezing, abrasive and adhesive wear (scoring and scuffing) limit the gear load capacity. These flank damages are parallel or complementary. For pitting development, it is necessary to start the crack and grow it up along with increased high stress cycle numbers.

The teeth on a gear represent a series of cams which transmit smooth motion from one member to its mate. These cams must have the proper shape in order to transmit this smooth motion. In addition, these cams must be capable of transmitting rather large forces. For this reason, the cam surfaces must have sufficient contact area to support this load. Most of the commonly used gears have theoretical line contact between mating surfaces. Under load, the Hertzian deformation will cause the contact to spread over a rectangular area.

For calculating surface strength, the necessary reference is the nominal tangential load on the reference pitch circle. Therefore, the nominal tangential load on the reference pitch circle should be equal to or below the allowable tangential load on the reference pitch circle, which is derived from calculating allowable Hertz stress. Wear is a general term which refers to the loss of material from the contacting surface of a gear. Wear is the process when layers of metal are removed from the surfaces that are in contact. Wear is related to Hertzian contact stresses and elasto-hydrodynamic (EHD) lubrication. Contact compressive stress should be within allowable limits to prevent failures due to wear like scuffing, scoring, welding, peening, rippling, ridging, and tooth breakage.

Contact compressive strength of spur gears and helical gears as per ISO 6336:2006-based JGMA 402-01 (1975) standard is given by,

$$\sigma_H = \sqrt{\frac{F_t(i \pm 1)}{9.81 d_w b i}} \frac{Z_H Z_M Z_\varepsilon Z_\beta}{K_{HL} Z_L Z_R Z_V Z_W K_{HX}} \sqrt{K_{H\beta} K_V K_O} S_H \quad \text{kgf/mm}^2$$

The "+" symbol in the above equation applies to two external gears in mesh, whereas the "–" symbol is used for an internal gear and an external gear pair in mesh. The above equation contains the following factors:

Zone factor, Z_H
Material factor, Z_M
Contact ratio factor, Z_ε
Helix angle factor, Z_β
Life factor, K_{HL}
Lubricant factor, Z_L
Surface roughness factor, Z_R
Lubrication speed factor, Z_V

Hardness ratio factor, Z_W
Size factor, K_{HX}
Longitudinal load distribution factor, $K_{H\beta}$
Dynamic load factor, K_V
Overload factor, K_O
Safety factor for pitting, S_H
For the sun gear–planet gear engagement, Hertz contact stress is given by,

$$\sigma_{H_1} = \sqrt{\frac{F_t(i+1)}{9.81 d_{w_2} bi} \frac{Z_H Z_M Z_\varepsilon Z_\beta}{K_{HL} Z_L Z_R Z_V Z_W K_{HX}}} \sqrt{K_{H\beta} K_V K_O} S_H \quad \text{kgf / mm}^2$$

$$i = i_1$$

$$F_{t_2} = F_{n_2}$$

Zone factor, Z_H, is given by,

$$Z_H = \sqrt{\frac{2\cos\beta_b \cos\alpha_{wt_1}}{\cos^2\alpha_t \sin\alpha_{wt_1}}} = \frac{1}{\cos\alpha_t}\sqrt{\frac{2\cos\beta_b}{\tan\alpha_{wt_1}}}$$

Material factor, Z_M, is given by,

$$Z_M = \left[\frac{1}{\pi\left\{\frac{1-\nu_1^2}{E_1} + \frac{1-\nu_3^2}{E_2}\right\}}\right]^{0.5}$$

For steel, $E_1 = 21000$ kgf/mm^2
Contact ratio factor, Z_ε, is given by the following:
For spur gear,

$$Z_\varepsilon = 1.0$$

For helical gear, when the overlap ratio $\varepsilon_\beta \le 1$,

$$Z_\varepsilon = \sqrt{1 - \varepsilon_\beta + \frac{\varepsilon_\beta}{\varepsilon_\alpha}}$$

When $\varepsilon_\beta > 1$,

$$Z_\varepsilon = \sqrt{\frac{1}{\varepsilon_\alpha}}$$

The helix angle factor, Z_β, is difficult to calculate and can be taken as 1.0.

The life factor, K_{HL}, depends on the duty cycle.

For duty cycle > 10^7 cycles,

$$K_{HL} = 1.0$$

Lubricant factor, Z_L, is based upon the lubricant's kinematic viscosity at 50°C, *cSt*.

From the standards,

$$Z_L = 1.0$$

The surface roughness factor, Z_R, is obtained on the basis of the average roughness R_{max} (μm).

The average roughness, R_{max}, is calculated by the equation given below using the surface roughness values of the pinion and gear, R_{max1} and R_{max2}, and the center distance a in mm.

$$R_{max} = \frac{R_{max1} + R_{max2}}{2} \left(\frac{100}{a'} \right)^{\frac{1}{3}}$$

For surface-hardened gear, and R_{max} = 10 microns,

$$Z_R = 0.92$$

Lubrication speed factor, Z_V, depends on the tangential speed at the pitch circle, v (m/s).

For surface-hardened gear and $v < 1$ m/s,

$$Z_V = 0.96$$

The hardness ratio factor, Z_W, applies only to the gear that is in mesh with a pinion which is quenched and ground. It is given by the equation,

$$Z_W = 1.2 - \frac{H_{B2} - 130}{1700}$$

Here, H_{B2} is in the range of the Brinell hardness of gear material.

$$130 \leq H_{B2} \leq 470$$

If a gear is out of this range, Z_W is assumed to be 1.0.

The size factor, K_{HX}, is usually taken as 1.0 because the conditions affecting this parameter are often unknown.

The longitudinal load distribution factor, $K_{H\beta}$, depends on the method of gear shaft support and the ratio, b / d_{w2}, of the gear face width b to the pitch diameter, d_{w2}. When tooth contact under load is good, the longitudinal load distribution factor can be taken to be in the range of 1.0–1.2. For the present case, let $K_{H\beta} = 1.0$.

The dynamic load factor, K_V, depends on the gear's precision grade and pitch circle tangential speed, v.

For speeds less than 1 m/s, the value of K_V is taken as 1.0.

The overload factor, K_O, depends on the impact load on the gears on the prime mover side and on the load side. For uniform load from the load side, this factor can be taken as 1.0.

The safety factor for pitting, S_H, is normally taken as 1.15 since the causes of pitting are difficult to precisely define because it involves many environmental factors.

For the planet gear–ring gear engagement, Hertz contact stress is given by,

$$\sigma_{H_2} = \sqrt{\frac{F_{t_1}(i-1)}{9.81 d_{w_1} bi} \frac{Z_H Z_M Z_\varepsilon Z_\beta}{K_{HL} Z_L Z_R Z_V Z_W K_{HX}}} \sqrt{K_{H\beta} K_V K_O} S_H \text{ ,kgf / mm}^2$$

$$i = i_2$$

$$F_{t_1} = F_{n_1}$$

Zone factor, Z_H, is given by,

$$Z_H = \sqrt{\frac{2\cos\beta_b \cos\alpha_{wt_2}}{\cos^2\alpha_t \sin\alpha_{wt_2}}} = \frac{1}{\cos\alpha_t}\sqrt{\frac{2\cos\beta_b}{\tan\alpha_{wt_2}}}$$

Material factor, Z_M, is given by,

$$Z_M = \left[\frac{1}{\pi\left\{\frac{1-\nu_1^2}{E_1} + \frac{1-\nu_3^2}{E_2}\right\}}\right]^{0.5}$$

For steel, $E_1 = 21000$ kgf/mm^2

Contact ratio factor, Z_ε, is given by the following:

For spur gear,

$$Z_\varepsilon = 1.0$$

For helical gear, When the overlap ratio $\varepsilon_\beta \le 1$,

$$Z_\varepsilon = \sqrt{1 - \varepsilon_\beta + \frac{\varepsilon_\beta}{\varepsilon_\alpha}}$$

When $\varepsilon_\beta > 1$,

$$Z_\varepsilon = \sqrt{\frac{1}{\varepsilon_\alpha}}$$

The helix angle factor, Z_β, is difficult to calculate and can be taken as 1.0.

The life factor, K_{HL}, depends on the duty cycle.

For duty cycle > 10^7 cycles,

$$K_{HL} = 1.0$$

Lubricant factor, Z_L, is based upon the lubricant's kinematic viscosity at 50°C, *cSt*.

From the standards,

$$Z_L = 1.0.$$

The surface roughness factor, Z_R, is obtained on the basis of the average roughness $R_{\max}$ (μm).

The average roughness, $R_{\max}$, is calculated by the equation given below using the surface roughness values of the pinion and gear, $R_{\max 1}$ and $R_{\max 2}$, and the center distance a in mm.

$$R_{\max} = \frac{R_{\max 1} + R_{\max 2}}{2}\left(\frac{100}{a'}\right)^{\frac{1}{3}}$$

For surface-hardened gear, and $R_{\max} = 10$ microns,

$$Z_R = 0.92$$

Lubrication speed factor, Z_V, depends on the tangential speed at the pitch circle, v (m/s).

For surface-hardened gear and $v < 1$ m/s,

$$Z_V = 0.96$$

The hardness ratio factor, Z_W, applies only to the gear that is in mesh with a pinion, which is quenched and ground. It is given by the equation,

$$Z_W = 1.2 - \frac{H_{B2} - 130}{1700}$$

Here, H_{B2} is in the range of the Brinell hardness of gear material.

$$130 \le H_{B2} \le 470$$

If a gear is out of this range, Z_W is assumed to be 1.0.

The size factor, K_{HX}, is usually taken as 1.0 because the conditions affecting this parameter are often unknown.

The longitudinal load distribution factor, $K_{H\beta}$, depends on the method of gear shaft support and the ratio, b / d_{w_1}, of the gear face width b to the pitch diameter, d_{w_1}. When tooth contact under load is good, the longitudinal load distribution factor can be taken to be in the range of 1.0–1.2. For the present case, let $K_{H\beta} = 1.0$.

The dynamic load factor, K_V, depends on the gear's precision grade and pitch circle tangential speed, v.

For speeds less than 1 m/s, the value of K_V is taken as 1.0.

The overload factor, K_O, depends on the impact load on the gears on the prime mover side and on the load side. For uniform load from the load side, this factor can be taken as 1.0.

The safety factor for pitting, S_H, is normally taken as 1.15 since the causes of pitting are difficult to precisely define because it involves many environmental factors.

3.6.14.4 Contact Ratio

The contact ratio for a given spur/helical gear mesh is the ratio of the length of contact along the line of action between the two addendum circles to the base pitch. In a similar way, the contact ratio of an internal gear pair drive is obtained by dividing the length of the path of contact between its intersection with the outer circle of the spur pinion by the base pitch of the series of involute curves that form the similar profiles of successive gear teeth. This condition holds good only when the inside circle of the internal gear is large enough to avoid involute interference with both the mating spur gear and the pinion-shaped cutter that is used to generate it.

The contact ratio is given by,

$$\varepsilon_\alpha' = \frac{\text{Path of contact}}{p_b}$$

where ε_α' is the contact ratio; p_b is the base pitch.

The base pitch, p_b, is defined as the distance from one tooth to the next tooth measured along the line of action. The base pitch is given by,

$$p_b = p_c \cos\alpha_t = \pi m_n \cos\alpha_t$$

where p_c is the circular pitch.

This contact ratio should be greater than 1.4 and is normally less than 2.0 for standard spur gears. For contact with an internal gear at low gear ratios and for nonstandard addenda and dedenda, it can exceed 2.0.

There are two types of contact ratio: one is transverse contact ratio (or radial contact ratio), and the other is overlap ratio. The transverse contact ratio is a value expressed by the length of the path of contact divided by the base pitch. For example, if a transverse contact ratio is 1.5, then during rotation, the engaging gear pair keeps tooth contact with a single tooth and with two teeth, alternately. If the transverse contact ratio is 2, then during rotation, the engaging gear pair always keeps tooth contact with two teeth. Overlap ratio is a value involving gears which have helix hands like helical gears, and is expressed by the face width divided by pitch in the direction of the face width.

The total contact ratio is the sum of the transverse contact ratio and overlap ratio. Normally, less noise is produced if the total contact ratio becomes an integer, such as 2 or 3. As for strength, a higher contact ratio leads to more strength as the load per tooth is dispersed.

The transverse contact ratio for the sun gear–planet gear pair engagement is given by,

$$\varepsilon_{\alpha_1}{}' = \frac{AB}{p_b} = \frac{AB}{\pi m_t \cos\alpha_t}$$

The transverse contact ratio for the planet gear–ring gear pair engagement is given by,

$$\varepsilon_{\alpha_2}{}' = \frac{EF}{p_b} = \frac{EF}{\pi m_t \cos\alpha_t}$$

$$\text{Overlap ratio } \varepsilon_\beta = \frac{b\sin\beta}{\pi m_n}$$

The total contact ratio for the sun gear–planet gear pair engagement is given by,

$$\varepsilon_{\alpha_1}{}' + \varepsilon_\beta \geq 1.4$$

The total contact ratio for the planet gear–ring gear pair engagement is given by,

$$\varepsilon_{\alpha_2}{}' + \varepsilon_\beta \geq 1.4$$

3.6.14.5 Top Land Thickness

The top land thickness is the tooth thickness on the tip circle. The top land thickness is normally referred to in terms of the percentage of the module. The ratio of top land thickness and the module is suggested to be between 0.2 and 0.4 depending on the material and heat treatment of the gear.

The top land thickness of the planet gear is given by,

$$S_{a_1} = d_{a_1}{}'\left(\frac{s_1}{d_1} + inv\alpha_n - \Omega_1\right)$$

where

$$S_1 = \frac{\pi m_n}{2} + 2x_{n_1} m_n \tan\alpha_n$$

$$\Omega_1 = \tan \alpha_{at_1} - \alpha_{at_1}\left(\frac{\pi}{180}\right)$$

Similarly, the top land thickness of the sun gear is given by,

$$S_{a_2} = d_{a_2}'\left(\frac{s_2}{d_2} + inv\alpha_n - \Omega_2\right)$$

where

$$S_2 = \frac{\pi m_n}{2} + 2x_{n_2} m_n \tan \alpha_n$$

$$\Omega_2 = \tan \alpha_{at_2} - \alpha_{at_2}\left(\frac{\pi}{180}\right)$$

Also, the top land thickness of the ring gear is given by,

$$S_{a_3} = d_{a_3}'\left(\frac{s_3}{d_3} - inv\alpha_n + \Omega_3\right)$$

where

$$S_3 = \frac{\pi m_n}{2} - 2x_{n_3} m_n \tan \alpha_n$$

$$\Omega_3 = \tan \alpha_{at_3} - \alpha_{at_3}\left(\frac{\pi}{180}\right)$$

The top land thickness should not be below 0.2 times the normal module, m_n, and correspondingly the constraint equations for the top land thickness of the planet gear, the sun gear, and the ring gear are as given below:

$$S_{a_1} \geq 0.2m_n$$
$$S_{a_2} \geq 0.2m_n$$
$$S_{a_3} \geq 0.2m_n$$

3.6.14.6 Operating Interference

To avoid operating interference on the two mating gears, the following conditions are to be met (refer to Figure 3.7):

1. The distance O_1M is greater than tip radius of the planet gear.
2. The distance O_2N is greater than tip radius of the sun gear.

Hence, the constraint equations are,

$$g_p\left(x_{n_1}, x_{n_2}\right) = \left(r_{b_1}{}^2 + \left(a' \sin \alpha_{wt_1}\right)^2\right)^{0.5} > r_{a_1}'$$

$$g_s\left(x_{n_1}, x_{n_2}\right) = \left(r_{b_2}{}^2 + \left(a' \sin \alpha_{wt_1}\right)^2\right)^{0.5} > r_{a_2}{}'$$

3.6.14.7 Constraints of Internal Gear

For proper tooth engagement and disengagement, involute interference (contact below the base circle) and secondary interference between the pinion gear tooth tip and the internal gear tooth tip for internal gearing must be avoided.

3.6.14.7.1 Involute Interference

If the profile of the internal involute gear extends below the point of tangency of the path of contact with the base circle of the mating spur pinion, or pinion-shaped cutter that is used to generate it, involute interference will exist. To avoid such involute interference, the inside radius of the internal gear must be made large enough to be outside this point.

Involute interference occurs when the addendum circle of one gear crosses the line of action past its point of tangency with the base circle on its mating gear. For standard tooth systems with equal addenda, the pinion will always be the gear on which contact occurs below the base circle. This is shown in Figure 3.7 for contact with an external gear and in Figure 3.8 for contact with an internal gear.

For contact with an internal gear, involute interference is avoided if,

$$KL \leq KE$$

that is,

$$a' \sin \alpha_{wt_2} \leq KE$$

Also, to avoid involute interference, the following condition must be satisfied (Buckingham, 1988):

$$\frac{z_1}{z_3} \geq 1 - \frac{\tan \alpha_{at_3}}{\tan \alpha_{wt_2}}$$

3.6.14.7.2 Tip Interference

The next constraint is related to tip interference. It is a type of kinematic interference for contact with an internal gear, in which the tip of the internal gear tooth contacts the tip of the pinion tooth as they come into engagement (Buckingham, 1988; Andersson, 1973). This interference is called fouling. For standard gearing geometry, this condition is not present for internal gear ratios greater than 2.5:1.

Even though the internal gear ratio is more than 2.5 in the present case, still this constraint is considered for the sake of completeness and also to check if the solution contains an internal gear ratio less than 2.5.

Tip interference is present when the trochoid of the path of the corner of the pinion teeth intersects the involute profile of the internal gear tooth.

To avoid tip interference, the following conditions must be satisfied (Buckingham, 1988):

$$\cos\varepsilon_1 = \frac{\left(r_{a_3}'^2 - r_{a_1}^2 - a'^2\right)}{2a' r_{a_1}'} - A$$

$$\therefore \varepsilon_1 = \cos^{-1} A$$

$$\varepsilon_2 = \left(\frac{z_1}{z_3}\right)\varepsilon_1$$

$$\theta_{t1} = \sin^{-1}\left[\frac{r_{a_1}' \sin\varepsilon_1}{r_{a_3}'}\right] - \varepsilon_2 = B$$

$$\theta_{t1r} = \frac{B \times \pi}{180} = C$$

$$\cos\alpha_{at_3} = \frac{r_{b_3}}{r_{a_3}'} = D$$

$$\cos\alpha_{at_1} = \frac{r_{b_1}}{r_{a_1}'} = E$$

$$\delta = \left(\frac{z_1}{z_3}\right)\left(inv\alpha_{at_1} - inv\alpha_t\right) + inv\alpha_t$$

$$inv\alpha_{at_1} = \tan\alpha_{at_1} - \frac{\alpha_{at_1} \times \pi}{180} = \tan\alpha_{at_1} - 0.0175\alpha_{at_1} = F$$

$$inv\alpha_t = \tan\alpha_t - 0.0175\alpha_t = 0.02881 = G$$

$$x_1 = inv\alpha_{at_3} = \tan\alpha_{at_3} - 0.0175\alpha_{at_3}$$

$$x_2 = \delta - \theta_{t1r} = \delta - C$$

Condition to avoid tip interference:
If $x_2 > x_1$, then there is no interference.
that is, $x_2 - x_1 \geq 0$

3.6.14.8 Condition for Proper Assembly

Additionally, the condition for proper assembly of planetary gearing is given below (Townsend, 2013):

$$\frac{z_2 + z_3}{n_w} = I$$

where I is an integer.

TABLE 3.3
PTO Planetary Gear Parameters

Data	Gear
Density of the material, ρ	7.86 × 106 kg/mm^3
Material of sun gear, planet gear, and ring gear	Steel 20 Mn5Cr5 (IS:4432-1988)
Design bending stress, σ_b	$400\mathrm{N} / \mathrm{mm}^2$
Design crushing stress, σ_c	$2000\mathrm{N} / \mathrm{mm}^2$
Input speed, N	104.65 rpm
Power transmitted, P	50 HP
Young's modulus of the material, E	$2.0 \times 10^5 \mathrm{kg} / \mathrm{mm}^2$
Pressure angle, α_n	25°
Ratio between face width and average module, ψ_m	13.225

3.6.15 Design Data for Optimization

The following are the input parameters considered for the planetary gear design based on the existing gear design as in Table 3.3.

3.6.16 Methodology of Optimization of Planetary Gear Using NSGA-II

This section describes how the NSGA-II algorithm is configured for solving the optimization problem of planetary gearing. This book utilizes the OPTIMO software package (Gokul Krishnan, 2024), which provides an efficient NSGA-II implementation in Python along with the complete problem set corresponding to this case study. Chapter 2 provides detailed instructions on how to obtain and use the software.[7]

3.6.16.1 Variable Bounds

In order to determine the values of design variables, such as module, number of teeth, face width, and helix angle, through the optimization procedure, the upper and lower bounds of the design variables are to be established first.

The following variables are considered, and their lower and upper bounds are fixed. Further, the type of variable – whether discrete (integer) or real-valued (float) – is also specified.

Number of planet gear teeth, z_1: 10 to 20 (integer)
Number of sun gear teeth, z_2: 20 to 30 (integer)
Number of ring gear teeth, z_3: 50 to 60 (integer)
Profile shift coefficient for planet gear, x_{n1}: 0.0 to 1.0 (float)
Profile shift coefficient for sun gear, x_{n2}: −0.5 to 0.5 (float)
Profile shift coefficient for ring gear, x_{n3}: 0.0 to 1.0 (float)
Module, m_n: 3 to 5 mm (float)
Number of planets, n_p: 2 to 6 (integer)

3.6.16.2 Input Parameters

The backbone of any system is its input data. For this model, the most important data inputs are the gear design parameters of the two mating gears. The decision maker is required to make this decision based on experience and design requirements.

The problem sets provided in this book contain Python code corresponding to the objective functions and design constraints formulated in Sections 3.6.9 and 3.6.14. The upper and lower bounds of design variables are also specified in the same file(s). The seed for the random number generator is fixed at 0.123 for reproducibility of results.

The problem definition corresponding to Chapter 3 can be found in *problem-sets/problem_ch3.py* file within the software repository. The optimization run is started by entering the command specified in Listing 3.1 from the repository root.

LISTING 3.1. COMMAND TO EXECUTE THE ALGORITHM RUN

```
# macOS / Linux
python3 run.py -a nsga2 -p ch3
# Windows
python.exe run.py -a nsga2 -p ch3
```

3.6.16.3 Decision-Making

The goal of decision-making is to select the most optimal solution from the Pareto solution set obtained from the optimization results. Selection of the single best solution from many conflicting non-dominated solutions from the Pareto-set is based on expertise along with any problem-specific requirements. Apart from the above, there are decision-making techniques such as the linear programming technique for multidimensional analysis of preference (LINMAP), Shannon Entropy (SE), and the technique of order preference by similarity to an ideal solution (TOPSIS), which can be used as a post-processing step to obtain a relative gradation of results.

The OPTIMO software makes use of a Common Objective Function, or COF, which assigns a scalar value to each solution in the Pareto-optimal set. This scalar value is essentially the normalized average of the set of objective function values. For a given solution i,

$$P_i = \left[p_1, p_2, \dots p_N \right]$$

$$S_i = COF\left(P_i\right)$$

where P_i is the vector of objective function values; N is the number of objectives; S_i is the scalar output assigned to the solution.

The solution set is then sorted on its COF value, such that the solution with the lowest COF value is considered to be, on average, the overall best solution (assuming minimization of objectives).

3.6.17 Interpretation of NSGA-II Results

In this section, the results obtained from NSGA-II are discussed and compared with the design of existing PTO planetary gearing.

Table 3.4 shows a sample of the NSGA-II output. Since constraints are specified in the form of ≥ 0 only, all the columns corresponding to the constraints shown in Table 3.4 indicate results obtained by modifying the constraint equations accordingly. Table 3.4 lists the values of the objective functions, constraints, and design variables from the generated solutions. The output represents a non-dominated Pareto-optimal solution set.

3.6.17.1 Objective Function Values

The first 14 columns represent the values of optimized objective functions.

- *Center distance (Column 1)*: The minimized center distance obtained.
- *Weight (Column 2)*: The corresponding minimized weight of the gearing.
- *Efficiency (Column 3)*: The maximized efficiency value.
- *Slide roll ratio (Column 4)*: The minimized value of slide roll ratio for the sun gear and planet gear engagement.
- *Specific sliding (Column 5)*: The difference in specific sliding ratios at SAP and EAP, respectively, of the planet gear and sun gear pair meshing. A minimized value in this column represents balanced or equalized values of sliding ratios at SAP and EAP of gear mesh.
- *Specific sliding (Column 6)*: The difference in specific sliding ratios at SAP and EAP, respectively, of the planet gear and ring gear pair meshing. A minimized value in this column represents balanced or equalized values of sliding ratios at SAP and EAP of gear mesh.
- *Almen product (Column 7)*: The difference in Almen product at SAP and EAP, respectively, of the planet gear and sun gear pair meshing. A minimized value in this column represents balanced or equalized values of Almen product at SAP and EAP of gear mesh.
- *Almen product (Column 8)*: The difference in Almen product at SAP and EAP, respectively, of the planet gear and ring gear pair meshing. A minimized value in this column represents balanced or equalized values of Almen product at SAP and EAP of gear mesh.
- *Flash temperature (Column 9 & 10)*: The minimized values of flash temperature at SAP and EAP, respectively, of the planet gear and sun gear pair

TABLE 3.4
Sample of NSGA-II Output

Objectives													
Center distance	Weight	Efficiency	Slide roll ratio	Specific sliding		Almen product			Flash temperature			Lambda ratio	
123.74	13.00	98.53	0.29	0.03	1.46	2.87	249.60	0.06	11.22	4.73	8.01	3.71	2.35
100.35	11.45	98.31	0.12	0.13	1.06	25.23	2156.54	9.49	14.03	2.90	10.97	2.63	2.43
98.70	11.34	98.40	0.15	0.34	1.80	22.16	1093.61	5.36	14.62	3.50	10.57	2.77	2.21
122.22	12.63	98.40	0.29	0.20	1.15	11.45	932.03	0.44	12.06	3.05	8.40	3.53	2.34
123.79	12.95	98.42	0.29	0.19	1.21	10.88	1443.48	0.40	11.86	2.40	8.24	3.60	2.36
114.82	9.99	98.46	0.24	0.18	1.62	32.91	127.21	2.57	15.80	9.59	12.79	3.26	2.23
123.17	12.87	98.38	0.29	0.32	1.06	23.62	2043.04	0.06	12.21	2.08	8.19	3.59	2.39
124.08	13.01	98.42	0.29	0.21	1.19	12.27	3138.41	0.38	11.87	1.64	8.22	3.60	2.37
70.00	**5.03**	**97.69**	**0.21**	**0.10**	**1.18**	**3.42**	**568.13**	**8.15**	**22.52**	**8.56**	**17.12**	**1.67**	**1.30**
97.19	10.57	98.07	0.12	1.07	1.01	90.75	1624.29	14.62	14.97	4.00	12.63	2.36	2.27
113.17	9.86	98.18	0.31	0.20	1.58	33.53	79.04	0.85	17.73	14.84	16.18	3.19	1.91
119.26	12.23	98.04	0.34	0.56	1.22	40.46	369.20	0.65	14.22	5.78	9.69	3.41	2.03
68.12	**4.86**	**97.70**	**0.24**	**0.08**	**1.39**	**17.50**	**201.99**	**6.96**	**22.44**	**13.93**	**18.96**	**1.67**	**1.20**
120.56	12.13	97.57	0.29	0.90	3.08	97.46	51.43	2.58	9.37	11.61	10.12	3.53	2.04
108.99	7.31	98.11	0.29	0.28	1.09	27.56	4223.74	0.47	25.45	3.61	17.29	2.80	1.86
115.48	8.20	98.16	0.29	0.28	1.10	27.35	6026.55	0.47	23.97	2.85	16.29	3.01	2.00
94.99	8.03	96.71	0.24	0.95	6.96	107.14	94.99	5.29	11.24	14.27	6.72	2.65	1.63
111.04	7.58	98.13	0.29	0.28	1.10	27.56	7938.66	0.46	24.96	2.59	16.96	2.87	1.91
101.55	8.22	98.15	0.12	0.11	1.05	33.18	9953.03	17.19	25.66	2.95	19.96	2.44	2.26
69.24	**5.36**	**97.20**	**0.12**	**2.46**	**3.26**	**220.31**	**29.34**	**24.43**	**13.17**	**24.94**	**24.27**	**1.65**	**1.35**

Constraints														
Interference		Top land thickness			Contact ratio		Bending stress		Contact stress		Involute interference		Tip interference	Proper assembly
20.08	10.10	1.18	1.50	1.96	0.07	1.97	328.29	360.80	942.01	1458.67	1.01	0.21	0.13	0.00
8.60	8.12	1.31	1.47	2.15	0.17	1.11	330.32	345.59	748.11	1402.99	0.05	0.19	0.13	0.00
12.25	9.34	1.38	1.08	2.22	0.06	1.32	323.31	343.11	807.80	1332.61	0.09	0.20	0.13	0.00
17.94	8.41	1.27	1.48	2.05	0.13	1.95	329.27	361.34	885.08	1468.03	0.17	0.18	0.12	0.00
18.32	8.62	1.28	1.49	2.09	0.13	1.96	330.88	362.22	902.36	1470.21	0.07	0.18	0.12	0.00
17.47	11.23	1.00	1.57	1.83	0.03	1.60	287.00	331.37	683.75	1299.00	2.42	0.26	0.14	0.00
18.68	8.47	1.36	1.41	2.02	0.13	2.01	330.04	361.76	901.40	1481.92	0.04	0.17	0.11	0.00
18.33	8.58	1.29	1.49	2.10	0.13	1.97	331.27	362.43	903.85	1472.14	0.02	0.18	0.12	0.00
7.08	**4.47**	**0.87**	**1.05**	**1.47**	**0.19**	**1.29**	**236.07**	**293.60**	**76.75**	**1085.25**	**0.39**	**0.15**	**0.10**	**0.00**
5.47	6.30	1.39	1.83	2.10	0.25	0.89	328.53	344.20	603.75	1373.62	0.09	0.17	0.12	0.00
18.62	9.46	0.87	1.61	1.87	0.03	1.38	276.22	334.76	581.23	1258.35	6.39	0.20	0.06	0.00
18.90	6.60	1.36	1.41	2.15	0.14	1.90	323.92	362.58	802.02	1430.42	1.26	0.13	0.07	0.00
7.88	**4.78**	**0.71**	**1.02**	**1.44**	**0.14**	**1.18**	**220.90**	**289.87**	**54.13**	**1033.42**	**1.72**	**0.16**	**0.08**	**0.00**
16.58	10.99	0.29	1.94	2.21	0.07	1.15	324.13	358.53	890.82	1351.65	7.98	0.41	0.13	0.00
16.25	7.44	1.18	1.28	1.81	0.13	1.98	221.89	302.65	236.47	1164.92	0.02	0.17	0.11	0.00
17.20	7.87	1.25	1.35	1.92	0.13	1.98	241.40	313.31	334.82	1211.33	0.01	0.17	0.11	0.00
12.22	9.81	0.22	1.62	1.82	0.05	1.05	290.52	333.51	668.87	1164.80	5.40	0.60	0.27	0.00
16.56	7.58	1.20	1.30	1.85	0.13	1.99	228.40	306.21	269.04	1179.84	0.00	0.17	0.11	0.00
8.73	8.20	1.33	1.48	2.17	0.17	1.11	263.89	293.72	250.81	1166.11	0.00	0.18	0.13	0.00
4.34	**7.26**	**0.20**	**1.40**	**1.62**	**0.13**	**0.29**	**249.93**	**282.82**	**188.46**	**952.03**	**4.94**	**0.41**	**0.15**	**0.00**

(Continued)

TABLE 3.4 (CONTINUED)
Sample of NSGA-II Output

Variables							
z_1	z_2	z_3	x_{n1}	x_{n2}	x_{n3}	m_n	n_p
20	30	60	0.45	0.49	0.41	4.95	6
20	21	51	0.16	0.08	0.57	4.90	6
19	21	51	0.28	0.48	0.08	4.94	6
20	30	60	0.26	0.33	0.54	4.89	6
20	30	60	0.27	0.34	0.48	4.95	6
20	27	57	0.59	0.46	0.50	4.89	4
20	30	60	0.24	0.41	0.67	4.93	6
20	30	60	0.26	0.34	0.48	4.96	6
19	**24**	**54**	**0.12**	**0.04**	**0.39**	**3.26**	**6**
20	21	51	0.08	–0.26	0.52	4.74	6
18	28	60	0.56	0.42	0.53	4.92	4
18	30	60	0.16	0.38	0.40	4.97	6
18	**24**	**54**	**0.26**	**0.12**	**0.39**	**3.24**	**6**
20	30	60	0.76	–0.01	0.01	4.82	6
20	30	60	0.24	0.37	0.61	4.36	3
20	30	60	0.24	0.37	0.60	4.62	3
20	27	51	0.81	0.03	0.08	4.04	6
20	30	60	0.24	0.37	0.60	4.44	3
20	21	51	0.15	0.08	0.57	4.95	3
20	**21**	**51**	**0.64**	**–0.40**	**0.13**	**3.38**	**6**

meshing. Balancing or equalization of these two values reduces noise and vibration during running.

- *Flash temperature (Column 11 & 12)*: The minimized values of flash temperature at SAP and EAP, respectively, of the planet gear and ring gear pair meshing. Balancing or equalization of these two values reduces noise and vibration during running.
- *Lambda ratio (Column 13 & 14)*: The lambda ratio values for the pair of planet gear/sun gear meshing and the pair of planet gear/ring gear meshing, respectively. A maximized value of the lambda ratio aimed at the EHL range (refer to Figure 3.12) will ensure smooth operation and enhance the overall efficiency of the gear system.

3.6.17.2 Constraint Values

The next 15 columns represent the value of constraint inequality functions. These functions are all specified in ≥ 0 format. Therefore, a negative value represents a violation of the constraint.

- *Interference (Column 15)*: The condition to avoid operating interference between the mating sun gear and the planet gear, that is, the condition of the distance O_1M > tip radius of the planet gear (refer to Figure 3.7).
- *Interference (Column 16)*: The condition to avoid operating interference between the mating sun gear and the planet gear, that is, the condition of the distance O_2N> tip radius of the sun gear (refer to Figure 3.7).
- *Top land thickness (Column 17, 18 & 19)*: The top land thickness conditions for the planet gear, sun gear, and ring gear. The values represent how much the top land thickness is more than the stipulated condition of 0.2 m_n.
- *Contact ratio (Column 20, 21)*: The contact ratio conditions for the sun gear/planet gear and the planet gear/ring gear pair engagement, respectively. The values represent how much the contact ratio is more than the stipulated condition value of 1.4.
- *Bending stress (Column 22,23)*: The bending stress conditions – how much the bending stress is less than the stipulated value of 400 N/mm^2 for the planet gear and sun gear, respectively.
- *Contact stress (Column 24,25)*: The contact stress conditions – how much the contact stress is less than the stipulated value of 2000 N/mm^2 for the planet gear and sun gear, respectively.
- *Involute interference (Column 26, 27)*: Any violation of 7the involute interference conditions between the ring gear and the planet gear.
- *Tip interference (Column 28)*: Any violation of the tip interference condition between the ring gear and the planet gear.
- *Proper assembly (Column 29)*: The condition for proper assembly of planetary gearing. A value of zero in this column means there is no violation of this constraint.

3.6.17.3 Design Variables

The last eight columns represent the values of the eight design variables corresponding to each optimized solution from the optimization problem.

3.6.17.4 Comparison of Results

In Table 3.5, a comparison has been made between the results that are obtained with NSGA-II and the design of the existing planetary gearing. The solutions corresponding to the number of teeth of the planet gear, namely, 19, 18, and 20, were selected from the results. The values of design variables and objective functions corresponding to each solution are indicated in Table 3.5. The total weight of planetary gearing in kg, the efficiency in percentage, and the center distance in mm were taken for comparison.

The comparison shows that the results obtained from NSGA-II are an improvement over the existing design by optimizing the 14 objective functions within the given constraints. The results were obtained with minimal computational effort and time. The three best results chosen from the solution set of NSGA-II output are analyzed for selecting a single best solution. The total weight of the existing planetary gearing was 6.11 kg, the center distance was 78 mm, and the efficiency was 96.95%.

Referring to Table 3.5, in the case of the first result, by selecting the number of teeth in the planet gear as 19, the weight has reduced to 5.03 kg, the efficiency has increased to 97.69%, and the center distance has reduced to 70.00 mm. There is a reduction of 17.68% in weight, an increase of 0.74% in efficiency, and a space reduction of 10.26%. Hence, there is an improvement in all three objective functions in the solution obtained with NSGA-II.

In the case of the second result, by selecting the number of teeth in the planet gear as 18, the weight has reduced to 4.86 kg, the efficiency has increased to 97.70%, and the center distance has reduced to 68.12 mm. There is a reduction of 20.46% in weight, an increase of 0.75% in efficiency, and a space reduction of 12.67%. Hence, there is an improvement in all three objective functions in the solution obtained with NSGA-II.

In the case of the third result, by selecting the number of teeth in the planet gear as 20, the weight has reduced to 5.36 kg, the efficiency has increased to 97.20%, and the center distance has reduced to 69.24 mm. There is a reduction of 12.27% in weight, an increase of 0.25% in efficiency, and a space reduction of 11.23%. Hence, there is an improvement in all the three objective functions in the solution obtained with NSGA-II.

3.6.17.5 Visualization of Results

The formation of PF from the non-dominated solutions containing 14 objective functions is depicted in the subsequent graphs in combinations of three objective functions for every graph. The graphs show the 3D scatter plot of the solution set generated by NSGA-II in various combinations of three objective functions.

TABLE 3.5
Optimized Design Parameters Using NSGA-II

	Variables				Objective Functions		
	Module, m (mm)	No. of Teeth (Planet Gear), Z_1	No. of Teeth (Sun Gear), Z_2	No. of Teeth (Ring Gear), Z_3	Total Weight, W (kg)	Efficiency, η	Center Distance (mm)
Existing planetary gearing design	4	15	24	56	6.11	96.95	78
NSGA-II design	3.26	19	24	54	5.03	97.69	70.00
	3.24	18	24	54	4.86	97.70	68.12
	3.38	20	21	51	5.36	97.20	69.24

Even though the graphs help in better visualization of the non-dominated Pareto-optimal solution set, one should be cautious of the fact that all the 14 objective functions put together to form a single set of Pareto-optimal solutions that cannot be visualized in totality in graphical format. Hence, for the sake of decision-making, the solutions better suited for the individual requirements of the user/decision maker must be given priority in the choice of objective function combinations taken for visualization.

As a result of optimization, a Pareto-optimal front is formed in the objective space of the solutions. In the scatter plot, the circular points represent non-dominated solutions with respect to all 14 objectives taken together, whereas the triangular points represent a non-dominated front formed with respect to the three objectives that are taken for visualization. It can be observed that the distribution of solutions along the front is uniform, which means the NSGA-II algorithm offers a high-quality solution set.

Referring to Figure 3.14(a), the solutions close to the upper right corner have higher total weight, higher center distance, and higher efficiency, which means that they have better efficiency but are heavier and less compact. On the contrary, the solutions near the lower left corner present better compactness and better total weight but lesser efficiency. The solutions lying in the lower right corner have the best value in terms of higher efficiency, lesser weight, and better compactness.

Referring to Figure 3.14(b), the solutions close to the lower right corner offer higher efficiency, better compactness, and lesser slide roll ratio values.

Referring to Figure 3.15(a), the solutions close to the lower right corner offer higher efficiency and lesser specific sliding values. A lower value of specific sliding indicates a balanced specific sliding at SAP and EAP of the gear tooth profile engagement during running.

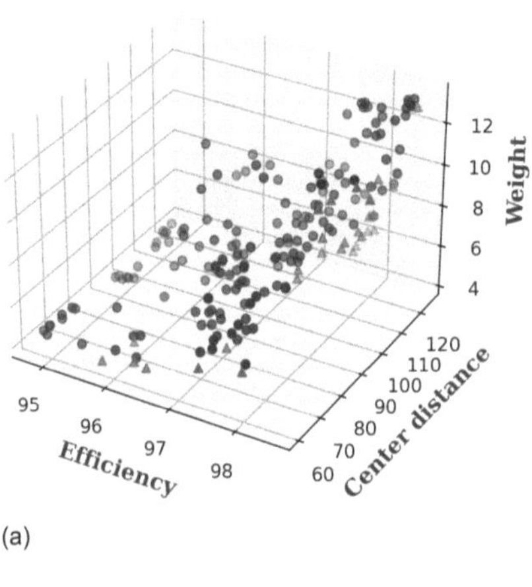

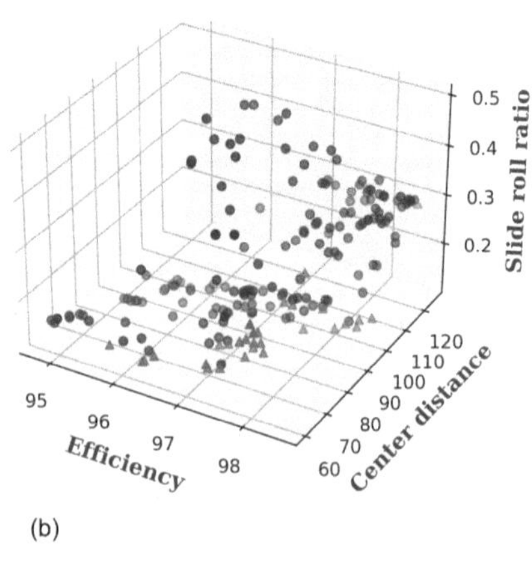

FIGURE 3.14. (a) and (b) 3D scatter plot of NSGA-II results.

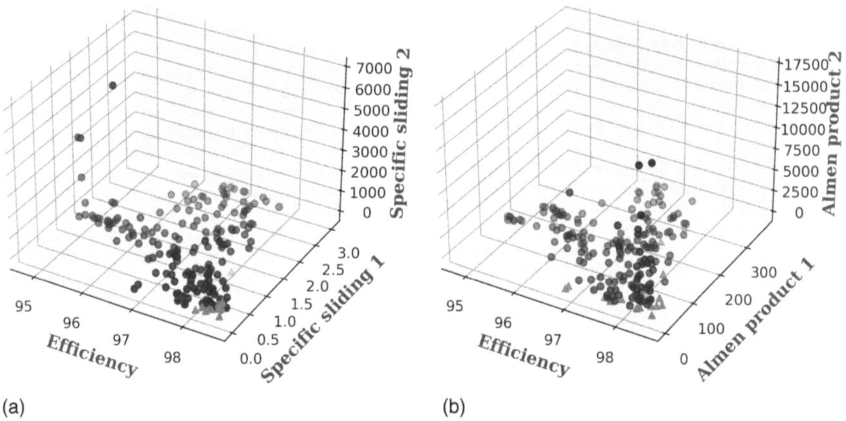

FIGURE 3.15 (a) and (b) 3D scatter plot of NSGA-II results.

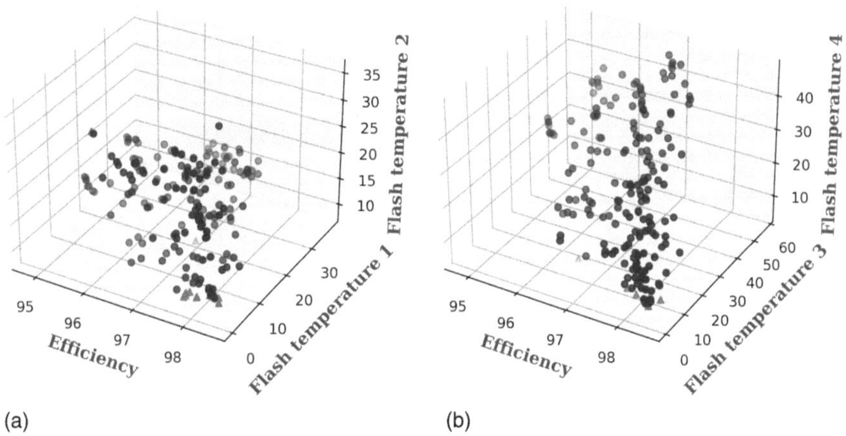

FIGURE 3.16 (a) and (b) 3D scatter plot of NSGA-II results.

Referring to Figure 3.15(b), the solutions close to the lower right corner offer higher efficiency and lower Almen product values at SAP and EAP of the gear tooth profile engagement during running.

Referring to Figure 3.16(a) and (b), the solutions close to the lower right corner offer higher efficiency and lower flash temperatures that are balanced at SAP and EAP of the gear tooth profile engagement during running.

Referring to Figure 3.17, the solutions close to the upper right corner offer higher efficiency and higher lambda ratios for the sun gear/planet gear engagement and the planet gear/ring gear engagement, respectively.

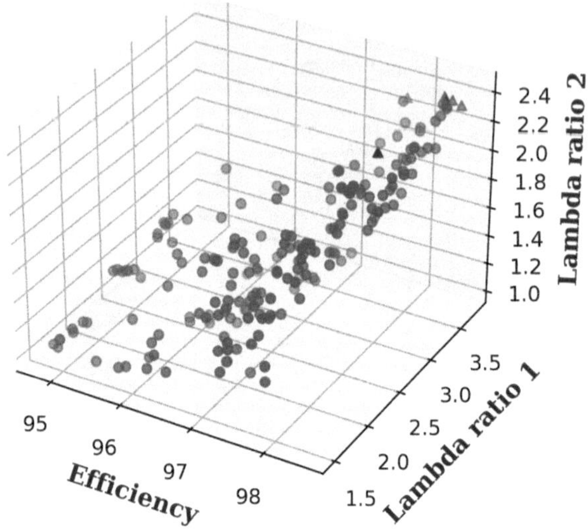

FIGURE 3.17 3D scatter plot of NSGA-II results.

3.6.17.6 Torque Ratio Comparison

The output torque of the spider arm for the three solutions of NSGA-II can be analyzed as follows:

For the first and second solutions,

Number of teeth of the planet gear, $Z_1 = 19/18$
Number of teeth of the sun gear, $Z_2 = 24$
Number of teeth of the ring gear, $Z_3 = 54$

Applying the equations given in Table 3.2,

Input speed to the sun gear = 104.651 rpm
Speed of the spider arm = 32.20 rpm

$$\text{Hence the torque ratio} = \frac{104.651}{32.20} = 3.25$$

Compared to the torque ratio of 3.33 of the existing planetary gearing, the above torque ratio is 2.5% less, which may not be acceptable.

For the third solution,

Number of teeth of the planet gear, $Z_1 = 20$
Number of teeth of the sun gear, $Z_2 = 21$
Number of teeth of the ring gear, $Z_3 = 51$

Applying the equations given in **Table 3.2**,

Input speed to the sun gear = 104.651 rpm
Speed of the spider arm = 30.52 rpm
Hence the torque ratio $= \frac{104.651}{30.52} = 3.429$

Compared to the torque ratio of 3.33 of the existing planetary gearing, the above torque ratio is 2.86% more, which is acceptable.

Therefore, the third solution is selected as the best solution.

3.6.17.7 Selection of Standard Module

From Table 3.5, it can be seen that the module for the best solution is 3.38 mm, which is a non-standard module. The nearest standard module is 3.5 mm. Hence, the final design can be modified to suit a standard module of 3.5 mm using the procedure given below.

The following design variables of the epicyclic gearing are made constant, based on the values from the best solution: (Table 3.4)

Number of teeth of the planet gear, $Z_1 = 20$
Number of teeth of the sun gear, $Z_2 = 21$
Number of teeth of the ring gear, $Z_3 = 51$
Number of planet gears, $n_p = 6$
Module, $m_n = 3.5$ mm (standard module)

The problem is reformulated with the three profile shift coefficients x_{n1}, x_{n2}, and x_{n3} taken as design variables, which will be computed again for the fixed parameters above. The upper and lower bounds for the profile shift coefficients are fixed based on the values from the best solution. (Table 3.4)

Profile shift coefficient for planet gear, x_{n1}: 0.6–0.7
Profile shift coefficient for sun gear, x_{n2}: −0.5—0.3
Profile shift coefficient for ring gear, x_{n3}: 0.1–0.2

The NSGA-II procedure is rerun for the modified problem formulation.
Table 3.6 shows a sample of the results generated for the standard module of 3.5 mm.

From the results generated, the solution listed in Table 3.7 is selected as the best solution.

From the above analysis, we can conclude that the solution from Table 3.7, obtained through the design optimization of the PTO planetary gearing using NSGA-II, can be considered the best solution.

TABLE 3.6
Sample of NSGA-II Output for Standard Module

Objectives

Center distance	Weight	Efficiency	Slide roll ratio	Specific sliding		Almen product		Flash temperature				Lambda ratio	
71.75	**5.78**	**97.60**	**0.12**	**1.99**	**2.66**	**186.17**	**3.78**	**21.32**	**13.54**	**20.35**	**21.16**	**1.74**	**1.45**
71.75	5.78	97.60	0.12	2.01	2.64	188.15	3.75	21.42	13.46	20.43	21.22	1.74	1.45
71.75	5.78	97.60	0.12	2.02	2.65	188.74	3.29	21.51	13.50	20.42	21.24	1.74	1.45
71.75	5.78	97.57	0.12	1.99	2.69	187.70	7.63	21.30	13.38	20.69	21.30	1.74	1.45
71.75	5.77	97.59	0.12	2.07	2.61	191.98	1.99	21.76	13.45	20.47	21.32	1.74	1.45

Constraints

Interference		Top land thickness			Contact ratio		Bending stress		Contact stress		Involute interference		Tip interference	Proper assembly
4.94	**7.47**	**0.30**	**1.40**	**1.64**	**0.13**	**0.45**	**260.32**	**290.93**	**266.69**	**1024.48**	**4.13**	**0.36**	**0.15**	**0.00**
4.93	7.50	0.29	1.40	1.64	0.13	0.45	260.23	290.86	267.27	1026.55	4.16	0.36	0.15	0.00
4.90	7.46	0.29	1.40	1.64	0.13	0.45	260.38	290.98	264.39	1025.15	4.13	0.36	0.14	0.00
4.99	7.57	0.28	1.40	1.64	0.12	0.44	259.97	290.66	271.86	1023.95	4.28	0.37	0.15	0.00
4.83	7.44	0.29	1.41	1.64	0.13	0.44	260.47	291.05	261.30	1026.90	4.12	0.36	0.14	0.00

Variables

x_{n1}	x_{n2}	x_{n3}
0.60	**–0.30**	**0.19**
0.61	–0.31	0.20
0.60	–0.31	0.19
0.62	–0.30	0.20
0.60	–0.32	0.20

TABLE 3.7
Optimized Design Parameters for Standard Module Using NSGA-II

	Variables				Objective Functions		
	Module, m (mm)	Number of Teeth (Planet Gear), Z_1	Number of Teeth (Sun Gear), Z_2	Number of Teeth (Ring Gear), Z_3	Total Weight, W (kg)	Efficiency, η	Center Distance (mm)
NSGA-II design	3.38	20	21	51	5.36	97.20	69.24
NSGA-II design for standard module	3.5	20	21	51	5.78	97.60	71.75

NOTES

1 Niches are stable subpopulations and niching techniques are used to promote the formation and maintenance of multiple final solutions.
2 Refer to Table 1.1.
3 The procedure is detailed in Section 2.2.1.
4 Refer to Section 3.6.9.4.
5 Refer to Section 3.6.9.10.
6 Refer to Section 3.6.10.3.
7 Refer to Section 2.9.

REFERENCES

AGMA 925–A03, "Effect of Lubrication on Gear Surface Distress."

Almen, J. O., 1950, "Surface Deterioration of Gear Teeth," *Mechanical Wear*, Cleveland, OH, USA, pp. 229–288.

Andersson, S. A. E., 1973, "On the Design of Internal Involute Spur Gears," *Transactions of Machine Elements Division*, Lund Technical University, Lund, Sweden.

Bhagubai, P. P. C., et al., "Design, Multi-Objective Optimization, and Prototyping of a 20 kW 8000 rpm Permanent Magnet Synchronous Motor for a Competition Electric Vehicle," *Energies* 2020, 13, 2465, https://doi.org/10.3390/en13102465, https://www.mdpi.com/journal/energies. Used under the terms and conditions of the Creative Commons Attribution (CC BY) license (http://creativecommons.org/licenses/by/4.0/)

Blok, H., June, 1937, "Les Températures de Surface dans les Conditions de Graissage sans Pression Extrême," *Second World Petroleum Congress*, Paris. Vol. 3, pp. 471–486.

Blok, H., "The Flash Temperature Concept," *Wear* 1963, 6(w63), 483–494.

Botka, I., "Egységes magyar homlokkerék fogazási rendszer." Mérnöki Továbbképzo Intézet Bp. 1953. (In Hungarian)

Buckingham, E., 1988, *Analytical Mechanics of Gears*, Dover Publications, New York, NY, USA.

Crowe, et al., "The case study approach." *BMC Medical Research Methodology* 2011, 11, 100. doi:10.1186/1471-2288-11-100, http://www.biomedcentral.com/1471-2288/11/100. Used under the terms and conditions of the Creative Commons Attribution License (http://creativecommons.org/licenses/by/2.0)

De Jong, K.A., 1975, "Analysis of Behavior of a Class of Genetic Adaptive Systems," PhD thesis, University of Michigan, Arm Arbor, MI.

Deb, K., Sindhya, K., and Okabe, T., 2007, "Self-adaptive simulated binary crossover for real-parameter optimization," *Proceedings of the 9th Annual Conference on Genetic and Evolutionary Computation, GECCO '07*, 1187–1194. ACM, New York, NY, USA, 2007. http://doi.acm.org/10.1145/1276958.1277190

Deb, K., et al., "A Fast and Elitist Multiobjective Genetic Algorithm: NSGA-II," *IEEE Transactions on Evolutionary Computation* 2002, 6(2), 182–197, https://doi.org/10.1109/4235.996017

Everitt, C.-M., Alfredsson, B., "The influence of gear surface roughness on rolling contact fatigue under thermal elastohydrodynamic lubrication with slip," *Tribology International* 2020, 151, https://doi.org/10.1016/j.triboint.2020.106394. Used under the terms and conditions of the Creative Commons Attribution CC BY license, (http://creativecommons.org/licenses/by/4.0/)

Holland, J. H., 1975, *Adaptation in Natural and Artificial Systems*, MIT Press, Ann Arbor.

JGMA 401-01, 1974, "Bending Strength of Spur Gears and Helical Gears," ISO6336:2006 based JGMA 401-01 (1974).

Krishnan, G., 2024, OPTIMO software – Algorithms for Multi-Objective optimization. https://github.com/epicyclic-book/optimo/. Used under the terms and conditions of the MIT license.

Martinsa, R. et al., "Friction coefficient in FZG gears lubricated with industrial gear oils: Biodegradable ester vs. mineral oil," *Tribology International* 2006, 39, 512–521.

Maximilian Prölß, et al., "Theoretical and Experimental Investigations on Transient Run-Up," *Procedures of Journal Bearings Including Mixed Friction Conditions, Lubricants* 2018, 6(4), 105; https://doi.org/10.3390/lubricants6040105. Used under the terms of the Creative Commons Attribution 4.0 International License (http://creativecommons.org/licenses/by/4.0/)

Ohlendorf, H., 1958, "Verlustleistung und Erwärmung von Stirnrädern." PhD thesis, THMünchen, pp. 9–16. (In German)

Rycerz, P., Kadiric, A., "The Influence of Slide–Roll Ratio on the Extent of Micropitting Damage in Rolling–Sliding Contacts Pertinent to Gear Applications," *Tribology Letters* 2019, 67, 63, https://doi.org/10.1007/s11249-019-1174-7. Used under the terms of the Creative Commons Attribution 4.0 International License (http://creativecommons.org/licenses/by/4.0/).

Schlenk, L., 1995, "Untersuchungen zur Freßtragfähigkeit von Großzahnrädern." PhD thesis, TU München, p 112.

Stribeck, R., "Die WesentlichenEigenschaften der Gleit und Rollenlager, Part 1," *Zeitschrift Ver Dtsch Ing* 1902a, 46(37), 1341–1348.

Stribeck, R., "Die WesentlichenEigenschaften der Gleit und Rollenlager, Part 2," *Zeitschrift Ver Dtsch Ing* 1902b 46(38), 1432–1438.

Stribeck, R., "Die WesentlichenEigenschaften der Gleit und Rollenlager, Part 3," *Zeitschrift Ver Dtsch Ing* 1902c 46(39), 1463–1470.

Townsend, D. P., 2013, *Dudley's Gear Handbook: The Design, Manufacture and Applications of Gears*, McGraw Hill, New York, NY, USA.

Yannibelli, V. et al., "A Comparative Analysis of NSGA-II and NSGA-III for Autoscaling Parameter Sweep Experiments in the Cloud," *Hindawi Scientific Programming* 2020, Article ID 4653204, 17, https://doi.org/10.1155/2020/4653204. Used under the terms and conditions of Creative Commons Attribution License (CC-BY).

4 Design Optimization of Wind Turbine Generator Gearbox Using Multi-Objective Particle Swarm Optimization Algorithm

Chandrasekaran G and Gokul Krishnan

4.1 INTRODUCTION

The following chapters focus on a selection of traditional single-objective optimization algorithms and their application in the multi-objective design optimization of epicyclic gearing. Specifically, the class of algorithms under consideration is non-classical, evolutionary, and stochastic optimization techniques, that is, methods of optimization that are based on the simulation of social behavior found in nature and its associated randomness. However, unlike genetic algorithms, there is no usage of operators such as crossover and mutation. Therefore, these algorithms are relatively simple to implement and run and could prove to be computationally less expensive depending on the use case.

Swarm intelligence comes under the category of artificial intelligence and is usually defined as the collective behavior exhibited by any decentralized and self-organized social system, either in nature or otherwise. Algorithms based on swarm intelligence use knowledge of collective entities (birds, insects, etc.) and their cooperative and group behavior in an attempt to find optimal solutions to a given problem. Two main swarm intelligence techniques widely used for the design and manufacturing optimization of epicyclic gearing are particle swarm optimization (PSO) and ant colony optimization (ACO).

4.2 PARTICLE SWARM OPTIMIZATION

The first such algorithm under consideration is particle swarm optimization (PSO). PSO was initially proposed by James Kennedy and Russell Eberhart (1995). The algorithm attempts to simulate the collective behavior observed in

DOI: 10.1201/9781003144465-4

nature among a group of individuals, such as a school of fish or a flock of birds. Each individual particle moves through the problem space and adjusts its course by following the current optimum particles.

PSO was modeled on the social behavior of birds and insects by considering every individual swarm member as a particle. The collective intelligence of these dispersed particles, owing to their cooperation and learning, is modeled for the purpose of understanding microscopic transformations and then finding a solution to an optimization problem within a search space for a given objective. Proper formulation and specification of the objective function or fitness function is a prerequisite for using meta-heuristic algorithms. The fitness function works by transforming a set of input variables into a fitness value that defines how far away the particles are from the ideal solution.

PSO treats each potential solution in the problem space as a *particle*. These particles move at random in search of an objective, such as the minimization of a function. In time, these particles learn from their own past trajectory and also from that of their neighboring particles and adjust their course of motion accordingly. Finally, these particles converge toward a solution (the minima or maxima of the objective under consideration). Each particle has an associated position, velocity, and fitness value.

Because PSO is a single-objective optimization algorithm, the technique cannot be directly applied to solve multi-objective problems. A fair amount of research has been done on extending PSO to solve multi-objective optimization problems and obtain a Pareto-optimal solution set.[1] These techniques fall under the category of multi-objective particle swarm optimization (MOPSO), and a few such techniques have been proposed by Parsopoulos and Vrahatis (2002) and Coello et al. (2004).

4.3 STANDARD PSO PROCEDURE

This section describes the steps involved in the standard PSO procedure that solves for the minimization of a single-objective, having one or more design variables and optionally a set of constraints. At the core of the algorithm is the velocity equation, which is visually depicted in Figure 4.1.

Initially, a set of N particles is generated at random positions within the problem space, where N is the population size. Each particle i has the following properties:

- The position vector X_i is defined by the independent design variables. The number of design variables defines the dimension of the problem space.
- The velocity vector V_i defines the instantaneous velocity of the particle along each dimension.
- The fitness value of the particle $P_i = F(X_i)$, where F is the fitness function.
- The position vector $(X_{best})_i$ corresponds to the best fitness value achieved by the particle so far.

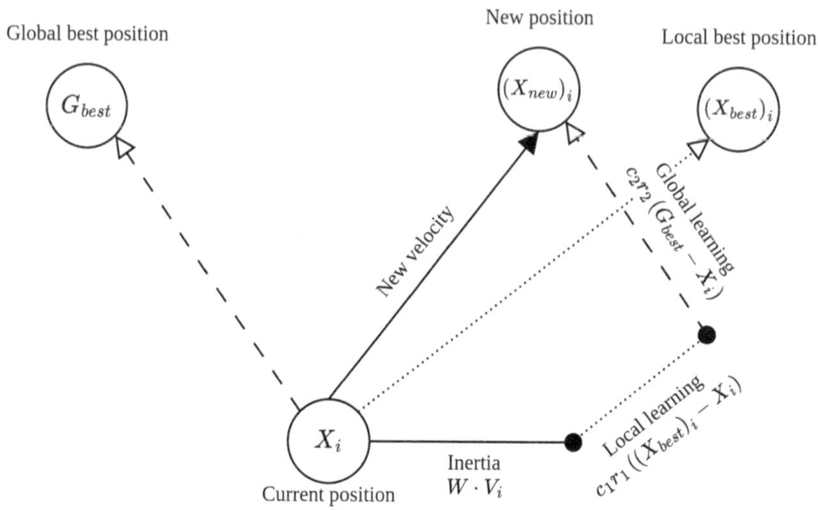

FIGURE 4.1 PSO algorithm procedure.

In addition, the following global properties are specified:

- The position vector G_{best} corresponds to the global best fitness value achieved so far.
- Upper and lower bounds for each design variable.
- Number of particles (population size).
- Number of iterations.
- Parameters that introduce random variation in the search space.

At each iteration, the velocity and position of each particle are updated based on two factors: (i) the particle's own best position achieved so far $(X_{best})_i$, and (ii) the global best position achieved so far G_{best}.

The particle velocity and position are updated based on the formula:

$$V_i = W \cdot V_i + c_1 r_1 \left(\left(X_{best} \right)_i - X_i \right) + c_2 r_2 \left(G_{best} - X_i \right)$$

$$X_i = X_i + V_i$$

where W is the inertial parameter; c_1 and c_2 are learning factors; r_1 and r_2 are randomly generated numbers in the 0–1 range.

Then, the best position of each particle and the global best position are updated based on whether the following conditions are satisfied:

1. The new position is an improvement over the previous best position. That is, in the case of minimization,

$P_i < F\left(\left(X_{best}\right)_i\right)$, for local best update
$P_i < F\left(G_{best}\right)$, for the global best update

2. The new position passes all the specified constraints, if any.

The process is repeated either until the required number of iterations is reached or based on any other termination conditions specified. In the end, position G_{best} represents the final optimum solution.

4.3.1 Algorithm Pseudo-code

The complete algorithm can be visualized through the flow chart in Figure 4.2 and pseudo-code representation in Listing 4.1.

LISTING 4.1 PSO ALGORITHM PSEUDO-CODE

```
FOR each particle i
    Initialize X_i to a random vector within bounds
    Initialize V_i to zero vector
    Initialize best position (X_best)_i to X_i
END FOR

Select random particle r
Initialize global best position G_best to X_r

WHILE iterations remaining
    FOR each particle i
        Compute fitness value P_i = F(X_i)
        IF P_i < F((X_best)_i)
            (X_best)_i = X_i
        END IF
        IF P_i < F(G_best)
            G_best = X_i
        END IF
    END FOR

    FOR each particle i
        Update velocity according to equation:
        V_i = W * V_i +
            c_1 * r_1 * ((X_best)_i - X_i) +
            c_2 * r_2 * ((G_best) - X_i)
        Update position according to equation:
        X_i = X_i + V_i
    END FOR
END WHILE
```

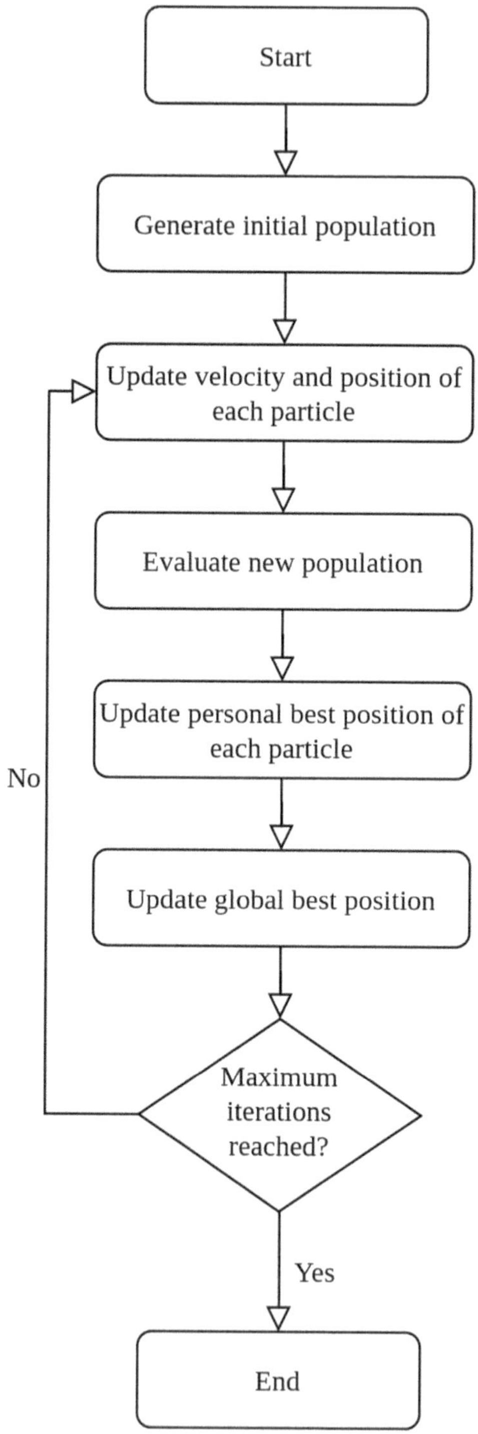

FIGURE 4.2 PSO algorithm flow chart.

4.4 MULTI-OBJECTIVE PSO (MOPSO)

This section describes the techniques and challenges involved in extending the PSO algorithm for solving constrained multi-objective optimization problems.

4.4.1 Fitness Vector

PSO deals with a single fitness value associated with each particle. In order to adapt the technique to solve problems containing multiple objectives, several approaches have been proposed. A simple approach is to use a weighted average of the fitness values from each objective function, that is,

$$P_{\text{combined}} = \frac{w_1 p_1 + w_2 p_2 + \ldots + w_N p_N}{N}$$

where N is the number of objectives.

Several other weighted aggregation techniques have been proposed (Jin et al., 2001).

However, an alternate approach involves modifying the scalar fitness value P into a fitness vector consisting of a separate value for each objective function, that is,

$$P = \left[p_1, p_2, \ldots, p_N\right]$$

where N is the number of objectives.

This approach has been selected in the implementation covered in this book.

4.4.2 Dominance Operator

While the use of a fitness vector is beneficial when dealing with multi-objective problems, it introduces the need for a new comparison operator that can be used to compare the fitness vectors of two particles. Therefore, the standard comparison operator is replaced with a *dominance* operator that checks for the Pareto-dominance of one solution by another. Consider two particles with fitness vectors P and Q.

Then, assuming minimization, P, is said to dominate Q if,

$P_i \leq Q_i$ for all values of i, and
$P_i < Q_i$ for at least one value of i,

where $i = \{0 \text{ to } N\}$, and N is the number of objectives.

4.4.3 Archive of Non-dominated Solutions

In order to extract a Pareto-optimal solution set, the approach considered is to maintain an archive of non-dominated solutions separate from the population set and update this archive at the end of each iteration. The procedure is described in detail in Section 4.4.9.

4.4.4 Population Diversity

In the standard PSO technique, all particles tend to follow a single global optimum solution, or G_{best}. Therefore, particles tend to bunch together after several iterations. When dealing with multi-objective optimization problems, we are interested in a set of non-dominated solutions which form the Pareto-optimal solution set. Therefore, there is a need to ensure a healthy diversity of Pareto-optimal solutions.

A number of approaches can be used in order to maintain the diversity of solutions. First, the single global optimum G_{best} is replaced by a global optimum $(G_{best})_i$ specific to each particle i. Each particle follows a global optimum particle chosen at random from the archive of non-dominated solutions. The following criterion is used in choosing the global optimum.

4.4.4.1 Crowding Distance

Each particle in the archive is assigned a crowding distance value that defines the density of particles surrounding it. Particles in denser areas are assigned a smaller crowding distance. The crowding distance is computed separately for each objective function, that is, along each dimension in the objective space, and then averaged together.

4.4.4.2 Roulette-Wheel Selection

For each particle, a global optimum solution is chosen at random from the archive of solutions through a roulette-wheel selection process with the weights defined by the crowding distance values. This ensures that there is a higher probability of particles chosen from less dense areas.

4.4.5 Handling Constraints

Constraints restrict the movement of particles by limiting the possible values that the design variables can assume. The feasibility of solutions is decided by one or more constraint functions defined as part of the problem. A solution is considered feasible if and only if all provided constraint tests pass. The restriction in the movement of particles limits the area of search space covered, and this reduces the quality of solutions provided by the algorithm.

In order to maintain a good quality of solutions and to ensure that particles cover a wide extent of the search space, the following approach is proposed. At every iteration, particles are allowed to move freely and assume any point in the search space irrespective of any constraint violation. However, a particle is only allowed to update its best position and best fitness $(P_{best})_i$ if all constraints pass. This allows particles to assume infeasible positions during their search for better solutions, whereas the fitness vector defined by $(P_{best})_i$ always holds the best feasible solution achieved so far.

Computation of dominance and crowding distance factors is performed based on the particle's best feasible position rather than the current actual position.

4.4.6 MOPSO Implementation

An implementation of the MOPSO algorithm is described along with the corresponding pseudo-code. We start with an overview of key functions used in the algorithm code.

4.4.6.1 RouletteWheelSelection

We employ a roulette-wheel selection process, which is essentially the randomized selection of members from a population, with each member assigned a weight that defines the probability of that member being selected. This can also be called a fitness-proportionate selection model. The function takes as input a list of weights and outputs the index of the member being selected. The weights array is normalized such that the elements add up to one (Listing 4.2).

LISTING 4.2 IMPLEMENTATION OF ROULETTE-WHEEL SELECTION

```
FUNCTION RouletteWheelSelection(weights):
    weights = normalize(weights)
    weights = generate_cumulative_sum(weights)
    r = rand() # random value in range [0,1]
    FOR EACH value in weights:
        IF value >= r:
            return i
        END IF
    END FOR
END FUNCTION
```

4.4.6.2 SelectGlobalOptimum

The global optimum or global best solution $(G_{\text{best}})_i$ for particle i is a non-dominated solution chosen from the archive at random that the particle will follow. To select a global optimum, a roulette-wheel selection is used with the particle crowding distances used as weights. Because the crowding distance value is inversely proportional to the density surrounding the particle, a larger crowding distance increases the probability of a global optimum being chosen from sparser regions. This ensures a healthy diversity in the population (Listing 4.3).

LISTING 4.3 IMPLEMENTATION OF GLOBAL OPTIMUM SELECTION

```
FUNCTION SelectGlobalOptimum(archive):
    weights = [c1, c2 ... cnum_archive]
    index = RouletteWheelSelection(weights)
    return index
END FUNCTION
```

4.4.6.3 SelectMemberToDelete

To select one among the non-dominated solutions to be removed from the archive, a roulette-wheel selection is used, with the inverse of the particle crowding distances used as weights. This increases the probability that particles from denser regions are chosen for removal (Listing 4.4).

LISTING 4.4 IMPLEMENTATION OF REDUNDANT MEMBER SELECTION

```
FUNCTION SelectMemberToDelete(archive):
    weights = [1/c1, 1/c2 ... 1/cnum_archive]
    index = RouletteWheelSelection(weights)
    return index
END FUNCTION
```

4.4.6.4 ComputeDominance

The function accepts a set of particles as input and computes the Pareto-dominance state of each particle based on the best feasible fitness achieved by the particle, or *best_fitness*. For each particle, the *is_dominated* property is set to *true* for all particles that are dominated by at least one other particle in the set.

4.4.6.5 ComputeCrowdingDistance

The function accepts a set of particles as input and computes the crowding distance factor of each particle, based on the best feasible fitness achieved by the particle, or *best_fitness*. The algorithm for crowding distance computation is adapted from the NSGA-II paper (K. Deb et al., 2002), with a few minor changes.

Because the crowding distance and inverse crowding distance are used as weighting factors in the roulette-wheel selection routines, the value is clamped within the half-closed range (0, 1) such that the crowding distance value is always nonzero and less than or equal to one.

The default value for crowding distance is chosen as 1.0, which ensures equal probability of each particle being selected.

4.4.6.6 FilterNonDominated

The function accepts a set of particles as input and removes each particle that has the *is_dominated* property set to *true*. It then returns the filtered non-dominated population set.

4.4.6.7 Fitness

This function is user-defined. It accepts an array of design variables as input and outputs an array of objective function values. Because we are dealing with a minimization problem, objective functions that are to be maximized must be negated.

4.4.6.8 Constraints

This function is user-defined. It accepts an array of design variables as input and outputs an array of constraint function values. Constraint functions are specified in such a way that they must produce a value greater than or equal to zero in order for the test to pass, that is, the inequality $C_i \geq N$ is rewritten as, $C_i - N \geq 0$.

4.4.7 Initialization

The following global properties are set (Listing 4.5):

- *num_vars*: number of design variables
- *bounds_min*: array of values representing minimum bounds for each design variable
- *bounds_max*: array of values representing maximum bounds for each design variable
- *num_iter*: number of iterations
- *num_pop*: population size
- *num_archive*: archive size for the non-dominated solution set
- w, c_1, c_2: intertial parameter and learning factors

LISTING 4.5 GLOBAL PROPERTIES

```
num_vars = 4
bounds_min = [x1, x2 ... xnum_vars]
bounds_max = [x1, x2 ... xnum_vars]
num_iter = 100
num_pop = 200
num_archive = 50
w = 0.5
c1 = 2
c2 = 2
```

Next, we allocate the required number of particles. Each particle is initialized at randomized positions within the provided minimum and maximum bounds (Listing 4.6).

LISTING 4.6 ALLOCATION OF PARTICLES

```
Population = [P₁, P₂ ... P_num_pop]
FOR EACH particle Pᵢ in Population:
    Pᵢ.position = [r₁, r₂ ... r_num_vars]
    Pᵢ.velocity = [v₁, v₂ ... v_num_vars]
    Pᵢ.fitness = Fitness(Pᵢ.position)
    Pᵢ.constraints = Constraints(Pᵢ.position)
    Pᵢ.best_position = Pᵢ.position
    Pᵢ.best_fitness = Pᵢ.fitness
    Pᵢ.is_dominated = false
    Pᵢ.crowding_dist = 1.0
END FOR
```

4.4.8 Iteration

Following the initialization, we reach the core of the algorithm where each iteration is processed. First, we loop through each particle in the population set (Listings 4.7).

LISTING 4.7 THE OVERALL ITERATION

```
FOR j = [1 to num_iter]:
    FOR i = [1 to num_pop]:
        Update each particle
    END FOR
    Update the archive
END FOR
```

For each particle, a global optimum solution is chosen from the archive that the particle will follow. A roulette-wheel selection process is utilized with the particle crowding distances used as weights. This prioritizes particles from sparser regions and thereby ensures a good diversity of solutions (Listings 4.8).

LISTING 4.8 STEP 1 – SELECT GLOBAL OPTIMUM

```
P_optimum = SelectGlobalOptimum(Archive)
```

We then compute the velocity of each particle based on the standard velocity equation and move the particle to its new position. It should be noted that the movement of particles is not restricted based on the provided constraints. The position is clamped to the provided bounds along each dimension (Listing 4.9).

LISTING 4.9 STEP 2 – UPDATE VELOCITY AND POSITION

```
Pi.velocity = w * Pi.velocity +
    c1 * r1 * (Pi.best_position - Pi.position) +
    c2 * r2 * (Poptimum.best_position - Pi.position)
Pi.position = Pi.position + Pi.velocity
Clamp(Pi.position, bounds_min, bounds_max)
```

Then, the fitness and constraint vectors are computed from the new position (Listing 4.10).

LISTING 4.10 STEP 3 – EVALUATE FITNESS AND CONSTRAINTS

```
Pi.fitness = Fitness(Pi.position)
Pi.constraints = Constraints(Pi.position)
```

In the case that all constraints are satisfied and the new position dominates the previous best position, the local best position and best fitness of each particle are updated to the new values.

On the other hand, in case the new solution does not dominate the previous best solution, the solution can still be accepted based on a probability value. This promotes a better diversity of solutions (Listing 4.11).

LISTING 4.11 STEP 4 – UPDATE LOCAL BEST

```
IF Pi.constraints >= 0:
    IF Pi.fitness dominates Pi.best_fitness:
        Pi.best_position = Pi.position
        Pi.best_fitness = Pi.fitness
    ELSE IF rand() > 0.5:
        Pi.best_position = Pi.position
        Pi.best_fitness = Pi.fitness
    END IF
END IF
```

4.4.9 Archiving

Archiving refers to the extraction of non-dominated solutions from the population set into a separately maintained archive. Once all particles have been processed in an iteration, the particles from the new population are merged with the pre-existing archive, which is initially an empty set. Then, the dominance factor is re-computed on this merged population. Solutions that are dominated by other solutions are filtered out, keeping only non-dominated solutions in the archive (Listing 4.12).

LISTING 4.12 ARCHIVING – STEP 1

```
Archive.append(Population)
ComputeDominance(Archive)
Archive = FilterNonDominated(Archive)
```

Then, the crowding distance factor is re-computed on the filtered solution set. The crowding distance is used as a weighting factor in the selection of global optima and in the elimination of redundant solutions from the archive.

In case the archive size exceeds the maximum allowed size, members are removed from the archive until the archive size requirement is met. This is done using a roulette-wheel selection process with the inverse crowding distance criterion. This prioritizes particles from denser regions for removal and ensures a stable archive size (Listing 4.13).

LISTING 4.13 ARCHIVING – STEP 2

```
ComputeCrowdingDistance(Archive)
WHILE size(Archive) > num_archive:
    r = SelectMemberToDelete(Archive)
    Archive.remove(r)
END WHILE
```

At the end of the final iteration, the particles in the archive represent the final non-dominated set of solutions. The *best_position* and *best_fitness* vectors hold the final feasible solution of each particle in the archive. The data can be output into a text file format, such as comma-separated values (CSV), or plotted to a graph for further processing and visualization. A good amount of diversity is preserved, which helps the decision-maker in selecting the most optimal solution for the problem at hand.

4.5 CASE STUDY: DESIGN OPTIMIZATION OF A WIND TURBINE GENERATOR (WTG) GEARBOX USING PARTICLE SWARM OPTIMIZATION

In a wind turbine, electricity is produced by converting the kinetic energy of air in motion. The blowing wind flows over the rotor blades, creating lift, thereby causing the blades to turn. The rotor, which is connected to the drive shaft, turns an electric generator, which generates electricity. In between the rotor shaft and the electric generator shaft, a gearbox is often used to increase the rotational speed from a low-speed rotor shaft to a high-speed shaft of the electric generator. Typically, the gearbox converts the 15 RPM turning speed of the rotor into the 1,800 RPM of the electric generator, the gear ratio being 120:1.

Modern wind power generation provides a significant and increasing portion of the global electrical energy requirement, with decreasing wind power production costs. The global wind power generation has crossed the threshold of 1 million megawatt in 2023, with 1,047,288 MW of installed capacity. Modern multi-megawatt wind turbines have an electric power generation capacity of 10 MW (on-shore) and 15 MW (off-shore).

4.5.1 Overview of WTG Gearbox

The wind turbine gearbox converts the high torque on the main shaft to the low torque on the high-speed shaft of the electric generator (refer to Figure 4.3). The wind turbine gearboxes may contain both the planetary and parallel axis gear stages depending on the capacity requirement of a wind turbine. Planetary gearing systems provide higher power densities than parallel axis gears and a large change in rpm within a small volume. These two qualities are very essential in the case of a wind turbine system since it reduces both the weight and the space occupied in the nacelle.

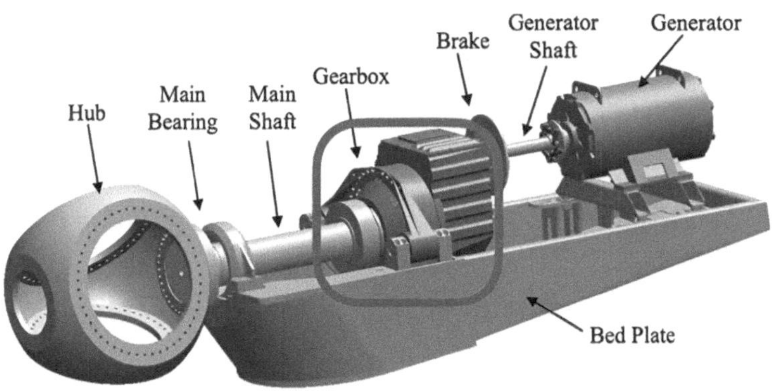

FIGURE 4.3 Wind turbine drive train (Marti-Puig et al., 2024).

In wind power generation, normally there is wind load uncertainty due to gusts of wind and their ever-changing direction for different locations and in different years. The wind turbine gearbox is a critical component that undergoes severe transient loading during start-ups, shut-downs, emergency stops, and grid connections. The wind turbine gearbox operates at unusual loading and environmental conditions, due to which special design considerations must be undertaken during its design, and specific international standards like IEC 61400-4:2012 (E) of The International Electrotechnical Commission, ANSI/AGMA 6006-B20 of the American standards, BS EN 50308:2004 of the British standards, etc., should be referred to.

4.5.2 Case Study of a 2.0 MW Wind Turbine Generator (WTG) Gearbox

The design optimization is attempted for the 2.0 MW WTG gearbox planetary gear train with the specifications listed in Table 4.1.

To find the speed of individual elements of the existing planetary gear train, let

- Number of teeth on the sun wheel, $z_2 = 20$
- Number of teeth on planet gears, $z_1 = 36$
- Number of teeth on the annular ring gear, $z_3 = 91$

The input to the planetary gearbox is given to the spider arm L. The ring gear is held stationary, and the output is taken from the sun gear. Hence, this arrangement comes under the second type of planetary gearing.[2] The sun gear shaft is connected to the electric generator through another set of gears.

The last row of Table 4.2 provides us with the equation for the speed of individual elements.

Input speed of WTG rotor to the spider arm shaft = 15 rpm (CW)

that is, $y = 15$

TABLE 4.1
WTG Planetary Gearing Specifications

	Sun Gear	Planet Gear	Ring Gear
Normal module (mm)	16	16	16
Number of teeth	20	36	91
Normal pressure angle (°)	20	20	20
Helix angle (°)	8	8	8
Individual shift in the basic profile X	0.4	0.3156	1.6429
Base diameter (mm)	303.307	545.95	1380.05
Pitch diameter (mm)	320	576	1456
Face width (mm)	310	310	310
Number of planets	–	3	–

TABLE 4.2
Speed Calculation for Planetary Elements

Operation	Spider Arm L	Sun Wheel Z_s	Planet Wheel Z_p	Annular Wheel Z_a
1. Arm L is locked and sun wheel S is given +1 revolution.	0	+1	$-\frac{Z_s}{Z_p}=-\frac{20}{36}$	$-\frac{Z_s}{Z_a}=-\frac{20}{91}$
2. Multiply by x	0	X	$-\frac{Z_s}{Z_p}x=-\frac{20}{36}x$	$-\frac{Z_s}{Z_a}x=-\frac{20}{91}x$
3. Add y to all columns	y	$y+x$	$y-\frac{Z_s}{Z_p}x=y-\frac{20}{36}x$	$y-\frac{Z_s}{Z_a}x=y-\frac{20}{91}x$

Since the annulus or ring gear is locked,

$$y-\frac{20}{91}x=0$$

$$\therefore x=68.25$$

$$\text{Speed of planets}=y-\frac{20}{36}x=15-\frac{20}{36}\times 68.25=-22.92\text{rpm}\,(\text{ACW})$$

$$\text{Speed of the sun gear}=y+x=83.25(\text{CW})$$

$$\text{Epicyclic speed ratio}\; i_p=\frac{83.25}{15}=5.55$$

$$\text{Power}=2.0\text{ MW}=2.0\times 1000\text{ kW}$$

4.5.3 Calculation of Gear Parameters

Gear parameters are calculated using the procedure explained in Chapter 3 for Power Take-Off (PTO) gearbox.[3]

4.5.4 Design Space

The existing gear design problem for gears made of a chosen material can be reduced to an eight-dimensional design problem where the eight free parameters are the number of teeth on the planet pinion z_1, the number of teeth on the sun gear z_2, the number of teeth on the ring gear z_3, the module m_n, the three profile shift coefficients x_{n1}, x_{n2}, and x_{n3} of the planet gear, the sun gear, and the ring gear, respectively, and the number of planets n_p. The inequality constraints which bound this design space are the bending stress, Hertz stress, contact ratio, top land thickness, operating interference of gears, and involute and tip interference of internal gears.

4.5.5 Objective Functions and Constraints

Fourteen objective functions for minimization/maximization with an array of inequality and equality constraints have been considered, and the problem can be stated as follows:

The following objective functions are considered for the design optimization of the planetary gear train:

$$f_{obj} = \begin{cases} Min.center\ distance\ between\ planetary\ gear\ pair \\ Min.weight\ of\ the\ planetary\ gear\ train \\ Max.efficiency\ of\ planetary\ gear\ train \\ Min.\ slide\ roll\ ratio \\ Balancing\ the\ specific\ sliding \\ Balancing\ the\ Almen\ product \\ Balancing\ the\ flash\ temperature \\ Balancing\ the\ lambda\ ratio \end{cases}$$

The independent design variables of the planetary gear train include a module, the number of teeth of the pinion and the three profile shift coefficients x_{n1}, x_{n2}, and x_{n3} of the planet gear, the sun gear, and the ring gear, respectively. The vector of design variables is given below:

$$x(i) = \begin{cases} planet\ gear\ no.of teeth, Z_1 \\ sun\ gear\ no.of teeth, Z_2 \\ ring\ gear\ no.of teeth, Z_3 \\ planet\ gear\ profile\ shift\ coefficient, x_{n_1} \\ sun\ gear\ profile\ shift\ coefficient, x_{n_2} \\ ring\ gear\ profile\ shift\ coefficient, x_{n_3} \\ normal\ module, m_n \\ number\ of\ planets, n_p \end{cases}$$

where $i = 1, 2, \ldots,$ the number of variables.

A number of constraints, which act as sub functions to restrict the objective function, are introduced into the objective function to be optimized and are used to provide suitable design choices. The very often encountered failure types in a transmission gear are tooth fracture, surface fatigue, abrasion, and plastic deformation. Thus, the following constraints are considered for the design optimization of the planetary gear train:

$$g(j)=\begin{cases}\textit{Bending stress constraint}\\ \textit{Surface contact compressive stress constraint}\\ \textit{Contact ratio constraint}\\ \textit{Top land thickness constraint}\\ \textit{Operating interference of planet gear constraint}\\ \textit{Operating interference of sun gear constraint}\\ \textit{Operating interference of ring gear constraint}\\ \textit{Involute interference of internal gear constraint}\\ \textit{Tip interference of internal gear constraint}\\ \textit{Proper assembly of planetary gearing constraint}\end{cases}$$

where $j = 1, 2, \ldots,$ the number of constraints.

4.5.6 Establishing a Mathematical Model

To establish a mathematical model for the planetary gear train design optimization, mathematical equations for the design variables, objective functions, and necessary constraints are to be developed. These sections are covered in more exhaustive detail in Chapter 3 for the PTO gearbox.[4]

4.5.7 Objective Functions for the Epicyclic Gearing

In the design of WTG gearbox planetary gearing, a number of load cases have to be considered for the independent gears, including non-torsional loads and dynamical effects that are transferred among the components of the drivetrain.

The dynamics of a WTG gearbox can be affected by several factors, including the wind speed changes, the rotor speed, the gear ratio, the stiffness of the components, the lubrication of the gears, and the applied forces (including the meshing forces, the centrifugal forces, and the gyroscopic forces), which can lead to fatigue damage (Llopis-Albert et al., 2024).

The main sources of the failures in WTG gearboxes and bearings encompass fatigue, wear, misalignment, corrosion, vibrations, and resonances in the drivetrain components, and harsh environments such as wind gusts, rain, snow, high temperatures, moisture, and dust. This leads to a significant cost of maintenance due to the high cost of spare parts, the need for specialized technicians, the WTG downtime costs, and the insurance costs to cover possible failures (Llopis-Albert et al., 2024).

In Chapter 3, we have seen that the sun gear is the input gear (driver) and the planet gear is the driven gear. In turn, the planet gear is the driver to the ring gear, which is the driven gear. However, in this chapter, we can see that the planet gear is the input gear (driver) since it receives input torque from the wind turbine rotor through the planet carrier. The sun gear and the ring gear have become the driven gears. Hence, the equations for the objective functions and constraints change accordingly, compared to those given in Chapter 3.

4.5.7.1 Minimization of Center Distance

The objective function for minimization of center distance is given by,

$$f_1 = \min a = \frac{m_n\left(z_2 + z_1\right)}{2}$$

4.5.7.2 Minimization of Weight

To develop a mathematical model for the second objective function of minimization of the weight of the planetary gear train, the second objective function may be written as,

Minimize f_2 = min total weight

$$\begin{aligned} f_2 &= W = W_1 + W_2 + W_3 \\ &= n_w \rho \pi \bar{d}_1 A_1 + \rho \pi \bar{d}_2 A_2 + \rho \pi \bar{d}_3 A_3 \end{aligned}$$

where W_1, W_2, and W_3 are the weight of the planet gear, the sun gear, and the ring gear, respectively, in kg; A is the swept area, mm^2; $\bar{d}$ is the diameter at C.G. mm; ρ is the material density, kg/mm^2.

Referring to Figure 3.6, assuming a rim thickness of 1 m_n for a gear.

For the sun gear and the planet gears,

$$\bar{d} = d_w - 2m_n$$

For the ring gear,

$$\bar{d} = d_w + 2m_n$$

The swept area is given by,

$$A = b \times 2m_n$$

4.5.7.3 Maximization of Efficiency

In the wind turbine gearbox, the rotor shaft drives the planet carrier shaft, which in turn drives the planet gears. Hence, the planet gear is the driver gear, and the sun gear is the driven gear.

The objective function for efficiency increase is given by,

Maximize

$$f_3 = \eta = \left[1 - \left(\mu_{m_1} Hv_1 + \mu_{m_2} Hv_2\right)\right]100$$

where Hv_1 and Hv_2 are the power loss factors.

The power loss factor Hv_1 for the planet gear and the sun gear pair is given by,

$$Hv_1 = \frac{\pi}{z_1 \cos\beta_b} \cdot \frac{1+i_1}{i_1}\left(1 - \varepsilon_{\alpha_1} + \varepsilon_1^{2} + \varepsilon_2^{2}\right)$$

where ε_{α_1}, the base contact ratio for the planet gear and sun gear pair, is given by,

$$\varepsilon_{\alpha_1} = \varepsilon_1 + \varepsilon_2$$

The power loss factor Hv_2 for the planet gear and ring gear pair is given by,

$$Hv_2 = \frac{\pi}{z_1 \cos\beta_b} \cdot \frac{1+i_2}{i_2}\left(1 - \varepsilon_{\alpha_2} + \varepsilon_1^{'2} + \varepsilon_3^{2}\right)$$

where ε_{α_2}, the base contact ratio for the planet gear and ring gear pair, is given by,

$$\varepsilon_{\alpha_2} = \varepsilon_1^{'} + \varepsilon_3$$

The base pitch and the base transverse pitch of the planet gear are given by,

$$p_b = \frac{\pi d_{b_1}}{z_1}$$

$$p_{b_t} = \frac{p_b}{\cos\beta}$$

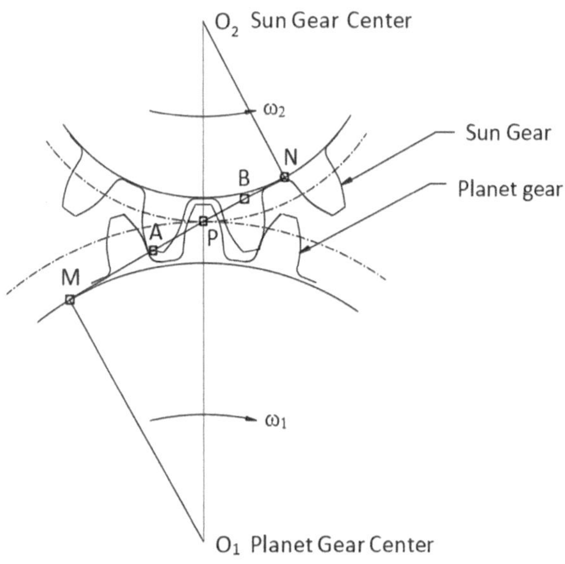

FIGURE 4.4 External gear mesh contact path.

Referring to Figure 4.4, for a clockwise rotation of the input planet gear, the contact path with the mating sun gear is along the common tangent to the respective base circles. The contact takes place along the line MN, and point A denotes the start of gear pair contact, and point B denotes the end of gear pair contact.

The addendum contact ratio for planet gear ε_1 and sun gear ε_2, in the planet-sun gear engagement, is given by,

$$\varepsilon_1 = \frac{AP}{p_{b_t}}$$

$$\varepsilon_2 = \frac{BP}{p_{b_t}}$$

Referring to Figure 4.5, for the planet gear and the ring gear pair engagement, the contact path is along the common tangent to the respective base circles. The contact takes place along the line KF, and point E denotes the start of gear pair contact and point F denotes the end of gear pair contact. As before, the pitch point P is

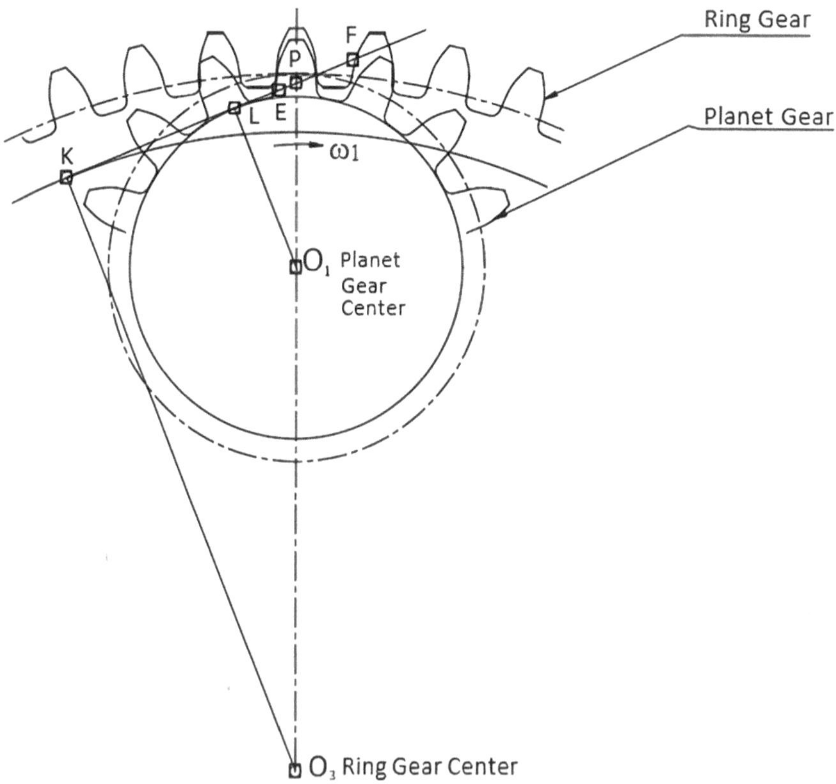

FIGURE 4.5 Internal gear mesh contact path.

where the pitch circles of the mating gear pair meet and also at the line connecting their centers of rotation.

Similarly, the addendum contact ratio for planet gear ε_1' and ring gear ε_3, in the planet ring gear engagement, is given by,

$$\varepsilon_1' = \frac{FP}{p_{b_t}}$$

$$\varepsilon_3 = \frac{EP}{p_{b_t}}$$

The base helix angle is given by,

$$\beta_b = \tan^{-1}\left(\frac{d_{b_1}\tan\beta}{\frac{m_n z_1}{\cos\beta}}\right)$$

Also,

$$\beta_b = \sin^{-1}\left(\sin\beta\cos\alpha_n\right)$$

The mean coefficient of friction is assumed to be constant for every contact point of the engagement region, even though it varies slightly along the path of contact. The approximate value of the mean coefficient of friction is given by,

$$\mu_{m_1} = 0.048\left(\frac{F_{t_{b1}}/b}{\gamma_{\varepsilon c_1}\rho_{redc_1}}\right)^{0.2}\eta_{oil_1}{}^{-0.05}R_{a_1}{}^{+0.25}X_{L_1}$$

where $F_{t_{b1}}$ is the circumferential force at the base circle (N); b is the face width for spur gears (mm); $\gamma_{\varepsilon c_1}$ is the sum velocity at the operating pitch circle (m/s); ρ_{redc_1} is the reduced radius of curvature at pitch point (mm); η_{oil} is the dynamic oil viscosity at oil temperature (mPa.s); R_a is the arithmetic mean roughness (μm); X_L is the lubricant correction factor.

The load-dependent losses are influenced by both gear geometry and lubricant properties.

The tangential force at the base circle for the planet gear and the sun gear engagement is,

$$F_{t_{b1}} = \frac{1000T_1}{r_{b_1}}$$

where T_1 is the torque for the planet gear and is calculated as,

$$T_1 = \frac{60000P}{2\pi N_1 n_p}$$

The power transmitted,

$$P = 2.0 \times 1000 Kw$$

The sum velocity at operating pitch circles for the planet gear and the sun gear pair is,

$$\gamma_{\varepsilon c_1} = 2 v_{t_1} \sin \alpha_{wt_1}$$

Here, the tangential velocity at the pitch circle is calculated as,

$$v_{t_1} = r_{w_1} \omega_1$$

The angular velocity of planet gear is,

$$\omega_1 = \frac{2\pi N_1}{60}$$

The curvature radius, or transverse profile radius, for planet gear (in the sun gear, planet gear pair) is,

$$\rho_1 = \frac{r_{w_1} \sin \alpha_{wt_1}}{\cos \beta_b}$$

The curvature radius for planet gear (in the ring gear, planet gear pair) is,

$$\rho_1' = \frac{r_{w_1} \sin \alpha_{wt_2}}{\cos \beta_b}$$

The curvature radius for sun gear (in the sun gear, planet gear pair) is,

$$\rho_2 = \frac{r_{w_2} \sin \alpha_{wt_1}}{\cos \beta_b}$$

The curvature radius for the ring gear is (in the ring gear, planet gear pair),

$$\rho_3 = \frac{r_{w_3} \sin \alpha_{wt_2}}{\cos \beta_b}$$

Then the reduced radius of curvature (in the sun gear, planet gear pair) is,

$$\rho_{redc_1} = \frac{\rho_1 \rho_2}{(\rho_1 + \rho_2) \cos \beta_b}$$

The lubricant correction factor X_{L_1} for the planet gear–sun gear pair is given by,

$$X_{L_1} = \frac{1}{\left(\frac{F_{tb_1}}{b}\right)^d}$$

Here, it is assumed that lubricating oil is mineral oil, for which the factor $d = 0.0651$ (Martinsa, R, 2006).

In a similar manner, the coefficient of friction for the planet gear and ring gear pair is given by,

$$\mu_{m_2} = 0.048\left(\frac{F_{t_{b2}} / b}{\gamma_{\varepsilon c_2}\rho_{redc_2}}\right)^{0.2}\eta_{oil_2}{}^{-0.05}R_{a_2}{}^{+0.25}X_{L_2}$$

The tangential force at the base circle for the planet gear and ring gear engagement is,

$$F_{t_{b2}} = F_{t_{b1}}$$

The sum velocity at operating pitch circles for the planet gear and ring gear pair is,

$$\gamma_{\varepsilon c_2} = 2v_{t_1}\sin\alpha_{wt_2}$$

The reduced radius of curvature (in the ring gear, planet gear pair) is,

$$\rho_{redc_2} = \frac{\rho_1^{'}\rho_3}{\left(\rho_3 - \rho_1^{'}\right)\cos\beta_b}$$

The lubricant correction factor X_{L_2} for the planet gear ring gear pair is given by,

$$X_{L_2} = \frac{1}{\left(\dfrac{F_{t_{b2}}}{b}\right)^{d}}$$

For the present case, the following values are taken from the standards:

$\eta_{oil} = 50$ mPa.s

$R_{a1} = R_{a2} = 0.8$ microns

4.5.7.4 Slide Roll Ratio (SRR)

Based on the principles and procedures already explained in Chapter 3, namely, sections starting from 3.6.9.4 to 3.6.9.6, the SRR is given by,

$$SRR = \frac{v_s}{v_e}$$

Hence, the objective function for SRR is given by,

$$\min f_4 = \min \text{SRR}$$

4.5.7.5 Specific Sliding

For balancing or equalization of specific sliding, the following equations are used, which form the next set of objective functions:

$$\min\ f_5 = \left|\gamma_{1\max} - \gamma_{2\max}\right|$$

$$\min\ f_6 = \left|\gamma_{3\max} - \gamma_1{}'_{\max}\right|$$

where $\gamma_{1\max}$ and $\gamma_{2\max}$ are the specific sliding ratios at SAP and EAP, respectively, for the planet gear and sun gear pair meshing.

To find out the values of $\gamma_{1\max}$ and $\gamma_{2\max}$, the following equations are used:

$$\gamma_{1\max} = \frac{\left(\tan\alpha_{at_1} - \tan\alpha_{wt_1}\right)}{\left(\left(i_1+1\right)\tan\alpha_{wt_1} - \tan\alpha_{at_1}\right)}\left(i_1+1\right)$$

$$\gamma_{2\max} = \frac{\left(\tan\alpha_{at_2} - \tan\alpha_{wt_1}\right)}{\left(\left(\frac{i_1+1}{i_1}\right)\tan\alpha_{wt_1} - \tan\alpha_{at_2}\right)}\left(\frac{i_1+1}{i_1}\right)$$

Here, α_{wt_1} is the working transverse pressure angle between the planet gear and sun gear pair meshing.

Similarly, $\gamma_{1\max}{}'$ and $\gamma_{3\max}$ are the specific sliding ratios at SAP and EAP, respectively, for the planet gear and ring gear pair meshing.

To find out the values of $\gamma_{1\max}{}'$ and $\gamma_{3\max}$, the following equations are used :

$$\gamma_{1\max}{}' = \frac{\left(\tan\alpha_{at_1} - \tan\alpha_{wt_2}\right)}{\left(\left(1+i_2\right)\tan\alpha_{wt_2} - \tan\alpha_{at_1}\right)}\left(i_2+1\right)$$

$$\gamma_{3\max} = \frac{\left(\tan\alpha_{at_3} - \tan\alpha_{wt_2}\right)}{\left(\left(\frac{i_2+1}{i_2}\right)\tan\alpha_{wt_2} - \tan\alpha_{at_3}\right)}\left(\frac{i_2+1}{i_2}\right)$$

Here, α_{wt_2} is the working transverse pressure angle between the planet gear and ring gear pair meshing.

Where i_1 and i_2 are the gear ratios between the planet gear and sun gear pair and the planet gear and ring gear pair, respectively, and are defined as below:

$$i_1 = \frac{z_2}{z_1}\quad i_2 = \frac{z_3}{z_1}$$

The addendum circle pressure angle of the planet gear is given by,

$$\alpha_{at_1} = \cos^{-1}\left(\frac{d_1\cos\alpha_t}{d_{a_1}{}'}\right)$$

The addendum circle pressure angle of the sun gear is given by,

$$\alpha_{at_2} = \cos^{-1}\left(\frac{d_2 \cos\alpha_t}{d_{a_2}{}'}\right)$$

The addendum circle pressure angle of the ring gear is given by,

$$\alpha_{at_3} = \cos^{-1}\left(\frac{d_3 \cos\alpha_t}{d_{a_3}{}'}\right)$$

The transverse pressure angle α_t is given by,

$$\alpha_t = \tan^{-1}\left(\frac{\tan\alpha_n}{\cos\beta}\right)$$

4.5.7.6 Almen Product

For balancing or equalization of the Almen product, the following equations are used, which form the next set of objective functions:

Minimize $f_7 = \left|\sigma_{H_{a1}} v_{s1_a} - \sigma_{H_{r1}} v_{s1_r}\right|$, for the planet gear and the sun gear pair meshing.
Minimize $f_8 = \left|\sigma_{H_{a2}} v_{s2_a} - \sigma_{H_{r2}} v_{s2_r}\right|$, for the planet gear and the ring gear pair meshing.

For the planet gear and sun gear pair meshing, the contact stress at SAP is given by,

$$\sigma_{H_{a1}} = Z_{E_1}\left[F_{n_a}{}'\left(\frac{1}{\rho_{1_a}} + \frac{1}{\rho_{2_a}}\right)\right]^{0.5}$$

where Z_{E_1} is the elasticity factor, *MPa*; $F_{n_a}{}'$ is the normal load per unit length at SAP, *N/mm*; ρ_{1_a} is the radius of curvature of the planet gear at SAP, *mm*; ρ_{2_a} is the radius of curvature of the sun gear at SAP, *mm*.

Similarly, for the planet gear and sun gear pair meshing, the contact stress at EAP is given by,

$$\sigma_{H_{r1}} = Z_{E_1}\left[F_{n_r}{}'\left(\frac{1}{\rho_{1_r}} + \frac{1}{\rho_{2_r}}\right)\right]^{0.5}$$

where Z_{E_1} is the elasticity factor, *MPa*; $F_{n_r}{}'$ is the normal load per unit length at EAP, *N/mm*; ρ_{1_r} is the radius of curvature of the planet gear at EAP, *mm*; ρ_{2_r} is the radius of curvature of the sun gear at EAP, *mm*.

The elasticity factor, or reduced modulus of elasticity, Z_E, depends on Young's moduli E_1, E_2, and Poisson's ratios ν_1, ν_2 of the planet gear and sun gear material according to the equation:

$$Z_{E_1} = \left[\frac{1}{\pi \left\{ \frac{1-\nu_1^2}{E_1} + \frac{1-\nu_2^2}{E_2} \right\}} \right]^{0.5}$$

where E_1, E_2 are Young's moduli of the planet gear and sun gear material, MPa; ν_1, ν_2 are Poisson's ratios of the planet gear and sun gear material.

Referring to Figure 4.4, the normal load per unit length at SAP is,

$$F_{n_a}' = \frac{1000T_1}{\left(r_{b_1}{}^2 + AM^2\right)^{0.5} \cos\alpha_t b}$$

The normal load per unit length at EAP is,

$$F_{n_r}' = \frac{1000T_1}{r_{a_1}' \cos\alpha_t b}$$

For the planet gear and the sun gear engagement,

The radius of curvature of the planet gear at SAP is,

$$\rho_{1_a} = \frac{MP - AP}{\cos\beta_b}$$

The radius of curvature of the planet gear at EAP is,

$$\rho_{1_r} = \frac{MP + PB}{\cos\beta_b}$$

The radius of curvature of the sun gear at SAP is,

$$\rho_{2_a} = \frac{NP + AP}{\cos\beta_b}$$

The radius of curvature of the sun gear at EAP is,

$$\rho_{2_r} = \frac{NP - PB}{\cos\beta_b}$$

Referring to Figure 4.4,

$$MP = r_{w_1} \sin\alpha_{w_{t1}}$$
$$NP = r_{w_2} \sin\alpha_{w_{t1}}$$

$$AP = \left(r_{a_2}'^2 - r_{b_2}^2\right)^{0.5} - r_{w_2}\sin\alpha_{w_{t1}}$$

$$PB = \left(r_{a_1}'^2 - r_{b_1}^2\right)^{0.5} - r_{w_1}\sin\alpha_{w_{t1}}$$

$$AM = MP - AP$$

Similarly, for the planet gear and the ring gear engagement, the contact stress at SAP is given by,

$$\sigma_{H_{a2}} = Z_{E_2}\left[F_{n_a}''\left(\frac{1}{\rho_{1_a}'} - \frac{1}{\rho_{3_a}}\right)\right]^{0.5}$$

where Z_{E_2} is the elasticity factor, MPa; F_{n_a}'' is the normal load per unit length at SAP, N/mm; ρ_{1_a}' is the radius of curvature of the planet gear at SAP, mm; ρ_{3_a} is the radius of curvature of the ring gear at SAP, mm.

Here, the elasticity factor is,

$$Z_{E_2} = \left[\frac{1}{\pi\left\{\frac{1-\nu_1^2}{E_1} + \frac{1-\nu_3^2}{E_3}\right\}}\right]^{0.5}$$

where E_3 is Young's modulus of the ring gear material, MPa; ν_3 is Poisson's ratio of the ring gear material.

Referring to Figure 4.5, the normal load per unit length at SAP is,

$$F_{n_a}'' = \frac{1000T_1}{\left(r_{b_1}^2 + LE^2\right)^{0.5}\cos\alpha_t b}$$

The contact stress at EAP is given by,

$$\sigma_{H_{r2}} = Z_{E_2}\left[F_{n_r}''\left(\frac{1}{\rho_{1_r}'} - \frac{1}{\rho_{3_r}}\right)\right]^{0.5}$$

where F_{n_r}'' is the normal load per unit length at EAP, N/mm.

The normal load per unit length at EAP is,

$$F_{n_r}'' = \frac{1000T_1}{r_{a_1}'\cos\alpha_t b}$$

The radius of curvature of the planet gear at SAP is,

$$\rho_{1_a}' = \frac{LE}{\cos\beta_b}$$

The radius of curvature of the planet gear at EAP is,

$$\rho_{1_r}' = \frac{LF}{\cos\beta_b}$$

The radius of curvature of the ring gear at SAP is,

$$\rho_{3_a} = \frac{KE}{\cos\beta_b}$$

The radius of curvature of the ring gear at EAP is,

$$\rho_{3_r} = \frac{KF}{\cos\beta_b}$$

Referring to Figure 4.5,

$$LF = \left(r_{a_1}'^2 - r_{b_1}^2\right)^{0.5}$$

$$KE = \left(r_{a_3}'^2 - r_{b_3}^2\right)^{0.5}$$

$$EF = LF - LE$$

$$EP = KP - KE$$

$$PF = LF - LP$$

$$KP = \left(r_{w_3}^2 - r_{b_3}^2\right)^{0.5}$$

$$LP = \left(r_{w_1}^2 - r_{b_1}^2\right)^{0.5}$$

$$LE = KE - KL$$

$$KL = a'\sin\alpha_{wt_2}$$

$$KF = KP + PF$$

The sliding velocity for the planet gear and the sun gear pair meshing at SAP is,

$$v_{s_{1a}} = AP(\omega_1 + \omega_2)$$

The sliding velocity for the planet gear and the sun gear pair meshing at EAP is,

$$v_{s_{1r}} = PB(\omega_1 + \omega_2)$$

The sliding velocity for the planet gear and the ring gear pair meshing at SAP is,

$$v_{s2a} = EP\left(\omega_2 - \omega_3\right)$$

The sliding velocity for the planet gear and the sun gear pair meshing at EAP is,

$$v_{s2r} = PF\left(\omega_2 - \omega_3\right)$$

The angular velocity of the planet gear and the sun gear is given by,

$$\omega_1 = \frac{2\pi N_1}{60}$$

$$\omega_2 = \frac{2\pi N_2}{60}$$

For the present case, the following material properties are taken from the standards:

Poisson's ratio and Young's modulus for the gear pair,

$$\nu_1 = \nu_2 = \nu_3 = 0.3$$
$$E_1 = E_2 = E_3 = 210GPa$$

Since the ring gear is stationary,

$$\omega_3 = 0$$

4.5.7.7 Flash Temperature

The flash temperature for a gear pair in engagement can be calculated for any local point on the involute flank along the path of contact using the following equation (AGMA925-A03):

$$\theta_{fl} = 31.62K\mu_m \frac{X_r F_n}{\left(b_H\right)^{0.5}} \frac{\left|v_{r1} - v_{r2}\right|}{B_{m1}\left(v_{r1}\right)^{0.5} + B_{m2}\left(v_{r2}\right)^{0.5}}$$

where θ_{fl} is the flash temperature, °C; K is 0.80, a numerical factor valid for a semi-elliptic (Hertzian) distribution of frictional heat over the instantaneous width, 2 b_H, of the rectangular contact band; μ_m is the coefficient of friction; X_r is the load-sharing factor; F_n is the normal unit load, *N/mm*; b_H is the semi-width of Hertzian contact band, *mm*; v_{r1} is the rolling tangential velocity of the pinion, *m/s*; v_{r2} is the rolling tangential velocity of the wheel, *m/s*; B_{m1} is the thermal contact coefficient of the pinion, N/(mm $s^{0.5}$K); B_{m2} is the thermal contact coefficient of the wheel, N/(mm $s^{0.5}$K).

The semi-width of the Hertzian contact band is given by,

$$b_H = \left[\frac{4F_n\rho_{redc}}{\pi E'}\right]^{0.5}$$

where F_n is the normal force at the contact point per unit width, N/mm; ρ_{redc} is the reduced radius of curvature at the contact point, mm; E' is the equivalent elastic modulus of gear pair materials, N/mm^2.

For balancing or equalization of flash temperature, the following equations are used, which form the next set of objective functions:

Minimize

$f_9 = \theta_{fla_1}$ for the planet gear and the sun gear pair meshing at SAP.
$f_{10} = \theta_{flr_1}$ for the planet gear and the sun gear pair meshing at EAP.
$f_{11} = \theta_{fla_2}$ for the planet gear and the ring gear pair meshing at SAP.
$f_{12} = \theta_{flr_2}$ for the planet gear and the ring gear pair meshing at EAP.

The flash temperature for the planet gear and the sun gear pair engagement at SAP is,

$$\theta_{fla_1} = 31.62 K \mu_{m_1} \frac{X_{r_1} F_{n_a}{}'}{\left(b_{H_{a1}}\right)^{0.5}} \frac{\left| v_{r_1 a}{}' - v_{r_2 a}{}' \right|}{B_{m1}\left(v_{r_1 a}{}'\right)^{0.5} + B_{m2}\left(v_{r_2 a}{}'\right)^{0.5}}$$

Here, the rolling tangential velocity of the planet gear at SAP is,

$$v_{r_{1a}}{}' = \rho_{1_a} \omega_1$$

The rolling tangential velocity of the sun gear at SAP is,

$$v_{r_{2a}}{}' = \rho_{2_a} \omega_2$$

The minimum Hertzian semi-contact width in one mesh cycle at SAP,

$$b_{H_{a1}} = \left[\frac{4 F_{n_a}{}' \rho_{redc_{1a}}}{\pi E_1'} \right]^{0.5}$$

The reduced radius of curvature at SAP (in the sun gear, planet gear pair) is,

$$\rho_{redc_{1a}} = \frac{\rho_{1_a} \rho_{2_a}}{\left(\rho_{1_a} + \rho_{2_a}\right) \cos \beta_b}$$

Here, E_1' is the equivalent or effective modulus of elasticity, which depends on the modulus of elasticity and Poisson ratio of the sun gear material and planet gear material.

$$E_1' = \frac{1}{2} \left[\frac{1 - \nu_1^2}{E_1} + \frac{1 - \nu_2^2}{E_2} \right]$$

Here, E_1 and E_2 are in N/mm^2.

The thermal contact coefficient B_{m1} for the planet gear is given by the following equation:

$$B_{m1} = \frac{\left(\lambda_m \rho_m c_m\right)^{0.5}}{10^3}$$

where λ_m is the heat conductivity of the planet gear material, W/(m.K); ρ_m is the planet gear material density, kg/m^3; c_m is the specific heat of planet gear material per unit of mass, J/(kg.K).

For the present case, the following material properties are taken from the standards:

$\lambda_m = 40$ W/(m.K)
$\rho_m = 7850$ kg/m^3
$c_m = 465$ J/(kg.K)

The thermal contact coefficient B_{m2} for the sun gear is given by the following equation:

$$B_{m2} = \frac{\left(\lambda_m \rho_m c_m\right)^{0.5}}{10^3}$$

where λ_m is the heat conductivity of the sun gear material, W/(m.K); ρ_m is the sun gear material density, kg/m^3; c_m is the specific heat of sun gear material per unit of mass, J/(kg.K).

The flash temperature for the planet gear and the sun gear pair engagement at EAP is,

$$\theta_{fl r_1} = 31.62 K \mu_{m_1} \frac{X_{r_1} F_{n_r}{}'}{\left(b_{H_{r1}}\right)^{0.5}} \frac{\left| v_{r_1 r}{}' - v_{r_2 r}{}' \right|}{B_{m1}\left(v_{r_1 r}{}'\right)^{0.5} + B_{m2}\left(v_{r_2 r}{}'\right)^{0.5}}$$

The minimum Hertzian semi-contact width in one mesh cycle at EAP,

$$b_{H_{r1}} = \left[\frac{4 F_{n_r}{}' \rho_{redc_{1r}}}{\pi E_1'}\right]^{0.5}$$

The reduced radius of curvature at EAP (in the sun gear, planet gear pair) is,

$$\rho_{redc_{1r}} = \frac{\rho_{1_r} \rho_{2_r}}{\left(\rho_{1_r} + \rho_{2_r}\right)\cos\beta_b}$$

The flash temperature for the planet gear and the ring gear pair engagement at SAP is,

$$\theta_{fla_2} = 31.62 K \mu_{m_2} \frac{X_{r_2} F_{n_a}{}''}{\left(b_{H_{a2}}\right)^{0.5}} \frac{\left| v_{r_1 a}{}'' - v_{r_3 a}{}'' \right|}{B_{m1}\left(v_{r_1 a}{}''\right)^{0.5} + B_{m2}\left(v_{r_3 a}{}''\right)^{0.5}}$$

The minimum Hertzian semi-contact width in one mesh cycle at SAP,

$$b_{H_{a2}} = \left[\frac{4 F_{n_a}{}'' \rho_{redc_{2a}}}{\pi E_2'}\right]^{0.5}$$

The flash temperature for the planet gear and the ring gear pair engagement at EAP is,

$$\theta_{flr_2} = 31.62K\mu_{m2}\frac{X_{r_2}F_{n_r}''}{\left(b_{H_{r2}}\right)^{0.5}}\frac{\left|v_{r_1r}''-v_{r_3r}''\right|}{B_{m1}\left(v_{r_1r}''\right)^{0.5}+B_{m2}\left(v_{r_3r}''\right)^{0.5}}$$

The minimum Hertzian semi-contact width in one mesh cycle at EAP,

$$b_{H_{r2}} = \left[\frac{4F_{n_r}''\rho_{redc2r}}{\pi E_2'}\right]^{0.5}$$

Here, E_2' is the equivalent or effective modulus of elasticity, which depends on the modulus of elasticity and Poisson ratio of the sun gear material and planet gear material.

$$E_2' = \frac{1}{2}\left[\frac{1-\nu_1^2}{E_1}+\frac{1-\nu_3^2}{E_3}\right]$$

Here, E_1 and E_3 are in N/mm^2.

The reduced radius of curvature at SAP (in the planet gear, ring gear pair meshing) is,

$$\rho_{redc2a} = \frac{\rho_{1_a}'\rho_{3_a}}{\left(\rho_{3_a}-\rho_{1_a}'\right)\cos\beta_b}$$

The reduced radius of curvature at EAP (in the planet gear, ring gear pair meshing) is,

$$\rho_{redc2r} = \frac{\rho_{1_r}'\rho_{3_r}}{\left(\rho_{3_r}-\rho_{1_r}'\right)\cos\beta_b}$$

Rolling tangential velocity of the planet gear at SAP (in the planet gear, ring gear pair meshing) is,

$$v_{r_1a}'' = \rho_{1_a}'\omega_2$$

The rolling tangential velocity of the ring gear at SAP (in the planet gear, ring gear pair meshing) is,

$$v_{r_3a}'' = 0$$

The rolling tangential velocity of the planet gear at EAP (in the planet gear, ring gear pair meshing) is,

$$v_{r_1r}'' = \rho_{1_r}'\omega_2$$

The rolling tangential velocity of the ring gear at EAP (in the planet gear, ring gear pair meshing) is,

$$v_{r_3r}'' = 0$$

4.5.7.8 Minimum Oil Film Thickness

Taking into account the unsteady speed and load conditions, the minimum oil film thickness for a gear pair engagement is given by,

$$h_{\min} = 1.714R_i\left(A\right)^{-0.128}\left(B\right)^{0.694}\left(C\right)^{0.568}$$

where $h_{\min}$ is the minimum oil film thickness, m; R_i is the radius of curvature of the interface between the mating gears, m.

Here, R_i is given by,

$$R_i = \frac{r_{w_1} r_{w_2} \sin \alpha_{wt_1}}{1000\left(r_{w_1} \pm r_{w_2}\right)}$$

Here, "+" sign holds good for external gear pair meshing, and the "–" sign holds good for internal gear pair meshing.

$$A = \frac{10^6 F_n}{bE'' R_i}$$

where F_n is the normal load at pitch point, N; b is the face width, m; E'' is the equivalent modulus of elasticity, N/m^2.

Here, the equivalent modulus of elasticity is given by,

$$E'' = \left[\frac{2}{\dfrac{1-v_1^2}{E_1} + \dfrac{1-v_2^2}{E_2}}\right]$$

Here, E_1 and E_2 are the modulus of elasticity of mating gears, in N/m^2.

$$B = \frac{\eta_o\left(v_1 \pm v_2\right)\cos\alpha_{wt_1}}{2\times 10^3 E'' R_i}$$

where η_o is the viscosity at working temperature; v_1, v_2 is the tangential velocity of mating gears at the pitch circle, m/s.

Here, "+" sign holds good for external gear pair meshing, and the "–" sign holds good for internal gear pair meshing.

$$C = \alpha' E''$$

where α' is the lubricant pressure viscosity coefficient, m^2/N.

The minimum oil film thickness for the planet gear and the sun gear pair engagement is given by,

$$h_{\min_1} = 1.714 R_{i_1} \left(A'\right)^{-0.128} \left(B'\right)^{0.694} \left(C'\right)^{0.568}$$

where

$$A' = \frac{10^6 F_{n_1}}{b E_1'' R_{i_1}}$$

where F_{n_1} is the normal load at pitch point, N; b is the face width, m

$$F_{n_1} = \left(\frac{1000 T_1}{r_{w1}} \right)$$

The equivalent modulus of elasticity in this case is given by,

$$E_1'' = \left[\frac{2}{\dfrac{1-\nu_1^2}{E_1} + \dfrac{1-\nu_2^2}{E_2}} \right]$$

Here, E_1 and E_2 are in N/m².

R_{i_1} is the radius of curvature of the interface between the planet gear and sun gear and is given by,

$$R_{i_1} = \frac{r_{w1} r_{w2} \sin \alpha_{wt_1}}{10^3 \left(r_{w_1} + r_{w2}\right)}$$

$$B' = \frac{\eta_o \left(v_1 + v_2\right) \cos \alpha_{wt_1}}{2 \times 10^3 E_1'' R_i}$$

where η_o is the viscosity at working temperature, mPa.s.

For the planet gear,

$$v_1 = \frac{r_{w1} \omega_1}{10^3} = \frac{v_{t1}}{10^3}$$

For the sun gear,

$$v_2 = \frac{r_{w2} \omega_2}{10^3} = \frac{v_{t2}}{10^3}$$

Here, v_1 and v_2 are in m/s.

$$C' = \alpha_1' E_1''$$

For the present case, the following values are taken from the standards:

$$\eta_o = 90 \text{ mPa.s}$$

$$\alpha_1' = 2 \times 10^{-8} m^2 / N$$

Similarly, the minimum oil film thickness for the planet gear and ring gear pair engagement is given by,

$$h_{min_2} = 1.714 R_{i_2} \left(A''\right)^{-0.128} \left(B''\right)^{0.694} \left(C''\right)^{0.568}$$

where

$$A'' = \frac{10^6 F_{n_2}}{b E_2'' R_{i_2}}$$

$$F_{n_2} = F_{n_1}$$

The equivalent modulus of elasticity in this case is given by,

$$E_2'' = \left[\frac{2}{\frac{1-\nu_1^2}{E_1} + \frac{1-\nu_3^2}{E_3}} \right]$$

$$R_{i_2} = \frac{r_{w1} r_{w3} \sin \alpha_{w_{t2}}}{1000\left(r_{w3} - r_{w1}\right)}$$

$$B'' = \frac{\eta_o \left(v_1 - v_3\right) \cos \alpha_{wt_2}}{2 \times 10^3 E_2'' R_i}$$

Since the ring gear is stationary,

$$v_3 = 0$$

$$C'' = \alpha_2' E_2''$$

Since both the sun gear-planet gear pair and the planet gear-ring gear pair operate in the same lubricant oil environment, the lubricant pressure viscosity coefficient will be the same for both pairs.

$$\alpha_1' = \alpha_2'$$

4.5.7.9 Lambda Ratio

Improvement in the lambda ratio value ensures better performance of the gear pair in engagement, and its value must be maximized. Hence, the following equations form the next set of objectives:

To maximize the lambda ratio,

$$Maximize \ f_{13} = \lambda_1 = \frac{h_{min_1}}{R_{a1}'}$$

$$Maximize\ f_{14} = \lambda_2 = \frac{h_{\min_2}}{R_{a2}'}$$

$$R_{a1}' = R_{a1} + R_{a2}$$

$$R_{a2}' = R_{a1} + R_{a3}$$

where R_{a1}, R_{a2}, and R_{a3} are the arithmetic mean roughness of the planet, sun, and ring gears, respectively (μ m).

For the present case,

$$R_{a1}' = R_{a2}' = 0.1$$

4.5.8 Design Variables and Constraints

The independent design parameters of a planetary gear drive include the module m_n, the number of teeth of planet gear z_1, the number of teeth of the sun gear z_2, the number of teeth of ring gear z_3, the three profile shift coefficients x_{n1}, x_{n2}, and x_{n3}, and the number of planets n_p.

Therefore, the design variables are,

$$X = \left[z_1, z_2, z_3, x_{n1}, x_{n2}, x_{n3}, m_n, n_p\right]^T = \left[x_1, x_2, x_3, x_4, x_5, x_6, x_7, x_8\right]^T$$

A number of constraints, which are introduced into the objective functions to be optimized, are used to provide suitable design choices. For an overview of types of gear failures, see Section 3.6.14.

4.5.8.1 Bending Strength

Maximum allowable bending stress is given by the bending strength of spur gears and helical gears (ISO 6336:2006-based JGMA 401-01 (1974) standard),

$$\sigma_F = F_t \frac{y_F y_\varepsilon y_\beta}{9.81 m_n b}\left(\frac{K_V K_O}{K_L K_{FX}}\right) S_F \ \text{kgf/mm}^2$$

where F_t is the tangential force at the working pitch circle, *kgf*; y_F is the tooth profile factor; y_ε is the load sharing factor; y_β is the helix angle factor; K_V is the dynamic load factor; K_O is the overload factor; K_L is the life factor; K_{FX} is the size factor of root stress; S_F is the safety factor for bending failure.

For the sun gear–planet gear engagement,

$$F_t = F_{n_1}$$

The tooth profile factor y_F is found by the equation (refer to Figure 3.13),

$$y_F = \frac{6\left(\frac{h_A}{m_n}\right)\cos\alpha_{wt}}{\left(\frac{S_A{}^2}{m_n{}^2}\right)\cos\alpha_n}$$

The load sharing factor y_ε is given by,

$$y_\varepsilon = \frac{1}{\varepsilon_{\alpha_1}{}'}$$

The helix angle factor y_β is given by,

$$y_\beta = 1 - \frac{\beta}{120}$$

The dynamic load factor, K_V, is based on the precision of the gear and the tangential speed of gears at the working pitch circle. The tangential speed of gears at the working pitch circle is,

$$v_{t_1} = \frac{\pi d_{w_1} N_1}{60000}$$

The dynamic load factor K_V is taken from the standards.

For tangential speed less than 1.0 m/s,

$K_V = 1.0$

The overload factor Ko is defined as,

$$K_O = \frac{\text{Actual tangential force}}{\text{Nominal tangential force } F_t}$$

Since the WTG gearbox is expected to be subjected to heavy input load, K_O is taken as 1.75 from the standards.

The life factor K_L depends on the number of cyclic repetitions or the total loaded meshing cycles during its lifetime. Its value is taken from the standards. For the number of cyclic repetitions of $>10^7$ cycles, K_L is taken as,

$$K_L = 1.0$$

The size factor of root stress K_{FX} is generally taken as 1.0.

S_F is the safety factor for bending failure. Since S_F is too complicated to be determined precisely, it is usually taken as at least 1.2.

4.5.8.2 Contact Compressive (Crushing) Strength

Contact compressive stress should be within allowable limits to prevent failures due to wear like scuffing, scoring, welding, peening, rippling, ridging, and tooth breakage.

Contact compressive strength of spur gears and helical gears as per ISO 6336:2006 based JGMA 402-01 (1975) standard is given by,

$$\sigma_H = \sqrt{\frac{F_t(i \pm 1)}{9.81 d_w b i} \frac{Z_H Z_M Z_\varepsilon Z_\beta}{K_{HL} Z_L Z_R Z_V Z_W K_{HX}}} \sqrt{K_{H\beta} K_V K_O} S_H \text{ kgf / mm}^2$$

The "+" symbol in the above equation applies to two external gears in mesh, whereas the "-" symbol is used for an internal gear and an external gear pair in mesh. The above equation contains the following factors:

Zone factor, Z_H
Material factor, Z_M
Contact ratio factor, Z_ε
Helix angle factor, Z_β
Life factor, K_{HL}
Lubricant factor, Z_L
Surface roughness factor, Z_R
Lubrication speed factor, Z_V
Hardness ratio factor, Z_W
Size factor, K_{HX}
Longitudinal load distribution factor, $K_{H\beta}$
Dynamic load factor, K_V
Overload factor, K_O
Safety factor for pitting, S_H

For the sun gear–planet gear engagement, Hertz contact stress is given by,

$$\sigma_{H_1} = \sqrt{\frac{F_t(i + 1)}{9.81 d_{w_2} b i} \frac{Z_H Z_M Z_\varepsilon Z_\beta}{K_{HL} Z_L Z_R Z_V Z_W K_{HX}}} \sqrt{K_{H\beta} K_V K_O} S_H \text{ kgf / mm}^2$$

$$i = i_1$$

$$F_t = F_{n_1}$$

Zone factor, Z_H, is given by,

$$Z_H = \sqrt{\frac{2\cos\beta_b \cos\alpha_{wt_1}}{\cos^2\alpha_t \sin\alpha_{wt_1}}} = \frac{1}{\cos\alpha_t}\sqrt{\frac{2\cos\beta_b}{\tan\alpha_{wt_1}}}$$

Material factor, Z_M, is given by,

$$Z_M = \left[\frac{1}{\pi\left\{\frac{1-\nu_1^2}{E_1} + \frac{1-\nu_2^2}{E_2}\right\}}\right]^{0.5}$$

For steel, $E_1 = 21000$ kgf/mm^2

Contact ratio factor, Z_ε, is given by the following:

For spur gear,

$$Z_\varepsilon = 1.0$$

For helical gear,

When the overlap ratio $\varepsilon_\beta \leq 1$,

$$Z_\varepsilon = \sqrt{1 - \varepsilon_\beta + \frac{\varepsilon_\beta}{\varepsilon_\alpha}}$$

When

$$\varepsilon_\beta > 1,$$

$$Z_\varepsilon = \sqrt{\frac{1}{\varepsilon_\alpha}}$$

The helix angle factor, Z_β, is difficult to calculate and can be taken as 1.0.

The life factor, K_{HL}, depends on the duty cycle.

For duty cycle > 10^7 cycles,

$$K_{HL} = 1.0$$

Lubricant factor, Z_L, is based on the lubricant's kinematic viscosity at 50°C, *cSt*.

From the standards,

$$Z_L = 1.0$$

The surface roughness factor, Z_R, is obtained on the basis of the average roughness R_{max} (μm).

The average roughness, R_{max}, is calculated by the below-given equation using the surface roughness values of the pinion and gear, R_{max1} and R_{max2}, and the center distance a in mm.

$$R_{max} = \frac{R_{max1} + R_{max2}}{2}\left(\frac{100}{a'}\right)^{\frac{1}{3}}$$

For surface-hardened gear, and R_{max} = 10 microns,

$$Z_R = 0.92$$

Lubrication speed factor, Z_V, depends on the tangential speed at the pitch circle, v (m/s).

For surface-hardened gear, and v = 1.0–3.0 m/s,

$$Z_V = 0.97$$

The hardness ratio factor, Z_W, applies only to the gear that is in mesh with a pinion, which is quenched and ground. It is given by the equation,

$$Z_W = 1.2 - \frac{H_{B2} - 130}{1700}$$

Here, H_{B2} is in the range of the Brinell hardness of gear material.

$$130 \leq H_{B2} \leq 470$$

If a gear is out of this range, Z_W is assumed to be 1.0.

The size factor, K_{HX}, is usually taken as 1.0 because the conditions affecting this parameter are often unknown.

The longitudinal load distribution factor, $K_{H\beta}$, depends on the method of gear shaft support and the ratio, b / d_{w_2}, of the gear face width b to the pitch diameter, d_{w_2}. When tooth contact under load is good, the longitudinal load distribution factor can be taken to be in the range of 1.0–1.2. For the present case, let $K_{H\beta} = 1.0$.

The dynamic load factor, K_V, depends on the gear's precision grade and pitch circle tangential speed, v.

For speeds less than 1 m/s, the value of K_V is taken as 1.0.

The overload factor, K_O, depends on the impact load on the gears on the prime mover side and on the load side. For uniform load from the load side, this factor can be taken as 1.0.

The safety factor for pitting, S_H, is normally taken as 1.15 since the causes of pitting are difficult to precisely define because it involves many environmental factors.

4.5.8.3 Contact Ratio

The contact ratio is given by,

$$\varepsilon_\alpha' = \frac{\text{Path of contact}}{p_b}$$

where ε_α' is the contact ratio; p_b is the base pitch.

The base pitch, p_b, is defined as the distance from one tooth to the next tooth measured along the line of action. The base pitch is given by,

$$p_b = p_c \cos \alpha_t = \pi m_n \cos \alpha_t$$

where p_c is the circular pitch.

This contact ratio should be greater than 1.4 and is normally less than 2.0 for standard spur gears. For contact with an internal gear at low ratios and for non-standard addenda and dedenda, it can exceed 2.0.

The transverse contact ratio for the sun gear–planet gear pair engagement is given by,

$$\varepsilon_{\alpha_1}{}' = \frac{AB}{p_b} = \frac{AB}{\pi m_t \cos\alpha_t}$$

The transverse contact ratio for the planet gear–ring gear pair engagement is given by,

$$\varepsilon_{\alpha_2}{}' = \frac{EF}{p_b} = \frac{EF}{\pi m_t \cos\alpha_t}$$

Overlap ratio $\varepsilon_\beta = \dfrac{b\sin\beta}{\pi m_n}$

The total contact ratio for the sun gear–planet gear pair engagement is given by,

$$\varepsilon_{\alpha_1}{}' + \varepsilon_\beta \geq 1.4$$

The total contact ratio for the planet gear–ring gear pair engagement is given by,

$$\varepsilon_{\alpha_2}{}' + \varepsilon_\beta \geq 1.4$$

4.5.8.4 Top Land Thickness

The top land thickness is the tooth thickness on the tip circle. The top land thickness is normally referred to in terms of the percentage of the module. The ratio of top land thickness and the module is suggested to be between 0.2 and 0.4 depending on the material and heat treatment of the gear.

The top land thickness of the planet gear is given by,

$$S_{a_1} = d_{a_1}{}'\left(\frac{s_1}{d_1} + inv\alpha_n - \Omega_1\right)$$

where

$$S_1 = \frac{\pi m_n}{2} + 2x_{n_1} m_n \tan\alpha_n$$

$$\Omega_1 = \tan\alpha_{at_1} - \alpha_{at_1}\left(\frac{\pi}{180}\right)$$

Similarly, the top land thickness of the sun gear is given by,

$$S_{a_2} = d_{a_2}{}'\left(\frac{s_2}{d_2} + inv\alpha_n - \Omega_2\right)$$

where

$$S_2 = \frac{\pi m_n}{2} + 2x_{n_2} m_n \tan \alpha_n$$

$$\Omega_2 = \tan \alpha_{at_2} - \alpha_{at_2}\left(\frac{\pi}{180}\right)$$

Also, the top land thickness of the ring gear is given by,

$$S_{a_3} = d_{a_3}{}'\left(\frac{S_3}{d_3} - inv\alpha_n + \Omega_3\right)$$

where

$$S_3 = \frac{\pi m_n}{2} - 2x_{n_3} m_n \tan \alpha_n$$

$$\Omega_3 = \tan \alpha_{at_3} - \alpha_{at_3}\left(\frac{\pi}{180}\right)$$

The top land thickness should not be below 0.2 times the normal module, m_n, and correspondingly the constraint equations for the top land thickness of the planet gear, the sun gear, and the ring gear are as given below:

$$S_{a_1} \geq 0.2m_n$$
$$S_{a_2} \geq 0.2m_n$$
$$S_{a_3} \geq 0.2m_n$$

4.5.8.5 Operating Interference

To avoid operating interference on the two mating gears, the following conditions are to be met (refer to Figure 4.4):

1. The distance O_2M is greater than the tip radius of the sun gear, and
2. The distance O_1N is greater than the tip radius of the planet gear.

Hence, the constraint equations are,

$$g_p\left(x_{n_1}, x_{n_2}\right) = \left(r_{b_1}{}^2 + \left(a' \sin \alpha_{wt_1}\right)^2\right)^{0.5} > r_{a_1}{}'$$

$$g_s\left(x_{n_1}, x_{n_2}\right) = \left(r_{b_2}{}^2 + \left(a' \sin \alpha_{wt_1}\right)^2\right)^{0.5} > r_{a_2}{}'$$

4.5.8.6 Constraints of Internal Gear

For proper tooth engagement and disengagement, involute interference (contact below the base circle) and secondary interference between the pinion tooth tip and the internal gear tooth tip for internal gearing must be avoided.

4.5.8.6.1 Involute interference

For contact with internal gear, involute interference is avoided if (refer to Figure 4.5),

$$KL \le KE$$

that is,

$$a' \sin \alpha_{wt_2} \le KE$$

Also, to avoid involute interference, the following condition must be satisfied (Buckingham, 1988):

$$\frac{z_1}{z_3} \ge 1 - \frac{\tan \alpha_{at_3}}{\tan \alpha_{wt_2}}$$

4.5.8.6.2 Tip Interference

To avoid tip interference, the following conditions must be satisfied (Buckingham, 1988):

$$\cos \varepsilon_1 = \frac{\left(r_{a_3}^{\ 2} - r_{a_1}^{\ 2} - a'^2\right)}{2a' r_{a_1}} = A$$

$$\therefore \varepsilon_1 = \cos^{-1} A$$

$$\varepsilon_2 = \left(\frac{z_1}{z_3}\right)\varepsilon_1$$

$$\theta_{t1} = \sin^{-1}\left[\frac{r_{a_1} \sin \varepsilon_1}{r_{a_3}}\right] - \varepsilon_2 = B$$

$$\theta_{t1r} = \frac{B \times \pi}{180} = C$$

$$\cos \alpha_{at_3} = \frac{r_{b_3}}{r_{a_3}} = D$$

$$\cos \alpha_{at_1} = \frac{r_{b_1}}{r_{a_1}} = E$$

$$\delta = \left(\frac{z_1}{z_3}\right)\left(inv\alpha_{at_1} - inv\alpha_t\right) + inv\alpha_t$$

$$inv\alpha_{at_1} = \tan \alpha_{at_1} - \frac{\alpha_{at_1} \times \pi}{180} = \tan \alpha_{at_1} - 0.0175\alpha_{at_1} = F$$

$$inv\alpha_t = \tan \alpha_t - 0.0175\alpha_t = 0.02881 = G$$

$$x_1 = inv\alpha_{at_3} = \tan\alpha_{at_3} - 0.0175\alpha_{at_3}$$
$$x_2 = \delta - \theta_{t1r} = \delta - C$$

Condition to avoid tip interference:

If $x_2 > x_1$, then there is no interference, that is, $x_2 - x_1 \geq 0$.

4.5.8.6.3 Condition for Proper Assembly

Additionally, the condition for proper assembly of planetary gearing is as given below (Townsend, 2013):

$$\frac{z_2 + z_3}{n_w} = I$$

where I is an integer.

4.5.9 Design Data for Optimization

Following are the input parameters considered for the planetary gear design based on the existing gear design as in Table 4.3.

4.5.10 Methodology of Optimization of Planetary Gear Using MOPSO

This section describes how the MOPSO algorithm is configured for solving the optimization problem of planetary gearing. This book utilizes the OPTIMO software package (Gokul Krishnan, 2024), which provides an efficient MOPSO implementation in Python along with the complete problem set corresponding to this case study. Chapter 2 provides detailed instructions on how to obtain and use the software.[5]

TABLE 4.3
WTG Gear Input Parameters

Data	Gear
Density of the material, ρ	7.86 × 106 kg/mm³
Material of sun gear, planet gear, and ring gear	Steel DIN 18 CrNiMo 7-6
Design bending stress, σ_b	$400\text{N}/\text{mm}^2$
Design crushing stress, σ_c	$2000\text{N}/\text{mm}^2$
Input speed, N	15 rpm
Power transmitted, P	2000 kW
Young's modulus of the material (E)	$2.0\times10^6\text{N}/\text{mm}^2$
Pressure angle, α_n	20°
Helix angle, β	8°

4.5.10.1 Variable Bounds

In order to determine the values of design variables, such as module, number of teeth, face width, and helix angle, through the optimization procedure, the upper and lower bounds of the design variables are to be established first.

The following variables are considered, and their lower and upper bounds are fixed. Further, the type of variable – whether discrete (integer) or real-valued (float) – is also specified.

Number of planet gear teeth, z_1: 30 to 40 (integer)
Number of sun gear teeth, z_2: 15 to 25 (integer)
Number Of ring gear teeth, z_3: 85 to 95 (integer)
Profile shift coefficient for planet gear, x_{n1}: 0.0 to 0.5 (float)
Profile shift coefficient for sun gear, x_{n2}: 0.0 to 0.5 (float)
Profile shift coefficient for ring gear, x_{n3}: 0.0 to 2.0 (float)
Module, m_n: 12 to 18 mm (float)
Number of planets, n_p: 2 to 6 (integer)

4.5.10.2 Input Parameters

For this model, the most important data inputs are the gear design parameters of the two mating gears. The decision-maker is required to make this decision based on experience and design requirements.

The problem sets provided in this book contain Python code corresponding to the objective functions and design constraints formulated in Sections 4.5.7 and 4.5.8. The upper and lower bounds of design variables are also specified in the same file(s). The seed for the random number generator is fixed at 0.123 for reproducibility of results.

The problem definition corresponding to Chapter 4 can be found in *problem-sets/problem_ch4.py* file within the software repository. The optimization run is started by entering the command specified in Listing 4.14 from the repository root.

LISTING 4.14 COMMAND TO EXECUTE THE ALGORITHM RUN

```
# macOS / Linux
python3 run.py -a pso -p ch4
# Windows
python.exe run.py -a pso -p ch4
```

4.5.10.3 Decision-Making

The goal of decision-making is to select the most optimal solution from the Pareto solution set obtained from the optimization results. Selection of the single best solution from many conflicting non-dominated solutions from the Pareto-set is based on expertise along with any problem-specific requirements. Apart from the above, there are decision-making techniques such as the linear programming technique for multidimensional analysis of preference (LINMAP), Shannon Entropy (SE), and the technique of order preference by similarity to an ideal solution (TOPSIS), which can be used as a post-processing step to obtain a relative gradation of results.

The OPTIMO software makes use of a Common Objective Function, or COF, which assigns a scalar value to each solution in the Pareto-optimal set. This scalar value is essentially the normalized average of the set of objective function values. For a given solution i,

$$P_i = \left[p_1, p_2, \dots p_N \right]$$

$$S_i = COF\left(P_i\right)$$

where P_i is the vector of objective function values; N is the number of objectives; S_i is the scalar output assigned to the solution.

The solution set is then sorted on its COF value, such that the solution with the lowest COF value is considered to be, on average, the overall best solution (assuming minimization of objectives).

4.5.11 Interpretation of MOPSO Results

In this section, the results obtained from MOPSO are discussed and compared with the design of existing WTG planetary gearing. For detailed information on interpreting the tables and plots in the subsequent sections, refer to Section 3.6.17.

Table 4.4 shows a sample of the MOPSO output. Since constraints are specified in the form of ≥ 0 only, all the columns corresponding to the constraints shown in Table 4.4 indicate results obtained by modifying the constraint equations accordingly. Table 4.4 lists the values of the objective functions, constraints, and design variables from the generated solutions. The output represents a non-dominated Pareto-optimal solution set.

4.5.11.1 Comparison of Results

In Table 4.5, a comparison has been made between the results that are obtained with MOPSO and the design of the existing planetary gearing. The solutions correspond to number of teeth of planet gear, namely, 40, 35, and 38 were selected from the results. The values of design variables and objective functions corresponding to each solution are indicated in Table 4.5. The total weight of planetary gearing in kg, the efficiency in percentage, and the center distance in mm were taken for comparison.

The comparison shows that the results obtained from MOPSO are an improvement over the existing design by optimizing the 14 objective functions within the given constraints. The results were obtained with minimal computational effort and time. The three best results chosen from the solution set of MOPSO output are analyzed for selecting a single best solution. The total weight of the existing

TABLE 4.4
Sample of MOPSO Output

			Objectives										
Center distance	Weight	Efficiency	Slide roll ratio	Specific sliding		Almen product			Flash temperature			Lambda ratio	
492.66	1474.42	98.97	0.27	0.99	0.96	201.03	207.67	58.81	24.74	19.66	21.30	6.63	6.88
507.16	1561.89	99.02	0.27	0.93	1.11	192.97	157.50	56.69	22.86	19.23	20.53	6.92	7.03
585.00	1780.42	99.23	0.39	0.01	1.18	19.62	134.26	63.99	6.61	18.53	19.75	9.21	7.27
535.78	1603.85	99.13	0.33	0.15	1.36	22.82	134.13	63.31	4.51	19.20	20.41	7.81	6.78
390.00	**787.29**	**98.98**	**0.39**	**0.01**	**1.87**	**19.62**	**19.04**	**97.36**	**10.06**	**30.94**	**30.79**	**5.53**	**4.08**
475.44	1168.37	98.75	0.23	1.16	0.68	261.66	266.41	85.43	48.01	38.31	40.40	5.84	6.62
510.49	1263.50	98.74	0.24	1.99	0.78	413.03	344.12	79.77	54.18	28.73	31.52	6.25	7.00
378.00	774.03	98.90	0.41	0.01	0.55	19.63	212.00	107.30	13.31	40.10	42.20	5.41	4.44
495.00	1489.34	98.68	0.15	8.39	0.60	1089.92	220.06	51.28	98.03	23.41	24.92	5.95	7.93
446.05	920.25	98.42	0.27	0.96	0.56	197.39	772.50	103.81	46.13	25.59	31.42	5.28	6.04
529.14	1288.10	99.10	0.39	0.04	1.88	2.90	26.22	82.74	7.33	26.62	26.49	7.91	5.84
475.18	1112.91	98.84	0.33	0.60	0.67	158.17	215.43	101.76	21.25	47.66	49.66	6.33	6.00
576.00	1741.22	99.16	0.40	0.01	1.87	19.60	15.24	67.14	7.64	22.03	21.96	9.10	6.54
576.00	1740.39	99.16	0.40	0.03	1.87	6.78	19.01	66.73	6.87	22.10	22.03	9.08	6.53
360.00	**731.30**	**98.86**	**0.44**	**0.00**	**0.69**	**20.14**	**158.11**	**125.26**	**19.19**	**56.16**	**58.03**	**5.21**	**3.86**
463.62	876.10	98.94	0.37	0.74	1.40	209.24	134.21	106.32	14.94	31.80	34.44	6.16	5.08
580.29	1125.20	98.81	0.39	0.57	0.87	252.96	2230.32	132.76	14.88	15.84	28.33	7.98	6.91
399.90	756.31	98.21	0.27	2.09	0.36	403.85	471.02	113.13	70.39	43.31	47.65	4.53	5.34
360.00	659.30	98.63	0.29	2.30	1.84	465.81	45.37	94.25	58.34	30.26	31.86	4.31	4.10
330.96	**721.95**	**98.67**	**0.24**	**1.72**	**1.56**	**321.96**	**62.03**	**84.80**	**54.86**	**32.36**	**33.15**	**3.97**	**4.04**

(Continued)

TABLE 4.4 (CONTINUED)
Sample of MOPSO Output

Constraints													
Interference		Top land thickness			Contact ratio		Bending stress	Contact stress		Involute interference		Tip interference	Proper assembly
13.16	68.10	8.06	7.23	10.90	1.08	2.01	311.83	819.17	1571.37	52.01	0.30	0.06	0.00
14.19	71.82	8.23	7.47	11.21	1.04	2.01	316.11	851.83	1568.95	54.05	0.32	0.07	0.00
32.52	74.02	9.05	7.63	11.29	0.96	2.00	321.40	1006.42	1566.07	56.22	0.31	0.07	0.00
22.38	68.93	8.71	7.07	11.54	1.03	2.04	316.54	915.92	1545.70	49.90	0.32	0.07	0.00
21.68	**49.35**	**6.04**	**5.09**	**7.99**	**1.34**	**2.31**	**223.14**	**574.88**	**1372.83**	**39.80**	**0.38**	**0.07**	**0.00**
10.62	67.86	8.39	6.75	11.27	1.05	1.44	267.41	485.59	1442.95	75.79	0.26	0.02	0.00
8.82	65.01	8.85	7.70	11.80	1.10	1.79	284.90	557.48	1488.57	58.33	0.26	0.04	0.00
22.10	46.12	5.98	5.12	8.24	1.33	1.75	213.02	524.02	1396.40	55.43	0.26	0.02	0.00
3.33	80.00	7.93	8.64	12.48	1.03	1.71	309.29	671.84	1585.24	73.58	0.31	0.05	0.00
9.11	49.38	8.17	6.19	10.22	1.26	2.24	246.23	379.30	1508.28	29.18	0.24	0.09	0.00
28.70	66.57	8.11	7.02	10.89	1.05	2.01	285.34	813.42	1441.09	54.41	0.38	0.07	0.00
16.74	61.41	7.98	7.16	11.15	1.06	1.25	254.42	509.51	1372.90	81.00	0.25	0.01	0.00
32.83	71.59	9.01	7.65	11.97	0.95	1.84	319.21	989.51	1497.20	61.53	0.37	0.06	0.00
32.23	71.28	8.94	7.75	12.01	0.96	1.83	319.54	988.68	1496.64	61.88	0.37	0.06	0.00
22.81	**41.35**	**5.88**	**5.17**	**7.80**	**1.32**	**1.42**	**195.52**	**437.43**	**1235.99**	**60.53**	**0.24**	**0.00**	**0.00**
16.66	52.54	6.92	6.96	10.00	1.27	2.27	230.53	497.11	1396.41	38.77	0.33	0.08	0.00
28.94	58.44	9.95	6.25	12.03	1.05	2.65	248.02	541.70	1463.87	5.78	0.24	0.15	0.00
6.16	42.62	7.29	6.11	10.22	1.40	1.69	212.46	225.79	1425.59	52.87	0.25	0.04	0.00
7.00	44.87	5.60	5.95	8.54	1.48	2.57	203.87	285.05	1376.09	28.86	0.39	0.12	0.00
7.06	**48.54**	**5.59**	**5.27**	**8.08**	**1.37**	**2.22**	**218.67**	**306.51**	**1375.31**	**38.40**	**0.35**	**0.06**	**0.00**

Variables							
z_1	z_2	z_3	x_{n1}	x_{n2}	x_{n3}	m_n	n_p
40	19	89	0.38	0.21	0.66	16.70	6
40	19	89	0.43	0.23	0.53	17.19	6
40	25	95	0.50	0.50	0.48	18.00	6
40	22	92	0.37	0.38	0.18	17.28	6
40	**25**	**95**	**0.50**	**0.50**	**0.00**	**12.00**	**6**
39	17	91	0.32	0.24	1.22	16.98	4
40	18	90	0.20	0.11	0.76	17.60	4
38	25	95	0.50	0.50	2.00	12.00	6
40	15	85	0.50	0.00	1.59	18.00	5
40	19	85	0.00	0.17	0.98	15.12	4
40	25	95	0.50	0.47	0.00	16.28	5
37	20	92	0.41	0.27	1.48	16.67	4
39	25	95	0.50	0.50	0.00	18.00	6
39	25	95	0.50	0.48	0.00	18.00	6
35	**25**	**95**	**0.50**	**0.50**	**1.69**	**12.00**	**6**
40	24	92	0.35	0.14	0.14	14.49	4
40	25	86	0.03	0.50	0.32	17.85	3
40	19	89	0.00	0.06	1.92	13.56	4
40	20	85	0.35	0.00	0.00	12.00	5
38	**17**	**85**	**0.44**	**0.17**	**0.23**	**12.04**	**6**

TABLE 4.5
Optimized Design Parameters Using MOPSO

	Variables				Objective Functions		
	Module, m (mm)	Number of Teeth (Planet Gear), Z_1	Number of Teeth (Sun Gear), Z_2	Number of Teeth (Ring Gear), Z_3	Total Weight, W (kg)	Efficiency, η	Center Distance (mm)
Existing planetary gearing design	16	36	20	91	870.24	98.70	448.0
MOPSO design	12.00	40	25	95	787.29	98.98	390.00
	12.00	35	25	95	731.30	98.86	360.00
	12.04	38	17	85	721.95	98.67	330.96

planetary gearing was 870.24 kg, the center distance was 448.0 mm, and the efficiency was 98.7%.

Referring to Table 4.5, in the case of the first result, by selecting the number of teeth in planet gear as 40, the weight has reduced to 787.29 kg, the efficiency has increased to 98.98%, and the center distance has reduced to 390.00 mm. There is a reduction of 9.53% in weight, an increase of 0.28% in efficiency, and a space reduction of 12.95%. Hence, there is an improvement in all three objective functions in the solution obtained with MOPSO.

In the case of the second result, by selecting the number of teeth in planet gear as 35, the weight has reduced to 731.30 kg, the efficiency has increased to 98.86%, and the center distance has reduced to 360.00 mm. There is a reduction of 15.97% in weight, an increase of 0.16% in efficiency, and a space reduction of 19.64%. Hence, there is an improvement in all three objective functions in the solution obtained with MOPSO.

In the case of the third result, by selecting the number of teeth in planet gear as 38, the weight has reduced to 721.95 kg, the efficiency has reduced to 98.67%, and the center distance has reduced to 330.96 mm. There is a reduction of 17.04% in weight, a reduction of 0.03% in efficiency, and a space reduction of 26.13%. Hence, there is an improvement in weight and center distance, along with marginal efficiency loss, in the solution obtained with MOPSO.

4.5.11.2 Visualization of Results

The formation of the Pareto front from the non-dominated solutions containing 14 objective functions is depicted in the subsequent graphs in combinations of three objective functions for every graph. The graphs show the 3D scatter plot of the solution set generated by MOPSO in various combinations of three objective functions.

Even though the graphs help in better visualization of the non-dominated Pareto-optimal solution set, one should be cautious of the fact that all the 14 objective functions put together to form a single set of Pareto-optimal solutions that cannot be visualized in totality in graphical format. Hence, for the sake of decision-making, the solutions better suited for the individual requirements of the user/decision-maker must be given priority in the choice of objective function combinations taken for visualization.

As a result of optimization, a Pareto-optimal front is formed in the objective space of the solutions. In the scatter plot, the circular points represent non-dominated solutions with respect to all 14 objectives taken together, whereas the triangular points represent a non-dominated front formed with respect to the three objectives that are taken for visualization. It can be observed that the distribution of solutions along the front is uniform, which means the MOPSO algorithm offers a high-quality solution set.

Referring to Figure 4.6(a), the solutions close to the upper right corner have higher total weight, higher center distance, and higher efficiency, which means they have better efficiency but are heavier and less compact. On the contrary, the solutions near the lower left corner present better compactness and better total weight but lesser efficiency. The solutions lying in the lower right corner have the best value in terms of higher efficiency, lesser weight, and better compactness.

Referring to Figure 4.6(b), the solutions close to the lower right corner offer higher efficiency, better compactness, and lesser slide roll ratio values.

Referring to Figure 4.7(a), the solutions close to the lower right corner offer higher efficiency and lesser specific sliding values. A lower value of specific sliding indicates a balanced specific sliding at SAP and EAP of the gear tooth profile engagement during running.

Referring to Figure 4.7(b), the solutions close to the lower right corner offer higher efficiency and lower Almen product values at SAP and EAP of the gear tooth profile engagement during running.

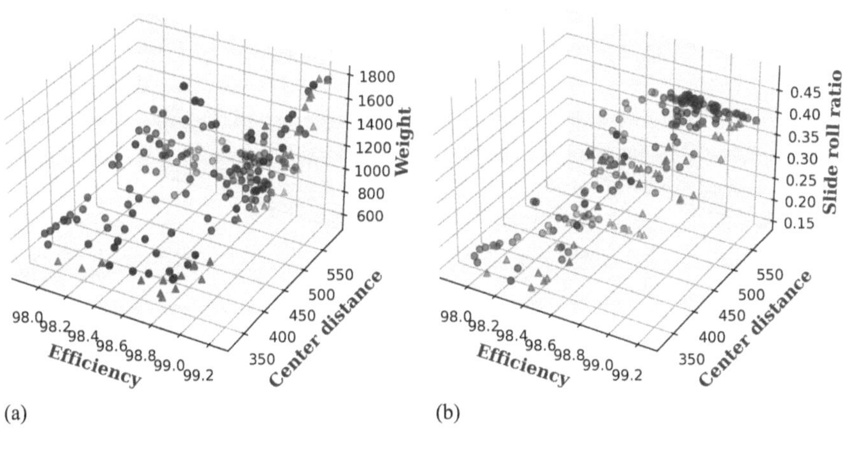

FIGURE 4.6 (a) and (b) 3D scatter plot of MOPSO results.

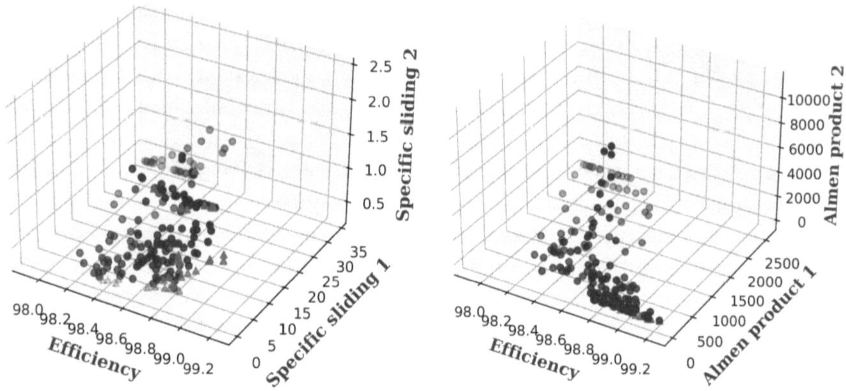

(a) (b)

FIGURE 4.7 (a) and (b) 3D scatter plot of MOPSO results.

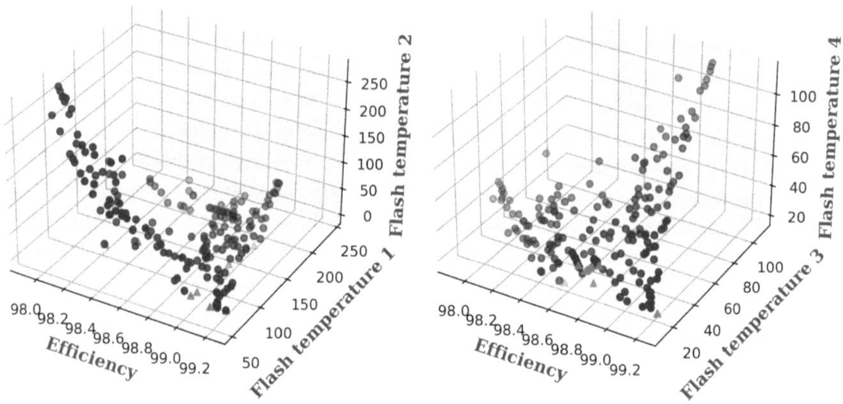

(a) (b)

FIGURE 4.8 (a) and (b) 3D scatter plot of MOPSO results.

Referring to Figure 4.8(a) and (b), the solutions close to the lower right corner offer higher efficiency and lower flash temperatures that are balanced at SAP and EAP of the gear tooth profile engagement during running.

Referring to Figure 4.9, the solutions close to the upper right corner offer higher efficiency and higher lambda ratios for the sun gear/planet gear engagement and the planet gear/ring gear engagement, respectively.

4.5.11.3 Speed Ratio Comparison

The output speed of the sun gear for the three solutions of MOPSO can be analyzed as follows:

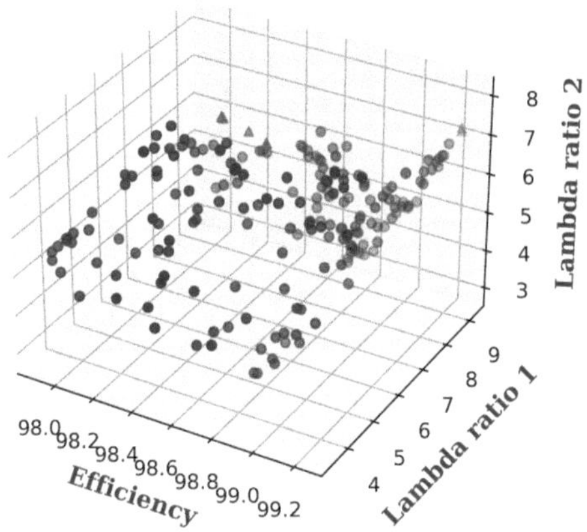

FIGURE 4.9 3D scatter plot of MOPSO results.

For the first and second solutions,
Number of teeth of the planet gear, Z_1 = 40/35
Number of teeth of the sun gear, Z_2 = 25
Number Of teeth of the ring gear, Z_3 = 95
Applying the equations given in Table 4.2,
Input speed of the rotor to the spider arm shaft = 15 rpm
Speed of the sun gear = 72 rpm

Hence, the speed ratio = 72 / 15 = 4.8.

Compared to the speed ratio of 5.55 of the existing planetary gearing, the above speed ratio is 13.51% less, which may not be acceptable.

For the third solution,

Number of teeth of the planet gear, Z_1 = 38
Number of teeth of the sun gear, Z_2 = 17
Number of teeth of the ring gear, Z_3 = 85
Applying the equations given in Table 4.2,
Input speed of the rotor to the spider arm shaft = 15 rpm
Speed of the sun gear = 90 rpm

Hence, the speed ratio = 90 / 15 = 6.00.

Compared to the speed ratio of 5.55 of the existing planetary gearing, the above speed ratio is 8.11% more, which is acceptable.

Therefore, the third solution is selected as the best solution.

4.5.11.4 Selection of Standard Module

From Table 4.5, it can be seen that the module for the best solution is 12.04 mm, which is a non-standard module. The nearest standard module is 12 mm. Hence,

TABLE 4.6
Sample of MOPSO Output for Standard Module

Objectives													
Center Distance	Weight	Efficiency	Slide roll Ratio	Specific Sliding		Almen Product		Flash Remperature				Lambda Ratio	
330.00	**718.61**	**98.69**	**0.24**	**1.49**	**1.41**	**281.90**	**105.45**	**86.35**	**51.78**	**31.77**	**33.07**	**3.95**	**4.08**
330.00	718.52	98.68	0.24	1.49	1.39	279.94	113.53	86.62	51.84	31.65	33.05	3.95	4.09
330.00	718.50	98.68	0.24	1.49	1.39	279.52	115.27	86.67	51.85	31.63	33.05	3.95	4.09
330.00	718.43	98.68	0.24	1.49	1.38	278.13	121.08	86.87	51.90	31.54	33.04	3.95	4.09
330.00	718.62	98.69	0.24	1.50	1.42	283.68	102.72	86.24	51.87	31.81	33.08	3.96	4.08

Constraints													
Interference		Top Land Thickness			Contact Ratio		Bending Stress	Contact Stress		Involute Interference		Tip Interference	Proper Assembly
7.38	**48.18**	**5.65**	**5.12**	**7.98**	**1.37**	**2.24**	**217.18**	**300.50**	**1387.49**	**37.24**	**0.33**	**0.06**	**0.00**
7.34	47.94	5.67	5.10	7.99	1.38	2.24	217.40	299.50	1388.94	36.92	0.33	0.06	0.00
7.33	47.89	5.68	5.09	7.99	1.38	2.24	217.45	299.29	1389.24	36.85	0.33	0.06	0.00
7.30	47.72	5.69	5.08	8.00	1.38	2.24	217.61	298.56	1390.25	36.61	0.32	0.06	0.00
7.37	48.25	5.65	5.13	7.98	1.37	2.24	217.14	300.79	1386.99	37.35	0.33	0.06	0.00

Variables		
x_{n1}	x_{n2}	x_{n3}
0.42	**0.20**	**0.30**
0.41	0.20	0.30
0.41	0.20	0.30
0.40	0.20	0.30
0.43	0.20	0.30

TABLE 4.7
Optimized Design Parameters for Standard Module Using MOPSO

	Variables				Objective Functions		
	Module, m (mm)	Number of Teeth (Planet Gear), Z_1	Number of Teeth (Sun Gear), Z_2	Number of Teeth (Ring Gear), Z_3	Total Weight, W (kg)	Efficiency, η	Center Distance (mm)
MOPSO design	12.04	38	17	85	721.95	98.67	330.96
MOPSO design for standard module	12.0	38	17	85	718.61	98.69	330.0

the final design can be modified to suit a standard module of 12 mm using the procedure given below.

The following design variables of the epicyclic gearing are made constant based on the values from the best solution (Table 4.4):

Number of teeth of the planet gear, $Z_1 = 38$
Number of teeth of the sun gear, $Z_2 = 17$
Number of teeth of the ring gear, $Z_3 = 85$
Number of planet gears, $n_p = 6$
Module, $m_n = 12$ mm (standard module)

The problem is reformulated with the three profile shift coefficients x_{n1}, x_{n2}, and x_{n3} taken as design variables, which will be computed again for the fixed parameters above. The upper and lower bounds for the profile shift coefficients are fixed based on the values from the best solution (Table 4.4).

Profile shift coefficient for planet gear, x_{n1}: 0.4 to 0.5
Profile shift coefficient for sun gear, x_{n2}: 0.1 to 0.2
Profile shift coefficient for ring gear, x_{n3}: 0.2 to 0.3

The MOPSO procedure is rerun for the modified problem formulation.
Table 4.6 shows a sample of the results generated for the standard module of 12 mm.

From the results generated, the solution listed in Table 4.7 is selected as the best solution.

From the above analysis, we can conclude that the solution from Table 4.7, obtained through the design optimization of the WTG planetary gearing using MOPSO, can be considered the best solution.

NOTES

1 For more information on Pareto-optimality, refer to Section 3.4.
2 Refer to Table 1.1.
3 Refer to Section 3.6.5.
4 Refer to Section 3.6.8 and onward.
5 Refer to Section 2.9.

REFERENCES

Buckingham, E., 1988, *Analytical Mechanics of Gears*, Dover Publications, New York, NY.
Coello, C. A. C., et al., 2004, "Handling multiple objectives with particle swarm optimization," *IEEE Transactions on Evolutionary Computation*. 8, 3, 256–279, https://doi.org/10.1109/TEVC.2004.826067

Deb, K., et al., 2002, "A fast and elitist multiobjective genetic algorithm: NSGA-II," *IEEE Transactions on Evolutionary Computation*, 6, 2, 182–197, https://doi.org/10.1109/4235.996017

Jin, Y., et al., 2001, "Dynamic Weighted Aggregation for evolutionary multi-objective optimization: why does it work and how," *Proceedings of the 3rd Annual Conference on Genetic and Evolutionary Computation (GECCO'01)*. Morgan Kaufmann Publishers Inc., San Francisco, CA, USA, 1042–1049.

Kennedy, J. and Eberhart, R., 1995, "Particle swarm optimization," *Proceedings of ICNN'95 - International Conference on Neural Networks*, Perth, WA, Australia, 1942–1948, 4, https://doi.org/10.1109/ICNN.1995.488968

Krishnan, G., 2024, OPTIMO software – Algorithms for Multi-Objective optimization. https://github.com/epicyclic-book/optimo/. Used under the terms and conditions of the MIT license.

Llopis-Albert, C., et al., 2024, "Digital twin-based approach for a multi-objective optimal design of wind turbine gearboxes," *Mathematics*, 12, 1383. https://doi.org/10.3390/math1209138. Used under Creative Commons Attribution 4.0 International License, http://creativecommons.org/licenses/by/4.0/

Marti-Puig, P. et al., (2024), "Wind turbine database for intelligent operation and maintenance strategies," *Scientific Data*, 11, 255, https://doi.org/10.1038/s41597-024-03067-9. https://www.nature.com/scientificdata. Used under Creative Commons Attribution 4.0 International License, https://creativecommons.org/licenses/by/4.0/

Parsopoulos, K. E. and Vrahatis, M. N., 2002, "Particle Swarm Optimization Method in Multiobjective Problems," *Proceedings of the 2002 ACM Symposium on Applied Computing (SAC '02)*, Association for Computing Machinery, New York, NY, USA, 603–607, https://doi.org/10.1145/508791.508907

Townsend, D. P., 2013, *Dudley's Gear Handbook – The design, manufacture and applications of gears*, McGraw Hill, New York, NY.

5 Design Optimization of Helicopter Gearbox Using Multi-Objective-Simulated Annealing Algorithm

Chandrasekaran G and Gokul Krishnan

5.1 INTRODUCTION

Another optimization technique inspired by real-world phenomena is simulated annealing (SA). It is based on the physical process of annealing, which involves the heating and subsequent cooling of metals and alloys. SA is a meta-heuristic optimization technique that starts from an initial solution and utilizes a local-search procedure in order to find better solutions. However, unlike other local-search-based algorithms, SA modifies the local-search heuristic such that the decision to accept locally better solutions is guided by a probability function. Unlike genetic algorithms, there is no usage of operators such as crossover and mutation. Therefore, it is relatively simple to implement and run and could prove to be computationally less expensive depending on the use case.

5.2 SIMULATED ANNEALING

Initially proposed by Kirkpatrick et al. (1983) and Černý (1985), SA is based on the process in metallurgy that involves heating a piece of metal and then gradually cooling it down until it reaches thermodynamic equilibrium or *ground state*. In a way, the algorithm is very similar to the Hill Climber algorithm, which also performs a local-search for the optimum solution among neighboring solutions. The primary drawback of the Hill Climber algorithm is that it uses a strict acceptance criterion that rejects any solution worse than the current solution. This can cause the system to get trapped inside a local optimum and fail to explore other regions of the problem space in the search for a global optimum.

SA, on the other hand, uses the concept of temperature to define the thermodynamic stability of the system. Initially, when the temperature of the system is high, the system is thermodynamically unstable, which means there is a higher

 DOI: 10.1201/9781003144465-5

probability of accepting solutions worse than the current solution. As the system cools down, the system attains thermodynamic stability. This results in an increased preference for solutions better than the current solution.

This ability of the system to accept worse solutions during the initial stages of optimization allows it to cover a wide area of the problem space. Then, as the system narrows down into the region of global optimum, the acceptance criterion is tightened so that the searching process remains focused within this region and ultimately converges to the point of global optimum.

Because SA is a single-objective optimization algorithm, the technique cannot be directly applied to solve multi-objective problems. Research has been conducted on extending SA to solve multi-objective problems and obtain a Pareto-optimal solution set.[1] An example of such a technique is Pareto-simulated annealing (PSA) (Czyżak and Jaszkiewicz, 1998).

5.3 STANDARD SA PROCEDURE

Unlike evolutionary algorithms and other stochastic optimization techniques like Particle Swarm Optimization, which are population based, SA maintains a single reference point, called the *candidate solution*, which is updated throughout the optimization procedure.

This section describes the steps involved in the standard SA algorithm used to solve for the minimization of a single-objective, having one or more design variables and optionally a set of constraints. The following global properties are specified:

- Starting temperature of the system.
- Cooling rate (the rate at which the system is cooled).
- Step size for traversal through the search space.
- Upper and lower bounds for each design variable.
- Number of iterations.

Initially, the candidate solution is placed at a random position within the search space bounds. The position vector X of the solution is defined by the set of design variables. The fitness value P of the solution is obtained by evaluating the objective function associated with the problem. The goal is assumed to be the minimization of the objective (i.e., a smaller fitness value is considered better).

The temperature of the system is initialized to the starting temperature, which also serves as the maximum allowable temperature. At each iteration of the algorithm, the following steps are performed:

- Generate a neighboring solution based on the current solution and step size:

$$X_{new} = X + \left[r_1, r_2, \ldots, r_M\right] \cdot S$$
$$P = F\left(X_{new}\right)$$

where X and X_{new} are the current and new positions, respectively; r_1, r_2, … r_M are randomly generated numbers in the range (−1, 1); M is the number of design variables; S is the step size; F is the fitness function.

- If the new solution has an improved fitness value compared to the existing solution, it is accepted directly. If the new solution is worse, on the other hand, compute the probability of acceptance based on, (i) the difference in fitness values and (ii) the current temperature of the system.

$$p = \exp\left(-\frac{\delta}{t}\right)$$

where $\delta = P - P_{best}$; p is the probability of acceptance; t is the current temperature.
This probability equation is called the *metropolis acceptance criterion.*

- For constrained optimization problems, there is an additional test for the feasibility of the solution. In the case that any of the provided constraints do not pass, the solution is rejected.
- Finally, the system is cooled down based on the chosen annealing schedule. A constant cooling rate is used here.

$$t = t(1 - R_c)$$

where R_c is the cooling rate.

These steps are repeated until the maximum number of iterations is reached, or the system is sufficiently cooled, that is, the temperature is very close to zero.

5.3.1 Algorithm Pseudo-code

The complete algorithm can be visualized through the flow chart in Figure 5.1 and pseudo-code representation in Listing 5.1.

LISTING 5.1 SA ALGORITHM PSEUDO-CODE

```
Initialize position X to a random vector within bounds
Initialize best fitness value Pbest to infinity
Initialize temperature t to starting temperature

WHILE iterations remaining
    Generate neighboring solution of X
    Compute fitness value P = F(X)

    IF P < Pbest
        Pbest = P
```

```
    ELSE
        diff = P - P_best
        p_acc = exp(-diff / t)
        IF rand() < p_acc
            P_best = P
        END IF
    END IF

    t *= 1 - cooling_rate
END WHILE
```

5.4 MULTI-OBJECTIVE-SIMULATED ANNEALING (MOSA)

This section describes the techniques and challenges involved in extending the SA algorithm for solving constrained multi-objective optimization problems.

5.4.1 Population of Solutions

SA maintains a single reference point, called the *candidate solution*, which is updated throughout the optimization process. MOSA attempts to convert the algorithm to a population-based algorithm, similar to evolutionary algorithms and particle swarm optimization.

Consider that there is a population of N candidates that are initialized at random locations within the search space. As the simulation is run, each candidate is updated independently according to the annealing technique used. At the end of the run, non-dominated solutions from this population are extracted into an external archive.

As a unique advantage of the population-based SA technique, considering each candidate is moved independently from the rest, the simulation run of the entire population can be parallelized effectively. This can result in significant speed-up on a multi-core CPU architecture.

5.4.2 Fitness Vector

SA deals with a single fitness value associated with each candidate solution. In order to adapt the technique to solve problems containing multiple objectives, several approaches have been proposed. A simple approach is to use a weighted average of the fitness values from each objective function, that is,

$$P_{\text{combined}} = \frac{w_1 p_1 + w_2 p_2 + \ldots + w_N p_N}{N}$$

where N is the number of objectives.

Several other weighted aggregation techniques have been proposed (Yaochu Jin et al., 2001).

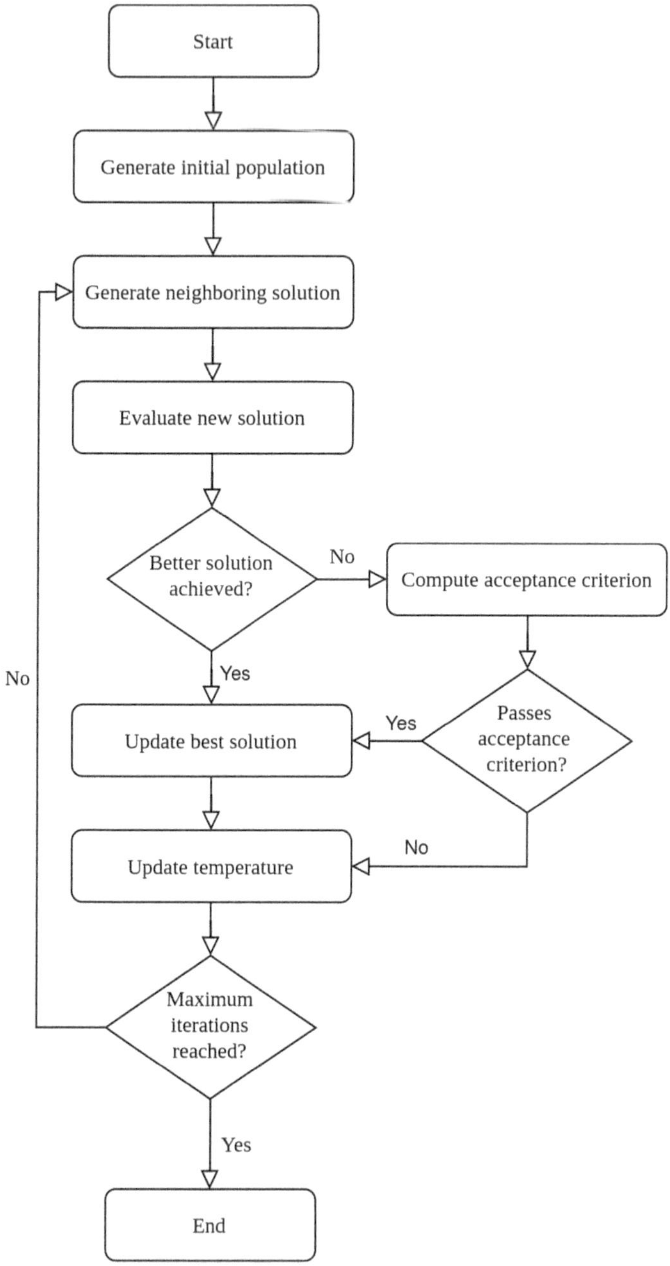

FIGURE 5.1 SA algorithm flow chart.

However, an alternate approach involves modifying the scalar fitness value P into a fitness vector consisting of a separate value for each objective function. That is,

$$P = \left[p_1, p_2, \ldots, p_N\right]$$

where N is the number of objectives.

This approach has been selected in the implementation covered in this book.

5.4.3 Dominance Operator

While the use of a fitness vector is beneficial when dealing with multi-objective problems, it introduces the need for a new comparison operator that can be used to compare the fitness vectors of two candidate solutions. Therefore, the standard comparison operator is replaced with a *dominance* operator that checks for the Pareto-dominance of one solution by another. Consider two candidates with fitness vectors P and Q.

Then, assuming minimization, P is said to dominate Q if,

$P_i \leq Q_i$ for all values of i, and
$P_i < Q_i$ for at least one value of i,

where $i = \{0 \text{ to } N\}$; N is the number of objectives.

5.4.4 Acceptance Function

If the fitness vector computed from the new solution dominates the existing fitness vector, the solution is accepted directly. In every other case, the decision whether to accept the solution is guided by an acceptance function, which is a probability function computed based on (i) the distance or delta between the fitness vectors and (ii) the current temperature of the system.

$$p = \exp\left(-\frac{\text{sum}(\Delta)}{t}\right)$$

where $\Delta = P - P_{best}$, the difference vector; p is the probability of acceptance; t is the current temperature.

This probability equation is based on the *metropolis acceptance criterion.* However, it uses the sum of individual elements of the distance vector in the acceptance calculation.

5.4.5 Archive of Non-dominated Solutions

In order to extract a Pareto-optimal solution set, the approach considered is to maintain an archive of non-dominated solutions separate from the population set and update this archive at the end of each iteration. The procedure is described in detail in Section 5.4.11.

5.4.6 Population Diversity

When dealing with multi-objective optimization problems, we are interested in a set of non-dominated solutions, which form the Pareto-optimal solution set. Therefore, there is a need to ensure a healthy diversity of Pareto-optimal solutions.

In order to prevent the candidate solutions from bunching together and to maintain the diversity of solutions, candidates that are closely bunched together are removed from the archive after each iteration and replaced by fresh candidates that further diversify the pool of solutions. The following criterion is used in choosing the candidates to eliminate.

5.4.6.1 Crowding Distance

Each solution in the archive is assigned a crowding distance value that defines the density of solutions surrounding it. Solutions in denser areas are assigned a smaller crowding distance. The crowding distance is computed separately for each objective function, that is, along each dimension in the objective space, and then averaged together.

5.4.6.2 Roulette-Wheel Selection

To select one among the non-dominated solutions to be removed from the archive, a roulette-wheel selection is used, with the weights defined by the inverse of the crowding distance values. This increases the probability that solutions from denser regions are chosen for removal.

5.4.7 Handling Constraints

Constraints restrict the movement of candidates by limiting the possible values that the design variables can assume. The feasibility of solutions is decided by one or more constraint functions defined as part of the problem. A solution is considered feasible if and only if all provided constraint tests pass. The restriction in the movement of candidates limits the area of search space covered, and this can reduce the quality of solutions provided by the algorithm.

In order to maintain a good quality of solutions and to ensure that candidates cover a wide extent of the search space, the following approach is proposed. At every iteration, candidates are allowed to move freely and assume any neighboring solution in the search space irrespective of constraint violation.

However, a solution is only accepted as the candidate's best solution achieved so far, P_{best}, if all constraints pass. This allows candidates to assume infeasible positions during their search for better solutions, whereas the fitness vector defined by P_{best} always holds the best feasible solution achieved so far.

Computation of dominance and crowding distance factors is performed based on the candidate's best feasible position rather than the current actual position.

5.4.8 MOSA Implementation

An implementation of the MOSA algorithm is described along with the corresponding pseudo-code. We start with an overview of key functions used in the algorithm code.

5.4.8.1 RouletteWheelSelection

We employ a roulette-wheel selection process, which is essentially the randomized selection of members from a population, with each member assigned a weight that defines the probability of that member being selected. This can also be called a fitness-proportionate selection model. The function takes as input a list of weights and outputs the index of the member being selected. The weights array is normalized such that the elements add up to one (Listing 5.2).

LISTING 5.2 IMPLEMENTATION OF ROULETTE-WHEEL SELECTION

```
FUNCTION RouletteWheelSelection(weights):
    weights = normalize(weights)
    weights = generate_cumulative_sum(weights)
    r = rand() # random value in range [0,1]
    FOR EACH value in weights:
        IF value >= r:
            return i
        END IF
    END FOR
END FUNCTION
```

5.4.8.2 SelectMemberToDelete

To select one among the non-dominated solutions to be removed from the archive, a roulette-wheel selection is used with the weights assigned based on the inverse of the candidate crowding distance. This increases the probability that candidates from denser regions are chosen for removal (Listing 5.3).

LISTING 5.3 IMPLEMENTATION OF REDUNDANT MEMBER SELECTION

```
FUNCTION SelectMemberToDelete(archive):
    weights = [1/c_1, 1/c_2 ... 1/c_num_archive]
    index =  RouletteWheelSelection(weights)
    return index
END FUNCTION
```

5.4.8.3 ComputeDominance

The function accepts a set of candidate solutions as input and computes the Pareto-dominance state of each candidate, based on the best feasible fitness achieved by the candidate, or *best_fitness*. For each candidate, the *is_dominated* property is set to *true* for all candidates that are dominated by at least one other candidate in the set.

5.4.8.4 ComputeCrowdingDistance

The function accepts a population as input and computes the crowding distance factor of each candidate, based on the best feasible fitness achieved by the candidate, or *best_fitness*. The algorithm for crowding distance computation is adapted from the NSGA-II paper (K. Deb, 2002), with a few minor changes.

Because the crowding distance and inverse crowding distance are used as weighting factors in the roulette wheel selection routines, the value is clamped within the half-closed range (0, 1) such that the crowding distance value is always nonzero and less than or equal to one.

The default value for crowding distance is chosen as 1.0, which ensures equal probability of each candidate being selected.

5.4.8.5 FilterNonDominated

The function accepts a set of candidate solutions as input and removes each solution that has the *is_dominated* property set to *true*. It then returns the filtered non-dominated population set.

5.4.8.6 Fitness

This function is user-defined. It accepts an array of design variables as input and outputs an array of objective function values. Because we are dealing with a minimization problem, objective functions that are to be maximized must be negated.

5.4.8.7 Constraints

This function is user-defined. It accepts an array of design variables as input and outputs an array of constraint function values. Constraint functions are specified in such a way that they must produce a value greater than or equal to zero in order for the test to pass, that is, the inequality $C_i \geq N$ is rewritten as, $C_i - N \geq 0$.

5.4.9 Initialization

The following global properties are set (Listing 5.4):

- *num_vars*: number of design variables
- *bounds_min*: array of values representing minimum bounds for each design variable
- *bounds_max*: array of values representing maximum bounds for each design variable

- *num_iter*: number of iterations
- *num_pop*: population size
- *num_archive*: archive size for the non-dominated solution set
- *step_size*: controls the step size for neighboring candidate generation
- *temp*: starting temperature of the system

LISTING 5.4 GLOBAL PROPERTIES

```
num_vars = 4
bounds_min = [x_1, x_2 ... x_num_vars]
bounds_max = [x_1, x_2 ... x_num_vars]
num_iter = 100
num_pop = 200
num_archive = 50
step_size = 1.0
temp = 100
```

Next, we allocate the required number of candidates. Each candidate is initialized at randomized positions within the provided minimum and maximum bounds (Listing 5.5).

LISTING 5.5 ALLOCATION OF CANDIDATES

```
Population = [P_1, P_2 ... P_num_pop]
FOR EACH candidate P_i in Population:
    P_i.position = [r_1, r_2 ... r_num_vars]
    P_i.fitness = Fitness(P_i.position)
    P_i.constraints = Constraints(P_i.position)
    P_i.best_position = P_i.position
    P_i.best_fitness = P_i.fitness
    P_i.is_dominated = false
    P_i.crowding_dist = 1.0
END FOR
```

5.4.10 Iteration

Following the initialization, we reach the core of the algorithm where each iteration is processed. First, we loop through each candidate in the population set. (Listing 5.6).

LISTING 5.6 THE OVERALL ITERATION

```
FOR j = [1 to num_iter]:
    FOR i = [1 to num_pop]:
        Update each particle
    END FOR
    Update the archive
END FOR
```

For each candidate, a neighboring solution is generated at a random location based on the specified step size. It should be noted that the movement of a candidate solution is not restricted based on the provided constraints. The position is clamped to the provided bounds along each dimension (Listing 5.7).

LISTING 5.7 STEP 1 – UPDATE POSITION

```
Pi.position = Pi.position + [r1, r2 ... rnum_vars] * step_size
Clamp(Pi.position, bounds_min, bounds_max)
```

Then, the fitness and constraint vectors are computed from the new position (Listing 5.8).

LISTING 5.8 STEP 2 – EVALUATE FITNESS AND CONSTRAINTS

```
Pi.fitness = Fitness(Pi.position)
Pi.constraints = Constraints(Pi.position)
```

In the case that all constraints are satisfied and the new solution dominates the previous best solution, the candidate is moved to the new position, and the best position and best fitness values are updated.

On the other hand, in case the new solution does not dominate the previous best solution, the acceptance probability is computed by applying the *metropolis acceptance criterion*, adapted to handle a multi-objective fitness vector.

This makes use of the current system temperature, which is computed as a fraction of the starting temperature, based on the number of iterations passed (Listing 5.9).

LISTING 5.9 STEP 3 – UPDATE LOCAL BEST

```
IF Population[i].constraints >= 0:
    IF Population[i].fitness dominates Population[i].
best_fitness:
        Population[i].best_position = Population[i].
position
        Population[i].best_fitness =
Population[i].fitness
    ELSE:
        T = temp * (1 - j / num_iter)
        diff = p.fitness - p.best_fitness
        P = exp(-sum(diff) / T)
        IF rand() < P:
            Population[i].best_position =
Population[i].position
            Population[i].best_fitness =
Population[i].fitness
        END IF
    END IF
END IF
```

5.4.11 Archiving

Archiving refers to the extraction of non-dominated solutions from the population set into a separately maintained archive. Once all candidates have been processed in an iteration, the solutions from the new population are merged with the pre-existing archive, which is initially an empty set. Then, the dominance factor is re-computed on this merged population. Solutions that are dominated by other solutions are filtered out, keeping only non-dominated solutions in the archive (Listing 5.10).

LISTING 5.10 ARCHIVING – STEP 1

```
Archive.append(Population)
ComputeDominance(Archive)
Archive = FilterNonDominated(Archive)
```

Then, the crowding distance factor is re-computed on the filtered solution set. The crowding distance is used as a weighting factor in the selection of global optima and in the elimination of redundant solutions from the archive.

In case the archive size exceeds the maximum allowed size, members are removed from the archive until the archive size requirement is met. This is done using a roulette-wheel selection process with the inverse crowding distance criterion. This prioritizes solutions from denser regions for removal and ensures a stable archive size (Listing 5.11).

LISTING 5.11 ARCHIVING – STEP 2

```
ComputeCrowdingDistance(Archive)
WHILE size(Archive) > num_archive:
 r = SelectMemberToDelete(Archive)
 Archive.remove(r)
END WHILE
```

At the end of the final iteration, the candidates in the archive represent the final non-dominated set of solutions. The *best_position* and *best_fitness* vectors hold the final feasible solution of each candidate in the archive. The data can be output into a text file format, such as CSV, or plotted on a graph for further processing and visualization. A good amount of diversity is preserved, which helps the decision-maker in selecting the most optimal solution for the problem at hand.

5.5 CASE STUDY: DESIGN OPTIMIZATION OF A HELICOPTER GEARBOX USING SIMULATED ANNEALING

An advantage of the planetary stage compared to a simple parallel shaft arrangement is that each planet gear must transmit only a part of the total torque. This load-sharing results in a smaller, lighter transmission.

5.5.1 Overview of a Helicopter Gearbox

The helicopter is a complex *rotary-wing* aircraft that obtains both lift and thrust from blades rotating about a vertical axis. The helicopter can have one or more engines, and it uses gearboxes connected to the engines by rotating shafts to transfer the power from the engines to the rotors (Figure 5.2) (Rotaru and Todorov, n.d.).

The helicopter transmission is responsible for providing the required torque to the main rotors and other accessory systems from the input power generated by the engine. The transmission system in helicopters is optimized for the demands of rotary-wing flight and designed to manage the varying power requirements and rotor speeds during flight. The helicopter transmission gearbox is designed for the precise control of rotor speeds and power output that is required during flight. The

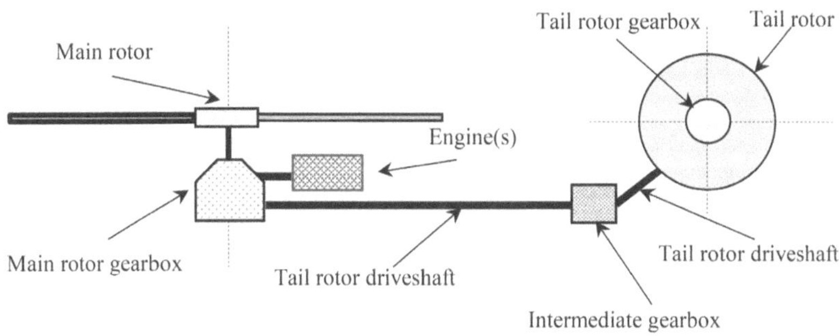

FIGURE 5.2 Helicopter driveline (Rotaru and Todorov, n.d.).

helicopter engine runs at thousands of rpm, which are converted into low speed of the rotor by the transmission system. The transmission also controls the speed of the tail rotor of the helicopter.

5.5.2 Case Study of a Helicopter Gearbox

The design optimization is attempted for the helicopter planetary gearing with the specifications listed in Table 5.1.

Power input to the planetary stage = 764 HP = 569.715 kW.

Input speed for the planetary stage $N = 3550$ rpm.

To find the speed of individual elements of the existing planetary gear train, let

Number of teeth on the sun wheel, $z_2 = 29$
Number of teeth on planet gears, $z_1 = 31$
Number of teeth on the annular ring gear, $z_3 = 91$

The engine power is transmitted to the rotor in a three-stage reduction, the first stage being a simple gear drive, the second stage being a bevel gear drive, and the third stage being a planetary gear drive. The input to the planetary gearbox is given to the sun gear. The ring gear is held stationary, and the output is taken from the spider arm L. Hence, this arrangement comes under the first type of epicyclic gearing, namely the planetary gearing.[2] The design optimization is attempted for the third stage of planetary gearing, with specifications as given in Table 5.1, using the SA technique.

The last row of Table 5.2 provides us with the equation for the speed of individual elements.

Input speed of sun gear = 3550 rpm, that is, $x + y = 3550$.

Since the annulus or ring gear is locked,

$$y - \frac{29}{91}x = 0$$
$$\therefore x = 2692.1$$

TABLE 5.1
Helicopter Planetary Gearing Specifications

	Sun Gear	Planet Gear	Ring Gear
Normal module (mm)	4.23	4.23	4.23
Number of teeth	29	31	91
Normal pressure angle (°)	25	25	25
Helix angle (°)	0	0	0
Individual shift in the basic profile X	0	0	0
Face width (mm)	38.862	38.862	38.862
Number of planets	–	5	–

TABLE 5.2
Speed Calculation for Planetary Elements

Operation	Spider Arm, L	Sun Wheel, Z_s	Planet Wheel, Z_p	Annular Wheel, Z_a
1. Arm L is locked and sun wheel S is given +1 revolution.	0	+1	$-\frac{Z_s}{Z_p}=-\frac{29}{31}$	$-\frac{Z_s}{Z_a}=-\frac{29}{91}$
2. Multiply by x	0	x	$-\frac{Z_s}{Z_p}x=-\frac{29}{31}x$	$-\frac{Z_s}{Z_a}x=-\frac{29}{91}x$
3. Add y to all columns	y	$y+x$	$y-\frac{Z_s}{Z_p}x=y-\frac{29}{31}x$	$y-\frac{Z_s}{Z_a}x=y-\frac{29}{91}x$

$$y = 857.9$$

$$\text{Speed of planets} = y-\frac{29}{31}x = 857.9-\frac{29}{31}\times 2692.1 = -1660.52\text{rpm}(\text{ACW})$$

$$\text{Output Speed of the spider arm} = y = 857.9\text{rpm}(\text{CW})$$

5.5.3 Calculation of Gear Parameters

Gear parameters are calculated using the procedure explained in Chapter 3 for Power Take-Off (PTO) gearbox.[3]

5.5.4 Design Space

The existing gear design problem for gears made of a chosen material can be reduced to an eight-dimensional design problem where the eight free parameters are the number of teeth on the planet pinion z_1, the number of teeth on the sun gear z_2, the number of teeth on the ring gear z_3, the module m_n, the three profile shift coefficients x_{n1}, x_{n2}, and x_{n3} of the planet gear, the sun gear, and the ring gear, respectively, and the number of planets n_p. The inequality constraints which bound this design space are the bending stress, Hertz stress, contact ratio, top land thickness, operating interference of gears, and involute and tip interference of internal gears.

5.5.5 Objective Functions and Constraints

Fourteen objective functions for minimization/maximization with an array of inequality and equality constraints have been considered, and the problem can be stated as follows:

The following objective functions are considered for the design optimization of the planetary gear train:

$$f_{obj} = \begin{cases} Min.center\ distance\ between\ planetary\ gear\ pair \\ Min.weight\ of\ the\ planetary\ gear\ train \\ Max.efficiency\ of\ planetary\ gear\ train \\ Min.slide\ roll\ ratio \\ Balancing\ the\ specific\ sliding \\ Balancing\ the\ Almen\ product \\ Balancing\ the\ flash\ temperature \\ Balancing\ the\ lambda\ ratio \end{cases}$$

The independent design variables of the planetary gear train include a module, the number of teeth of the pinion and the three profile shift coefficients x_{n1}, x_{n2}, and x_{n3} of the planet gear, the sun gear, and the ring gear, respectively. The vector of design variables is given below:

$$x(i) = \begin{cases} planet\ gear\ no.ofteeth, Z_1 \\ sun\ gear\ no.ofteeth, Z_2 \\ ring\ gear\ no.ofteeth, Z_3 \\ planet\ gear\ profile\ shift\ coefficient, x_{n_1} \\ sun\ gear\ profile\ shift\ coefficient, x_{n_2} \\ ring\ gear\ profile\ shift\ coefficient, x_{n_3} \\ normal\ module, m_n \\ number\ of\ planets, n_p \end{cases}$$

where $i = 1, 2, \ldots$, the number of variables.

A number of constraints, which act as sub functions to restrict the objective function, are introduced into the objective function to be optimized and are used to provide suitable design choices. The very often encountered failure types in a transmission gear are tooth fracture, surface fatigue, abrasion, and plastic deformation. Thus, the following constraints are considered for the design optimization of the planetary gear train:

$$
g(j)=\begin{cases}
\textit{Bending stress constraint}\\
\textit{Surface contact compressive stress constraint}\\
\textit{Contact ratio constraint}\\
\textit{Top land thickness constraint}\\
\textit{Operating interference of planet gear constraint}\\
\textit{Operating interference of sun gear constraint}\\
\textit{Operating interference of ring gear constraint}\\
\textit{Involute interference of internal gear constraint}\\
\textit{Tip interference of internal gear constraint}\\
\textit{Proper assembly of planetary gearing constraint}
\end{cases}
$$

where $j = 1, 2, \ldots,$ the number of constraints.

5.5.6 Establishing a Mathematical Model

To establish a mathematical model for the planetary gear train design optimization, mathematical equations for the design variables, objective functions, and necessary constraints are to be developed.

5.5.7 Objective Functions for the Epicyclic Gearing

Since the problem formulation is identical to that of the PTO gearbox in Chapter 3, the objective functions are formulated based on the principles and procedures explained in Section 3.6.9 for the PTO gearbox.

5.5.7.1 Lambda Ratio

In the equations for maximization of lambda ratio, for the present case,

$$R_{a1}' = R_{a2}' = 0.2$$

5.5.8 Design Variables and Constraints

The independent design parameters of the planetary gear drive include the module m_n, the number of teeth of planet gear z_1, the number of teeth of the sun gear z_2, the number of teeth of ring gear z_3, the three profile shift coefficients x_{n1}, x_{n2}, and x_{n3}, and the number of planets n_p.

Therefore, the design variables are,

$$X=\left[z_1,z_2,z_3,x_{n1},x_{n2},x_{n3},m_n,n_p\right]^T=\left[x_1,x_2,x_3,x_4,x_5,x_6,x_{7,}x_8\right]^T$$

A number of constraints, which are introduced into the objective function to be optimized, are used to provide suitable design choices. The constraints are formulated based on the principles and procedures explained in Section 3.6.14 for the PTO gearbox.

5.5.8.1 Bending Strength and Contact Compressive Strength

The dynamic load factor, K_V, and lubrication speed factor, Z_V, depend on the gear's tangential speed at the pitch circle, v (m/s).

As per the standards,

For the sun gear–planet gear pair engagement,
For surface-hardened gear, and $v < 25.0$ m/s,

$$K_V = 1.5$$
$$Z_V = 1.02$$

For the planet gear–ring gear pair engagement,
For surface-hardened gear, and $v = 10.7$ m/s,

$$K_V = 1.2$$
$$Z_V = 1.0$$

5.5.9 Design Data for Optimization

The following are the input parameters considered for the planetary gear design based on the existing gear design as in Table 5.3.

TABLE 5.3
Helicopter Planetary Gear Parameters

Data	Gear
Density of the material, ρ	7.86×10^6 kg/mm^3
Material of sun gear, planet gear, and ring gear	Steel AISI 9310
Design bending stress, σ_b	$400\text{N}/\text{mm}^2$
Design crushing stress, σ_c	$2000\text{N}/\text{mm}^2$
Input speed, N	3550 rpm
Power transmitted, P	764 HP
Young's modulus of the material (E)	$2.0\times10^5\text{kg}/\text{mm}^2$
Pressure angle, α_n	25°

5.5.10 Methodology of Optimization of Planetary Gear Using MOSA

This section describes how the MOSA algorithm is configured for solving the optimization problem of planetary gearing. This book utilizes the OPTIMO software package (Gokul Krishnan, 2024), which provides an efficient MOSA implementation in Python along with the complete problem set corresponding to this case study. Chapter 2 provides detailed instructions on how to obtain and use the software.[4]

5.5.10.1 Variable Bounds

In order to determine the values of design variables, such as module, number of teeth, face width, and helix angle, through the optimization procedure, the upper and lower bounds of the design variables are to be established first.

The following variables are considered, and their lower and upper bounds are fixed. Further, the type of variable – whether discrete (integer) or real-valued (float) – is also specified.

Number of planet gear teeth, z_1: 20 to 40 (integer)
Number of sun gear teeth, z_2: 20 to 40 (integer)
Number of ring gear teeth, z_3: 80 to 100 (integer)
Profile shift coefficient for planet gear, x_{n1}: −1.0 to 1.0 (float)
Profile shift coefficient for sun gear, x_{n2}: −1.0 to 1.0 (float)
Profile shift coefficient for ring gear, x_{n3}: −1.0 to 1.0 (float)
Module, m_n: 2 to 6 *mm* (float)
Number of planets, n_p: 2 to 8 (integer)

5.5.10.2 Input Parameters

For this model, the most important data inputs are the gear design parameters of the two mating gears. The decision-maker is required to make this decision based on experience and design requirements.

The problem sets provided in this book contain Python code corresponding to the objective functions and design constraints formulated in Sections 5.5.7 and 5.5.8. The upper and lower bounds of design variables are also specified in the same file(s). The seed for the random number generator is fixed at 0.123 for reproducibility of results.

The problem definition corresponding to Chapter 5 can be found in *problem-sets/problem_ch5.py* file within the software repository. The optimization run is started by entering the command specified in Listing 5.12 from the repository root.

LISTING 5.12 COMMAND TO EXECUTE THE ALGORITHM RUN

```
# macOS / Linux
python3 run.py -a sa -p ch5

# Windows
python.exe run.py -a sa -p ch5
```

5.5.10.3 Decision-Making

The goal of decision-making is to select the most optimal solution from the Pareto solution set obtained from the optimization results. Selection of the single best solution from many conflicting non-dominated solutions from the Pareto-set is based on expertise along with any problem-specific requirements. Apart from the above, there are decision-making techniques like the linear programming technique for multidimensional analysis of preference (LINMAP), Shannon Entropy (SE), the technique of order preference by similarity to an ideal solution (TOPSIS), etc., which can be used as a post-processing step to obtain a relative gradation of results.

The OPTIMO software makes use of a Common Objective Function, or COF, which assigns a scalar value to each solution in the Pareto-optimal set. This scalar value is essentially the normalized average of the set of objective function values. For a given solution *i*,

$$P_i = \left[p_1, p_2, \ldots p_N\right]$$

$$S_i = COF\left(P_i\right)$$

where P_i is the vector of objective function values; N is the number of objectives; S_i is the scalar output assigned to the solution.

The solution set is then sorted on its COF value, such that the solution with the lowest COF value is considered to be, on average, the overall best solution (assuming minimization of objectives).

5.5.11 Interpretation of MOSA Results

In this section, the results obtained from MOSA are discussed and compared with the design of existing helicopter planetary gearing. For detailed information on interpreting the tables and plots in the subsequent sections, refer to Section 3.6.17.

Table 5.4 shows a sample of the MOSA output. Since constraints are specified in the form of ≥ 0 only, all the columns corresponding to the constraints shown in Table 5.4 indicate results obtained by modifying the constraint equations accordingly. Table 5.4 lists the values of the objective functions, constraints, and design variables from the generated solutions. The output represents a non-dominated Pareto-optimal solution set.

5.5.11.1 Comparison of Results

In Table 5.5, a comparison has been made between the results that are obtained with MOSA and the design of the existing planetary gearing. The solutions correspond to the number of teeth of the planet gear, namely, 38, 40, and 34 were selected from the results. The values of design variables and objective functions corresponding to each solution are indicated in Table 5.5. The total weight of

TABLE 5.4
Sample of MOSA Output

						Objectives								
Center distance	Weight	Efficiency	Slide roll ratio	Specific sliding		Almen product			Flash temperature			Lambda ratio		
197.75	21.16	99.74	0.26	0.28	0.81	802.11	368.42	3.31	12.79	9.51	10.38	33.62	26.91
151.57	13.04	99.68	0.28	0.31	0.94	968.59	253.30	0.45	12.34	10.01	11.41	24.01	19.43
225.80	24.36	99.74	0.36	0.07	0.91	151.70	145.83	4.46	11.24	7.58	8.31	42.03	27.26
186.87	18.10	99.71	0.24	0.05	1.07	297.54	325.18	0.58	14.82	17.61	18.16	28.69	23.39
209.40	16.77	99.68	0.30	0.03	1.27	57.94	135.77	1.13	16.92	15.38	16.90	31.04	22.25
181.99	24.11	99.68	0.28	0.14	1.02	458.88	776.85	1.32	9.98	19.67	18.88	32.00	23.06
126.45	**7.94**	**99.67**	**0.27**	**0.02**	**0.79**	**70.03**	**763.42**	**2.84**	**22.52**	**19.38**	**21.20**	**18.40**	**14.79**
147.63	17.23	99.65	0.27	0.26	1.04	776.67	349.94	0.14	12.48	20.13	19.75	24.12	18.30
191.41	16.16	99.66	0.33	0.04	1.30	35.35	118.06	1.47	15.71	14.12	15.60	28.79	19.65
159.72	10.56	99.60	0.18	0.40	0.83	1150.14	2647.93	4.83	31.21	27.01	30.55	20.16	19.29
207.00	20.46	99.58	0.49	0.22	0.94	595.23	1317.79	4.09	13.26	13.89	14.86	35.85	17.70
80.72	4.52	99.41	0.26	0.71	0.73	1694.85	2348.77	7.66	28.53	15.50	19.01	10.93	9.00
68.83	3.54	99.40	0.13	0.78	0.91	1830.15	2454.75	2.65	37.70	21.89	27.33	7.96	8.25
81.12	**3.38**	**99.53**	**0.20**	**0.49**	**0.73**	**1559.30**	**4867.38**	**6.16**	**39.15**	**15.15**	**23.37**	**10.26**	**9.81**
102.42	4.65	99.41	0.33	0.12	1.30	505.02	2002.67	7.54	27.67	21.14	22.16	14.92	10.33
189.00	27.04	99.29	0.29	1.33	0.43	2710.31	1786.00	3.94	14.01	14.42	15.48	30.86	24.14
186.60	21.29	99.52	0.49	0.34	0.87	557.73	1171.17	4.32	12.73	14.79	15.61	32.68	16.07
172.34	10.57	99.56	0.47	0.01	1.05	131.37	191.21	13.31	30.18	27.80	28.42	28.22	13.92
167.04	13.45	99.50	0.54	0.12	1.11	264.05	350.25	10.99	22.25	25.30	25.42	29.87	12.15
69.02	**3.40**	**99.49**	**0.32**	**0.10**	**0.98**	**349.51**	**232.79**	**4.30**	**28.84**	**49.43**	**48.74**	**9.31**	**6.31**

Constraints														
Interference		Top land thickness			Contact ratio		Bending stress		Contact stress		Involute interference		Tip interference	Proper assembly
24.14	30.99	2.17	1.31	1.77	0.03	1.47	368.05	381.32	1351.26	1747.33	15.51	0.37	0.14	0.00
15.32	23.07	1.09	1.54	1.55	0.12	1.51	354.70	374.32	1184.10	1698.17	7.70	0.45	0.24	0.00
32.72	31.94	2.14	1.90	1.75	0.02	2.06	371.03	386.45	1410.11	1773.39	14.54	0.38	0.16	0.00
16.78	26.52	1.80	1.67	2.50	0.16	0.71	363.92	377.89	1206.51	1681.14	18.73	0.38	0.11	0.00
17.26	21.47	2.11	1.98	3.37	0.31	1.30	360.37	378.81	1121.03	1651.72	13.68	0.37	0.13	0.00
21.33	29.67	1.62	1.92	1.74	0.02	0.11	373.43	385.08	1390.99	1742.38	31.26	0.38	0.04	0.00
13.67	**18.78**	**1.19**	**1.10**	**1.33**	**0.10**	**1.25**	**313.17**	**350.21**	**855.30**	**1571.65**	**9.62**	**0.36**	**0.14**	**0.00**
15.39	23.10	1.28	1.68	1.74	0.06	0.31	364.47	379.69	1237.93	1692.59	20.26	0.36	0.07	0.00
16.25	18.50	1.98	1.88	3.18	0.31	1.33	362.15	380.78	1128.42	1656.29	12.24	0.35	0.12	0.00
11.13	19.82	1.95	1.22	2.55	0.24	0.81	327.21	348.93	746.85	1522.98	11.85	0.29	0.10	0.00
26.32	16.41	1.64	2.25	2.66	0.23	1.46	367.12	388.85	1203.90	1727.07	18.21	0.23	0.04	0.00
9.31	10.78	1.02	0.29	0.96	0.09	1.73	269.24	322.53	575.21	1462.70	4.17	0.29	0.13	0.00
5.66	10.16	0.93	0.21	1.11	0.12	1.16	236.79	273.87	216.06	1299.16	3.14	0.32	0.16	0.00
8.83	**13.11**	**0.99**	**0.37**	**0.83**	**0.04**	**1.96**	**197.50**	**264.69**	**281.54**	**1370.22**	**1.38**	**0.37**	**0.25**	**0.00**
13.22	15.64	0.85	1.02	0.89	0.03	1.77	245.14	321.36	597.72	1437.69	6.11	0.51	0.25	0.00
20.53	17.98	2.61	0.54	2.57	0.16	0.46	377.97	387.92	1346.22	1770.31	29.97	0.17	0.01	0.00
24.92	13.39	1.94	1.78	2.58	0.23	1.07	371.28	390.10	1242.03	1733.78	24.17	0.17	0.01	0.00
28.42	18.90	1.54	1.65	1.69	0.05	1.70	298.14	363.35	803.15	1544.89	17.23	0.28	0.05	0.00
32.50	15.57	1.72	1.59	1.66	0.01	1.33	331.19	379.37	991.34	1619.22	25.01	0.20	0.02	0.00
7.71	**9.25**	**0.62**	**0.68**	**0.84**	**0.12**	**0.29**	**221.03**	**308.18**	**291.50**	**1312.07**	**10.88**	**0.31**	**0.03**	**0.00**

(Continued)

TABLE 5.4 (CONTINUED)
Sample of MOSA Output

Variables							
z_1	z_2	z_3	x_{n1}	x_{n2}	x_{n3}	m_n	n_p
40	32	94	0.64	1.00	0.62	5.49	6
40	33	86	0.85	0.33	0.71	4.15	7
38	38	94	0.87	1.00	0.71	5.94	6
38	29	91	0.42	0.36	–0.05	5.58	5
40	35	97	–0.06	0.03	–1.00	5.58	4
36	30	98	0.96	0.67	0.57	5.51	8
38	**31**	**89**	**0.63**	**0.60**	**0.77**	**3.67**	**5**
33	27	85	0.82	0.45	0.67	4.92	8
38	35	95	–0.06	0.01	–1.00	5.24	5
39	26	91	–0.17	0.32	–0.42	4.91	3
29	40	92	0.32	0.02	0.33	6.00	6
38	30	90	0.08	1.00	0.18	2.37	8
38	23	82	–0.08	0.74	–0.31	2.26	7
40	**28**	**82**	**0.40**	**1.00**	**0.59**	**2.39**	**5**
38	35	85	1.00	0.74	0.46	2.81	5
34	29	99	–0.46	0.86	0.83	6.00	8
28	38	98	0.05	0.30	0.11	5.65	8
30	39	93	0.71	0.81	0.49	5.00	3
25	39	96	0.63	1.00	0.48	5.22	5
34	**31**	**97**	**0.60**	**0.45**	**0.31**	**2.12**	**8**

planetary gearing in kg, the efficiency in percentage, and the center distance in mm were taken for comparison.

The comparison shows that the results obtained from MOSA are an improvement over the existing design by optimizing the 14 objective functions within the given constraints. The results were obtained with minimal computational effort and time. The three best results chosen from the solution set of MOSA output are analyzed for selecting a single best solution. The total weight of the existing planetary gearing was 9.1 kg, the center distance was 126.9 mm, and the efficiency was 99.48%.

Referring to Table 5.5, in the case of the first result, by selecting the number of teeth in the planet gear as 38, the weight has reduced to 7.94 kg, the efficiency has increased to 99.67%, and the center distance has reduced to 126.45 mm. There is a reduction of 12.75% in weight, an increase of 0.19% in efficiency, and a space reduction of 0.35%. Hence, there is an improvement in all three objective functions in the solution obtained with MOSA.

In the case of the second result, by selecting the number of teeth in the planet gear as 40, the weight has reduced to 3.38 kg, the efficiency has increased to 99.53%, and the center distance has reduced to 81.12 mm. There is a reduction of 62.86% in weight, an increase of 0.05% in efficiency, and a space reduction of 36.08%. Hence, there is an improvement in all three objective functions in the solution obtained with MOSA.

In the case of the third result, by selecting the number of teeth in the planet gear as 34, the weight has reduced to 3.40 kg, the efficiency has increased to 99.49%, and the center distance has reduced to 69.02 mm. There is a reduction of 62.64% in weight, an increase of 0.01% in efficiency, and a space reduction of 45.61%. Hence, there is an improvement in all three objective functions in the solution obtained with MOSA.

5.5.11.2 Visualization of Results

The formation of the Pareto front from the non-dominated solutions containing 14 objective functions is depicted in the subsequent graphs in combinations of three objective functions for every graph. The graphs show the 3D scatter plot of the solution set generated by MOSA in various combinations of three objective functions.

Even though the graphs help in better visualization of the non-dominated Pareto-optimal solution set, one should be cautious of the fact that all the 14 objective functions put together to form a single set of Pareto-optimal solutions that cannot be visualized in totality in graphical format. Hence, for the sake of decision-making, the solutions better suited for the individual requirements of the user/decision-maker must be given priority in the choice of objective function combinations taken for visualization.

As a result of optimization, a Pareto-optimal front is formed in the objective space of the solutions. In the scatter plot, the circular points represent non-dominated solutions with respect to all 14 objectives taken together, whereas the

TABLE 5.5
Optimized Design Parameters Using MOSA

	Variables				Objective Functions		
	Module, m (mm)	Number of Teeth (Planet Gear), Z_1	Number of Teeth (Sun Gear), Z_2	Number of Teeth (Ring Gear), Z_3	Total Weight, W (kg)	Efficiency, η	Center Distance (mm)
Existing planetary gearing design	4.23	31	29	91	9.1	99.48	126.9
MOSA design	3.67	38	31	89	7.94	99.67	126.45
	2.39	40	28	82	3.38	99.53	81.12
	2.12	34	31	97	3.40	99.49	69.02

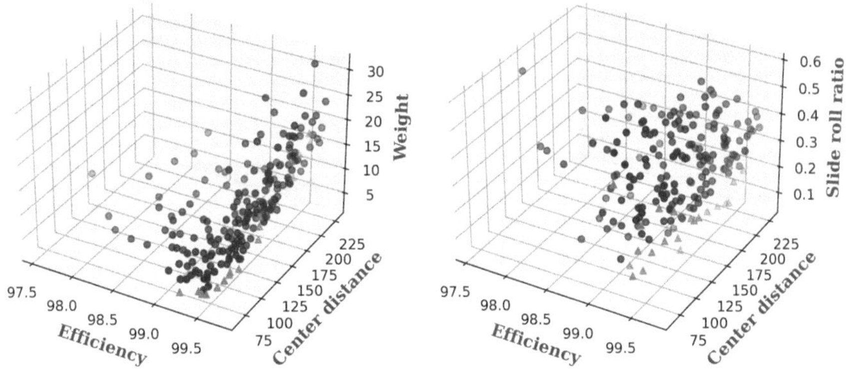

(a) (b)

FIGURE 5.3 (a) and (b) 3D scatter plot of MOSA results.

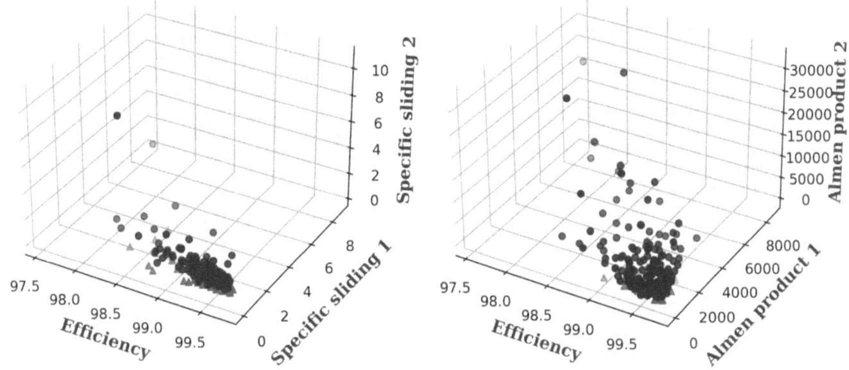

(a) (b)

FIGURE 5.4 (a) and (b) 3D scatter plot of MOSA results.

triangular points represent a non-dominated front formed with respect to the three objectives that are taken for visualization. It can be observed that the distribution of solutions along the front is uniform, which means the MOSA algorithm offers a high-quality solution set.

Referring to Figure 5.3(a), the solutions close to the upper right corner have higher total weight, higher center distance, and higher efficiency, which means they have better efficiency but are heavier and less compact. On the contrary, the solutions near the lower left corner present better compactness and better total weight but lesser efficiency. The solutions lying in the lower right corner have the best value in terms of higher efficiency, lesser weight, and better compactness.

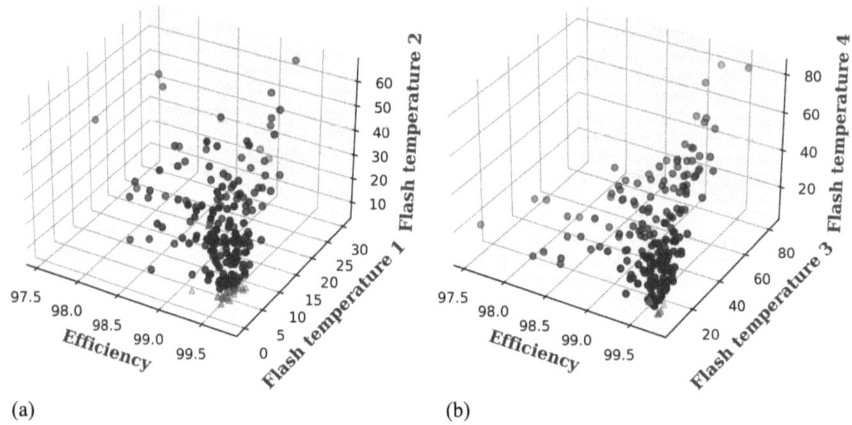

FIGURE 5.5 (a) and (b) 3D scatter plot of MOSA results.

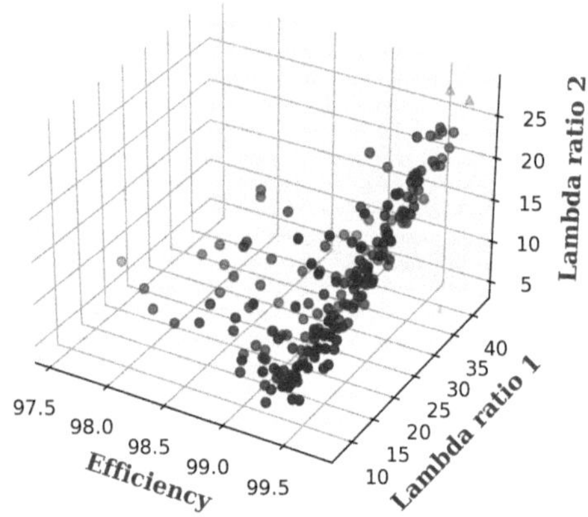

FIGURE 5.6 3D scatter plot of MOSA results.

Referring to Figure 5.3(b), the solutions close to the lower right corner offer higher efficiency, better compactness, and lesser slide roll ratio values.

Referring to Figure 5.4(a), the solutions close to the lower right corner offer higher efficiency and lesser specific sliding values. A lower value of specific sliding indicates a balanced specific sliding at SAP and EAP of the gear tooth profile engagement during running.

Referring to Figure 5.4(b), the solutions close to the lower right corner offer higher efficiency and lower Almen product values at SAP and EAP of the gear tooth profile engagement during running.

Referring to Figure 5.5(a) and (b), the solutions close to the lower right corner offer higher efficiency and lower flash temperatures that are balanced at SAP and EAP of the gear tooth profile engagement during running.

Referring to Figure 5.6, the solutions close to the upper right corner offer higher efficiency and higher lambda ratios for the sun gear/planet gear engagement and the planet gear/ring gear engagement, respectively.

5.5.11.3 Torque Ratio Comparison

The output torque of the Spider arm for the three solutions of MOSA can be analyzed as follows:

For the first solution,

Number of teeth of the planet gear, $Z_1 = 38$
Number of teeth of the sun gear, $Z_2 = 31$
Number of teeth of the ring gear, $Z_3 = 89$

Applying the equations given in Table 5.2,

Input speed of sun gear = 3550 rpm
Speed of the spider arm = 917.09 rpm

Hence, the torque ratio = $3550/917.09 = 3.87$

Compared to the torque ratio of 4.14 of the existing planetary gearing, the above torque ratio is 6.45% less, which may not be acceptable.

For the second solution,

Number of teeth of the planet gear, $Z_1 = 40$
Number of teeth of the sun gear, $Z_2 = 28$
Number of teeth of the ring gear, $Z_3 = 82$

Applying the equations given in Table 5.2,

Input speed of sun gear = 3550 rpm
Speed of the Spider arm = 903.64 rpm

Hence, the torque ratio = $3550/903.64 = 3.93$

Compared to the torque ratio of 4.14 of the existing planetary gearing, the above torque ratio is 5.06% less, which may not be acceptable.

For the third solution,

Number of teeth of the planet gear, $Z_1 = 34$
Number of teeth of the sun gear, $Z_2 = 31$
Number of teeth of the ring gear, $Z_3 = 97$

Applying the equations given in Table 5.2,

Input speed of sun gear = 3550 rpm
Speed of the spider arm = 859.5 rpm

TABLE 5.6
Sample of MOSA Output for Standard Module

Objectives

Center distance	Weight	Efficiency	Slide roll ratio	Specific sliding		Almen product			Flash temperature			Lambda ratio	
73.13	**3.82**	**99.52**	**0.32**	**0.06**	**0.95**	**254.62**	**148.07**	**4.53**	**27.40**	**45.45**	**44.96**	**10.04**	**6.83**
73.13	3.82	99.52	0.32	0.06	0.95	254.62	148.81	4.53	27.40	45.45	44.96	10.04	6.83
73.13	3.82	99.52	0.32	0.06	0.95	256.64	148.13	4.52	27.40	45.48	44.98	10.04	6.83
73.13	3.82	99.52	0.32	0.06	0.95	254.62	150.06	4.53	27.40	45.45	44.96	10.04	6.83
73.13	3.82	99.52	0.32	0.06	0.95	260.55	148.26	4.50	27.39	45.53	45.03	10.04	6.83

Constraints

Interference		Top land thickness			Contact ratio		Bending stress		Contact stress		Involute interference		Tip interference	Proper assembly
8.32	**9.85**	**0.68**	**0.71**	**0.87**	**0.11**	**0.32**	**239.82**	**317.82**	**393.64**	**1357.16**	**11.52**	**0.31**	**0.03**	**0.00**
8.32	9.85	0.68	0.71	0.87	0.11	0.32	239.82	317.82	393.64	1357.10	11.51	0.31	0.03	0.00
8.32	9.85	0.68	0.71	0.87	0.11	0.32	239.84	317.83	393.50	1357.14	11.52	0.31	0.03	0.00
8.32	9.85	0.68	0.71	0.87	0.11	0.32	239.82	317.82	393.64	1357.02	11.51	0.31	0.03	0.00
8.31	9.85	0.67	0.71	0.87	0.11	0.32	239.87	317.85	393.24	1357.11	11.52	0.31	0.03	0.00

Variables		
x_{n1}	x_{n2}	x_{n3}
0.60	**0.50**	**0.40**
0.60	0.50	0.40
0.60	0.50	0.40
0.60	0.50	0.40
0.60	0.50	0.40

TABLE 5.7
Optimized Design Parameters for Standard Module Using MOSA

	Variables				Objective Functions		
	Module, m (mm)	Number of Teeth (Planet Gear) Z_1	Number of Teeth (Sun Gear) Z_2	Number of Teeth (Ring Gear) Z_3	Total Weight, W (kg)	Efficiency, η	Center Distance (mm)
MOSA design	2.12	34	31	97	3.40	99.49	69.02
MOSA design for standard module	2.25	34	31	97	3.82	99.52	73.13

Hence, the torque ratio = 3550 / 859.77 = 4.13.

Compared to the torque ratio of 4.14 of the existing planetary gearing, the above torque ratio is 0.22% less, which is acceptable.

Therefore, the third solution is selected as the best solution.

5.5.11.4 Selection of Standard Module

From Table 5.5, it can be seen that the module for the best solution is 2.12 mm, which is a non-standard module. The nearest standard module is 2.25 mm. Hence, the final design can be modified to suit a standard module of 2.25 mm using the procedure given below.

The following design variables of the epicyclic gearing are made constant, based on the values from the best solution (Table 5.4):

Number of teeth of the planet gear, $Z_1 = 34$
Number of teeth of the sun gear, $Z_2 = 31$
Number of teeth of the ring gear, $Z_3 = 97$
Number of planet gears, $n_p = 8$
Module, $m_n = 2.25$ mm (standard module)

The problem is reformulated with the three profile shift coefficients x_{n1}, x_{n2}, and x_{n3} taken as design variables, which will be computed again for the fixed parameters above. The upper and lower bounds for the profile shift coefficients are fixed based on the values from the best solution (Table 5.4).

Profile shift coefficient for planet gear, x_{n1}: 0.6 to 0.7
Profile shift coefficient for sun gear, x_{n2}: 0.4 to 0.5
Profile shift coefficient for ring gear, x_{n3}: 0.3 to 0.4

The MOSA procedure is rerun for the modified problem formulation. Table 5.6 shows a sample of the results generated for the standard module of 2.25 mm.

From the results generated, the solution listed in Table 5.7 is selected as the best solution.

From the above analysis, we can conclude that the solution from Table 5.7, obtained through the design optimization of the helicopter planetary gearing using MOSA, can be considered the best solution.

NOTES

1 For more information on Pareto-optimality, refer to Section 3.4.
2 Refer to Table 1.1.
3 Refer to Section 3.6.5.
4 Refer to Section 2.9.

REFERENCES

Černý, V., 1985, "Thermodynamical approach to the traveling salesman problem: An efficient simulation algorithm," *The Journal of Optimization Theory and Applications*, 45, 41–51. https://doi.org/10.1007/BF00940812

Czyzżak, P., & Jaszkiewicz, A., 1998, "Pareto simulated annealing: A metaheuristic technique for multiple-objective combinatorial optimization," *Journal of Multi-criteria Decision Analysis*, 7, 34–47.

Deb, K., et al., 2002, "A fast and elitist multiobjective genetic algorithm: NSGA-II," *IEEE Transactions on Evolutionary Computation*, 6, 2, 182–197, https://doi.org/10.1109/4235.996017

Jin, Y., et al., 2001, "Dynamic Weighted Aggregation for evolutionary multi-objective optimization: why does it work and how," *Proceedings of the 3rd Annual Conference on Genetic and Evolutionary Computation (GECCO'01)*. Morgan Kaufmann Publishers Inc., San Francisco, CA, USA, 1042–1049.

Kirkpatrick, et al., 1983, "Optimization by Simulated Annealing," *Science*, 220, 671–680. https://doi.org/10.1126/science.220.4598.671

Krishnan, G., 2024, OPTIMO software – Algorithms for Multi-Objective optimization. https://github.com/epicyclic-book/optimo/. Used under the terms and conditions of the MIT license.

Rotaru, C., & Todorov, M., n.d. "Helicopter Flight Physics," https://doi.org/10.5772/intechopen.71516. Used under the terms and conditions of the Creative Commons Attribution license (https://creativecommons.org/licenses/by/3.0)

6 Design Optimization of Electric Vehicle Gearbox Using Multi-Objective Ant Colony Optimization Algorithm

Chandrasekaran G and Gokul Krishnan

6.1 INTRODUCTION

The third algorithm in the category of evolutionary and stochastic optimization techniques under consideration is the ant colony optimization (ACO) algorithm. The algorithm is based on the social behavior of ants in a colony as they forage for food. This places ACO in the same category as particle swarm optimization, which is also based on the application of collective behavior and swarm intelligence in the optimization of systems.

However, there are key differences in ACO that set it apart from other stochastic optimization techniques like particle swarm optimization and simulated annealing. First, ACO is a path-finding algorithm, and second, it deals with a discrete set of values as opposed to continuous real-valued functions. Therefore, it finds applications in *combinatorial optimization problems* such as finding the shortest path on a graph, the most common example being the travelling salesman problem. The goal in optimization problems such as these is to figure out the combination of values, visualized as nodes on a graph, out of a discrete set of interconnected nodes, which leads to the most optimum path within the system.

6.2 ANT COLONY OPTIMIZATION

The ACO algorithm was initially proposed by Marco Dorigo (1992) and further developed in the years that followed (Dorigo and Di Caro, 1999). The system is visualized as a graph consisting of nodes, with some or all of the nodes being connected by edges. The nodes represent discrete values, and the edges represent a path connecting two nodes (i.e., an allowed sequence or combination of values).

DOI: 10.1201/9781003144465-6

Ants traverse through this graph in search of food, initially selecting random edges to traverse upon. In doing so, they mark the path that was traversed using a special chemical called *pheromone*. These pheromone trails are used by successive ants in selecting the edges that must be followed in order to reach the food source within the shortest time. This reinforces certain paths through the graph which contain stronger pheromone trails. Eventually, this leads to the deduction of the most optimum path, which is a path composed of edges with the highest pheromone density.

Being a path-finding algorithm dealing with discrete values, ACO cannot be applied directly in the optimization of continuous real-valued functions, as is the case with the optimization of epicyclic gearing. The other major challenge is in extending the technique to solve multi-objective optimization problems and obtain a Pareto-optimal solution set.[1] A few such techniques have been proposed by Inès Alaya et al., (2007) and Liqiang Liu et al., (2014).

6.3 STANDARD ACO PROCEDURE

This section explores the standard ACO algorithm, and the subsequent sections discuss techniques available for extending this algorithm to solve for the constrained, multi-objective design optimization of epicyclic gearing.

Although ACO is a population-based algorithm, it is in fact a combinatorial optimization technique involving discrete variables, which means it cannot be applied directly to solve for the minimization/maximization of continuous objective functions. Instead, ACO finds application in path and routing-based optimization problems, the most common example being the travelling salesman problem. Given a set of interconnected nodes on a graph, the objective is to find the shortest possible path (i.e., sequence of nodes) that passes through each node exactly once.

Consider a graph consisting of M nodes. For each pair of nodes i and j, there may or may not exist an edge e_{ij} connecting the two nodes. Each edge is assigned a distance value d_{ij}, pheromone value t_{ij} and probability value p_{ij}. The following global properties are specified:

- Number of ants N.
- Number of iterations.
- Control parameters: *alpha* and *beta*.
- Pheromone weight Q.
- Pheromone evaporation factor ρ.

Initially, a population of N ants is generated and placed at randomly chosen nodes on the graph. Then, at each iteration of the algorithm, the following steps are performed.

6.3.1 Building the Solution

Each ant is made to traverse through the graph by selecting one of the edges based on the following criteria: assuming the ant is situated on node i, the edge is one of e_{ik}, where $k \in M$, and k is not a node already traversed by the ant during this iteration.

Among the set of such valid nodes, one is selected at random based on the computed probability of each edge. The edge probability is defined by the following equation:

$$p_{ij} = t_{ij}^{\alpha} \cdot \frac{1}{d_{ij}^{\beta}}$$

where α and β are control parameters.

For a given ant situated on node i, the probability value is normalized by dividing it by the sum of all the probabilities of all the valid edges emanating from node i.

$$p_{ij} = \frac{t_{ij}^{\alpha} \cdot \frac{1}{d_{ij}^{\beta}}}{\sum_{k=1}^{M} p_{ik}}$$

Once no further unvisited nodes are available, the ant is considered to have completed the traversal. The sequence of nodes visited by the ant forms the candidate solution. The sum of all traversed edge distances forms the total fitness of the solution P.

6.3.2 Pheromone Update

The next step involves updating the pheromone quantity on each edge. For an edge e_{ij}, there is a contribution of pheromone from each ant that has traversed through that edge. Therefore, for an ant a, assuming the edge e_{ij} was traversed by the ant, the change in pheromone is calculated as:

$$\left(\Delta t_{ij}\right)_a = \frac{Q}{P_a}$$

where Q is a predefined pheromone weight; P_a is the total fitness of solution for the ant a.

The pheromone contribution from all such ants is added to the existing pheromone value for edge e_{ij},

$$t_{ij} = t_{ij} + \sum_{a=1}^{N} \left(\Delta t_{ij}\right)_a$$

6.3.3 Pheromone Evaporation

Finally, the pheromone on each edge undergoes a natural evaporation process. This promotes diversity of edge selection by reducing dependency on edges traversed earlier in the search process. Pheromone evaporation is modeled according to the following equation:

$$t_{ij} = (1-\rho)t_{ij}$$

where ρ is the pheromone evaporation factor.

These steps are repeated until the maximum number of iterations is reached. At the end of the final iteration, the path defined by the set of edges with the highest pheromone density forms the best solution, which is the solution with the lowest total distance covered.

6.3.4 Algorithm Pseudo-code

The entire algorithm can be visualized through the flow chart in Figure 6.1 and pseudo-code representation in Listing 6.1.

LISTING 6.1 ACO ALGORITHM PSEUDO-CODE

```
WHILE iterations remaining
    FOR each ant a
        Initialize ant at random node N_a
        WHILE unvisited nodes remaining
            FOR each unvisited node n
                Compute probability p_an
            END FOR
            Select best unvisited node
        END WHILE
        Compute fitness P_a
    END FOR

    FOR each edge e_ij
        Evaporate pheromone from t_ij
        FOR each ant a
            Compute pheromone contribution (Δt_ij)_a
            Add pheromone contribution (Δt_ij)_a to t_ij
        END FOR
    END FOR
END WHILE
```

6.4 ACO in the Continuous Domain

In order to apply ACO for the minimization of continuous real-valued objective functions, it is necessary to re-purpose the problem in the form of a graph

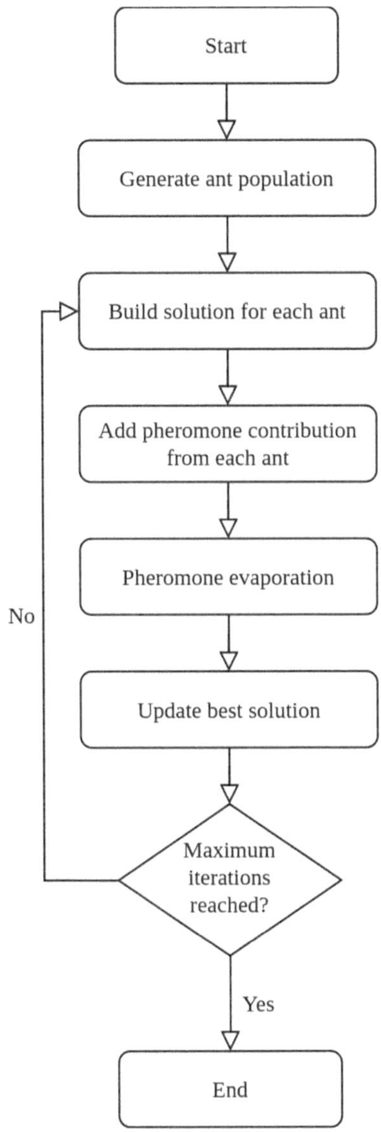

FIGURE 6.1 ACO algorithm flow chart.

containing nodes and edges. This is done through segmentation and by making use of a multi-layered approach, which is explained below:

- Consider a continuous search problem containing *N* design variables. Each design variable takes values within the provided minimum and maximum bounds, which define the range of the variable.

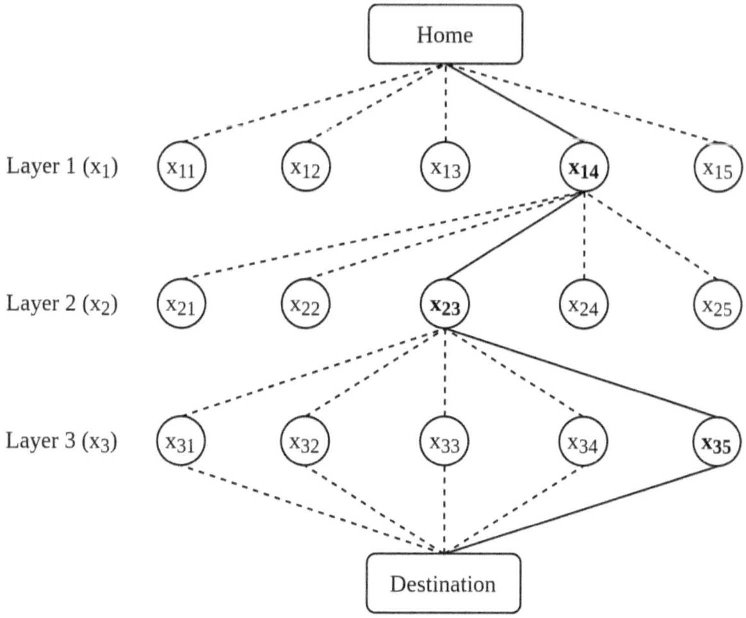

FIGURE 6.2 Sample ACO graph.

- For each design variable, the variable range is equally subdivided into M discrete values, with each value assigned to a distinct node on the graph. This results in a set of M disjoint nodes corresponding to a particular design variable, which is called a *layer*.
- Each node belonging to a layer l is connected to every node belonging to the subsequent layer $l + 1$. There exists a special *home* node, where ants are spawned, and which is directly connected to nodes of the first layer. Likewise, nodes of the last layer are connected to a special end node that marks the *food source*.
- Edges that connect nodes are unidirectional. Each edge has a start node, end node, and pheromone level assigned to it.

With the above setup, a problem composed of N design variables and M subdivisions results in a graph containing N layers, $NM+2$ nodes and $(N-1)M^2+2M$ edges. Figure 6.2 shows a sample ACO graph with three layers and five subdivisions.

6.4.1 Building the Solution

Initially, N ants are generated and spawned at the home node, where N is the population size. From there, each ant selects one of the forward edges based on the pheromone value on the edges. The edge probability is defined by the equation:

$$p_{ij} = t_{ij}^{\alpha}$$

where t_{ij} is the pheromone value on the edge connecting nodes i and j; α is the pheromone scaling factor.

A roulette-wheel selection process is used to select one of the edges extending outward from the current node. The ant traverses the edge to reach a node belonging to the next layer. The process is repeated until the ant reaches the final layer. The sequence of nodes traversed then forms a candidate solution.

The variable values associated with the sequence of nodes can then be plugged into the objective function to compute the fitness P of the solution.

6.4.2 Pheromone Update

Once all the ants have obtained candidate solutions, each ant performs a local pheromone update on the edges that belong to the path traversed by the ant. Therefore, for an edge e_{ij} and ant a, the change in pheromone is calculated as:

$$\left(\Delta t_{ij}\right)_a = \frac{Q}{P_a}$$

where Q is a predefined pheromone weight; P_a is the total fitness of solution for the ant a.

The pheromone contribution from all such ants adds up to the existing pheromone value in edge e_{ij},

$$t_{ij} = t_{ij} + \sum_{a=1}^{N} \left(\Delta t_{ij}\right)_a$$

6.4.3 Pheromone Evaporation

In the global update stage, a portion of pheromone is evaporated on all the edges throughout the system. This promotes diversity of edge selection by reducing dependency on edges traversed earlier in the search process. Pheromone evaporation is modeled according to the following equation:

$$t_{ij} = \left(1 - \rho\right) t_{ij}$$

where ρ is the pheromone evaporation factor.

The steps are repeated until the maximum number of iterations is reached. At the end of the final iteration, the path defined by the set of edges with the highest pheromone density forms the best solution, which, in the case of minimization, is the solution with the lowest fitness value.

6.5 MULTI-OBJECTIVE ANT COLONY OPTIMIZATION (MOACO)

This section describes the techniques and challenges involved in extending the ACO algorithm for solving constrained multi-objective optimization problems.

6.5.1 Fitness Vector

ACO deals with a single fitness value associated with each ant. In order to adapt the technique to solve problems containing multiple objectives, several approaches have been proposed. A simple approach is to use a weighted average of the fitness values from each objective function, that is,

$$P_{\text{combined}} = \frac{w_1 p_1 + w_2 p_2 + \ldots + w_N p_N}{N}$$

where N is the number of objectives.

Several other weighted aggregation techniques have been proposed (Jin et al., 2001).

However, an alternate approach involves modifying the scalar fitness value P into a fitness vector consisting of a separate value for each objective function. That is,

$$P = \left[p_1, p_2, \ldots, p_N \right]$$

where N is the number of objectives.

This approach has been selected in the implementation covered in this book.

6.5.2 Dominance Operator

While the use of a fitness vector is beneficial when dealing with multi-objective problems, it introduces the need for a new comparison operator that can be used to compare the fitness vectors of two candidate solutions. Therefore, the standard comparison operator is replaced with a *dominance* operator that checks for Pareto-dominance of one solution by another. Consider two candidates with fitness vectors P and Q.

Then, assuming minimization, P is said to dominate Q if,

$P_i \le Q_i$ for all values of i, and
$P_i < Q_i$ for at least one value of i,

where $i = \{0 \text{ to } N\}$, and N is the number of objectives.

6.5.3 Pheromone Update

The pheromone contribution for an edge e_{ij} traversed by an ant a is defined by,

$$\left(\Delta t_{ij}\right)_a = \frac{Q}{P_a}$$

where Q is a predefined pheromone weight; P_a is the total fitness of solution for the ant a.

When the fitness value is replaced with a fitness vector, the equation is modified to use a summation of the objective function components.

$$\left(\Delta t_{ij}\right)_a = \frac{Q}{sum\left(P_a\right)}$$

6.5.4 Archive of Non-Dominated Solutions

In order to extract a Pareto-optimal solution set, the approach considered is to maintain an archive of non-dominated solutions separate from the population set and update this archive at the end of each iteration. The procedure is described in detail in Section 6.5.10.

6.5.5 Population Diversity

When dealing with multi-objective optimization problems, we are interested in a set of non-dominated solutions which form the Pareto-optimal solution set. Therefore, there is a need to ensure a healthy diversity of Pareto-optimal solutions.

In order to prevent the candidate solutions from bunching together and to maintain the diversity of solutions, candidates that are closely bunched together are removed from the archive after each iteration and replaced by fresh candidates that further diversify the pool of solutions. The following criterion is used in choosing the candidates to delete.

6.5.5.1 Crowding Distance

Each solution in the archive is assigned a crowding distance value that defines the density of solutions surrounding it. Solutions in denser areas are assigned a smaller crowding distance. The crowding distance is computed separately for each objective function, that is, along each dimension in the objective space, and then averaged together.

6.5.5.2 Roulette-Wheel Selection

To select one among the non-dominated solutions to be removed from the archive, a roulette-wheel selection is used, with the weights defined by the inverse of the crowding distance values. This increases the probability that particles from denser regions are chosen for removal.

6.5.6 Handling Constraints

Constraints restrict the movement of candidates by limiting the possible values that the design variables can assume. The feasibility of solutions is decided by one or more constraint functions defined as part of the problem. A solution is considered feasible if and only if all provided constraint tests pass.

A solution is only accepted as the candidate's best solution achieved so far if all constraints pass. Otherwise, the solution is not considered for the local pheromone update process. The computation of dominance and crowding distance factors is performed based on the candidate's best feasible solution achieved so far.

6.5.7 MOACO Implementation

An implementation of the MOACO algorithm is described along with the corresponding pseudo-code. We start with an overview of key functions used in the algorithm code.

6.5.7.1 RouletteWheelSelection

We employ a roulette-wheel selection process, which is essentially the randomized selection of members from a population, with each member assigned a weight that defines the probability of that member being selected. This can also be called a fitness-proportionate selection model. The function takes as input a list of weights and outputs the index of the member being selected. The weights array is normalized such that the elements add up to one.

LISTING 6.2 IMPLEMENTATION OF ROULETTE-WHEEL SELECTION

```
FUNCTION RouletteWheelSelection(weights):
    weights = normalize(weights)
    weights = generate_cumulative_sum(weights)
    r = rand() # random value in range [0,1]
    FOR EACH value in weights:
        IF value >= r:
            return i
        END IF
    END FOR
END FUNCTION
```

6.5.7.2 SelectMemberToDelete

To select one among the non-dominated solutions to be removed from the archive, a roulette-wheel selection is used with the weights assigned based on the inverse of the candidate crowding distance. This increases the probability that candidates from denser regions are chosen for removal.

LISTING 6.3 IMPLEMENTATION OF REDUNDANT MEMBER SELECTION

```
FUNCTION SelectMemberToDelete(archive):
    weights = [1/c_1, 1/c_2 ... 1/c_num_archive]
    index =  RouletteWheelSelection(weights)
    return index
END FUNCTION
```

6.5.7.3 SelectEdge

To select an edge for traversal from among all the edges emanating from a specific node, a roulette-wheel selection is used with the weights assigned based on the pheromone value on the edges. This increases the probability that edges with denser pheromone quantity are preferred more.

LISTING 6.4 IMPLEMENTATION OF EDGE SELECTION

```
FUNCTION SelectEdge(node):
    weights = [t_1, t_2 ... t_num_edges]
    index =  RouletteWheelSelection(weights)
    return index
END FUNCTION
```

6.5.7.4 ComputeDominance

The function accepts a set of candidate solutions as input and computes the Pareto-dominance state of each candidate, based on the best feasible fitness achieved by the candidate, or *best_fitness*. For each candidate, the *is_dominated* property is set to *true* for all candidates that are dominated by at least one other candidate in the set.

6.5.7.5 ComputeCrowdingDistance

The function accepts a population as input and computes the crowding distance factor of each candidate, based on the best feasible fitness achieved by the candidate, or *best_fitness*. The algorithm for crowding distance computation is adapted from the NSGA-II paper (K. Deb, 2002), with a few minor changes.

Because the crowding distance and inverse crowding distance are used as weighting factors in the roulette wheel selection routines, the value is clamped within the half-closed range (0, 1) such that the crowding distance value is always nonzero and less than or equal to one.

The default value for crowding distance is chosen as 1.0, which ensures equal probability of each particle being selected.

6.5.7.6 FilterNonDominated

The function accepts a set of candidate solutions as input and removes each solution that has the *is_dominated* property set to *true*. It then returns the filtered non-dominated population set.

6.5.7.7 Fitness

This function is user-defined. It accepts an array of design variables as input and outputs an array of objective function values. Because we are dealing with a minimization problem, objective functions that are to be maximized must be negated.

6.5.7.8 Constraints

This function is user-defined. It accepts an array of design variables as input and outputs an array of constraint function values. Constraint functions are specified in such a way that they must produce a value greater than or equal to zero in order for the test to pass; that is, the inequality $C_i \geq N$ is rewritten as $C_i - N \geq 0$.

6.5.8 Initialization

The following global properties are set:

- *num_vars*: number of design variables
- *bounds_min*: array of values representing minimum bounds for each design variable
- *bounds_max*: array of values representing maximum bounds for each design variable
- *num_iter*: number of iterations
- *num_pop*: population size
- *num_archive*: archive size for non-dominated solution set
- *divisions*: number of allowed discrete values for each design variable
- *alpha*, *q*: pheromone scaling factors
- *rho*: pheromone evaporation factor

LISTING 6.5 GLOBAL PROPERTIES

```
num_vars = 4
bounds_min = [x1, x2 ... xnum_vars]
bounds_max = [x1, x2 ... xnum_vars]
num_iter = 100
num_pop = 200
num_archive = 50
divisions = 100
alpha = 1.0
q = 1.0
rho = 0.1
```

Next, we allocate the required number of nodes and edges. Each node is assigned the corresponding discrete design variable value. Each edge is initialized with a default pheromone value.

6.5.8.1 Allocation of Layers and Nodes

LISTING 6.6 ALLOCATION OF NODES

```
Layers = new Array()
Home = new Node()
FOR i = [1 to num_vars]:
    Li = new Array()
    FOR j = [1 to divisions]:
        node = new Node()
        node.value = bounds_mini +
            (bounds_maxi - bounds_mini) * j / divisions
        Li.append(node)
    END FOR
    Layers.append(Li)
END FOR
```

6.5.8.2 Allocation of Edges

LISTING 6.7 ALLOCATION OF EDGES

```
Layers = [L1, L2 ... Lnum_vars]
Edges = new Array()

Home.edges = new Array()
FOR EACH end_node in L1:
    edge = Edge()
    edge.end_node = end_node
    edge.pheromone = 1.0
    Home.edges.append(edge)
    Edges.append(edge)
END FOR

FOR i = [1 to num_vars-1]:
    FOR EACH node in Li:
        node.edges = new Array()
        FOR EACH end_node in Li+1:
            edge = Edge()
            edge.end_node = end_node
```

```
            edge.pheromone = 1.0
            node.edges.append(edge)
            Edges.append(edge)
        END FOR
    END FOR
END FOR
```

6.5.8.3 Allocation of Ants

LISTING 6.8 ALLOCATION OF ANTS

```
Population = [P1, P2 ... Pnum_pop]
FOR EACH ant Pi in Population:
    Pi.path = []
    Pi.fitness = []
    Pi.constraints = []
    Pi.best_path = []
    Pi.best_fitness = []
    Pi.is_dominated = false
    Pi.crowding_dist = 1.0
END FOR
```

6.5.9 Iteration

Following the initialization, we reach the core of the algorithm where each iteration is processed. First, we loop through each ant in the population set.

LISTING 6.9 THE OVERALL ITERATION

```
FOR j = [1 to num_iter]:
    FOR i = [1 to num_pop]:
        Update each ant
    END FOR
    Update pheromone
    Update the archive
END FOR
```

Each ant is placed at the home node. From here, an edge is selected, based on the probability rule discussed in Section 6.4.1, and the ant is moved to the new node. The process is repeated until the final node is reached. Each traversed edge is appended to the ant's path vector. This outputs a path made up of a sequence of N edges, where N is the number of layers, which is also the number of design variables.

LISTING 6.10 STEP 1 – PATH TRAVERSAL

```
current_node = Home
Pi.path = []
WHILE current_node.edges is not empty:
    r = SelectEdge(current_node)
    edge = current_node.get_edge(r)
    Pi.path.append(edge)
    current_node = edge.next_node
END FOR
```

Each edge that is part of the ant's path vector has an end node, and each such end node holds the associated design variable value. These design variable values make up the ant's candidate solution.

LISTING 6.11 STEP 2 – SOLUTION EXTRACTION

```
solution = []
FOR EACH edge in Pi.path:
    node = edge.next_node
    solution.append(node.value)
END FOR
```

Then, the fitness and constraint vectors are computed as usual from the extracted solution.

LISTING 6.12 STEP 3 – EVALUATE FITNESS AND CONSTRAINTS

```
Pi.fitness = Fitness(solution)
Pi.constraints = Constraints(solution)
```

In the case that all constraints are satisfied and the new solution dominates the previous best solution, the candidate path is accepted as the ant's best path achieved so far.

On the other hand, in case the new solution does not dominate the previous best solution, the solution can still be accepted based on a probability value. This promotes a better diversity of solutions.

LISTING 6.13 STEP 4 – UPDATE LOCAL BEST

```
IF Pi.constraints >= 0:
    IF Pi.fitness dominates Pi.best_fitness:
        Pi.best_path = Pi.path
        Pi.best_fitness = Pi.fitness
    ELSE IF rand() > 0.5:
        Pi.best_path = Pi.path
        Pi.best_fitness = Pi.fitness
    END IF
END IF
```

Once all the ants in the colony have completed their traversal, the pheromone update process is performed. This is done in two stages. The first is the global pheromone evaporation stage, where pheromone quantity from every edge throughout the graph undergoes dampening or evaporation, according to the specified evaporation factor. Additionally, clamping is performed in order to restrict the pheromone range.

LISTING 6.14 STEP 5 – GLOBAL PHEROMONE EVAPORATION

```
FOR EACH edge in Edges:
    edge.pheromone *= (1.0 - rho)
    Clamp(edge.pheromone, 1.0, pheromone_max)
END FOR
```

The second stage is the local pheromone update stage, wherein each ant deposits pheromone on the edges traversed by it. The amount of pheromone deposited by each ant follows the pheromone delta equation discussed in Section 6.5.3. Further, an ant can deposit pheromone only if the solution reached by the ant is a feasible solution.

LISTING 6.15 STEP 6 – LOCAL PHEROMONE UPDATE

```
FOR EACH ant Pi in Population:
    IF Pi.constraints >= 0:
        delta = q / sum(Pi.fitness)
        FOR EACH edge in Pi.path:
```

```
            edge.pheromone += delta
        END FOR
    END IF
END FOR
```

6.5.10 Archiving

Archiving refers to the extraction of non-dominated solutions from the population set into a separately maintained archive. Once all candidates have been processed in an iteration, the solutions from the new population are merged with the pre-existing archive, which is initially an empty set. Then, the dominance factor is re-computed on this merged population. Solutions that are dominated by other solutions are filtered out, keeping only non-dominated solutions in the archive.

LISTING 6.16 ARCHIVING – STEP 1

```
Archive.append(Population)
ComputeDominance(Archive)
Archive = FilterNonDominated(Archive)
```

Then, the crowding distance factor is re-computed on the filtered solution set. The crowding distance is used as a weighting factor in the selection of global optima and in the elimination of redundant solutions from the archive.

In case the archive size exceeds the maximum allowed size, members are removed from the archive until the archive size requirement is met. This is done using a roulette-wheel selection process with the inverse crowding distance criterion. This prioritizes solutions from denser regions for removal and ensures a stable archive size.

LISTING 6.17 ARCHIVING – STEP 2

```
ComputeCrowdingDistance(Archive)
WHILE size(Archive) > num_archive:
    r = SelectMemberToDelete(Archive)
    Archive.remove(r)
END WHILE
```

At the end of the final iteration, the candidates in the archive represent the final non-dominated set of solutions. The *best_path* and *best_fitness* vectors hold the final feasible solution of each candidate in the archive. The data can be output into a text file format, such as CSV, or plotted on a graph for further processing and visualization. A good amount of diversity is preserved, which helps the decision maker in selecting the most optimal solution for the problem at hand.

6.6 CASE STUDY: DESIGN OPTIMIZATION OF AN ELECTRIC VEHICLE 2-SPEED TRANSMISSION USING ANT COLONY OPTIMIZATION

In this chapter, a case study on the design optimization of an electric vehicle 2-speed transmission using ACO algorithm is undertaken.

The spread of cars using environmentally friendly alternative energy is increasing rapidly, with the growth of electric vehicles being the most noticeable. Electric vehicles are eco-friendly vehicles that use electric motors and batteries. In general, most electric vehicles currently do not use transmissions. Due to the power characteristics of the electric motor, which is the power source of the batteries, a single-gear ratio reducer is used instead of a transmission (Han et al., 2021).

6.6.1 Overview of Electric Vehicle Gearbox (Han, J.-O et al., 2021)

Figure 6.3(a) shows the power characteristics of the internal combustion engine. The maximum torque is generated in proportion to the rotational speed by generating a relatively constant maximum torque up to a certain speed. In general, a car needs a large torque to accelerate from a standstill and reach a certain speed. To maintain high speeds, only the torque needed to maintain the speed is required, and this can be achieved with a relatively small torque. The issue here is that internal combustion engines have some difficulty reconciling these two requirements in terms of power characteristics.

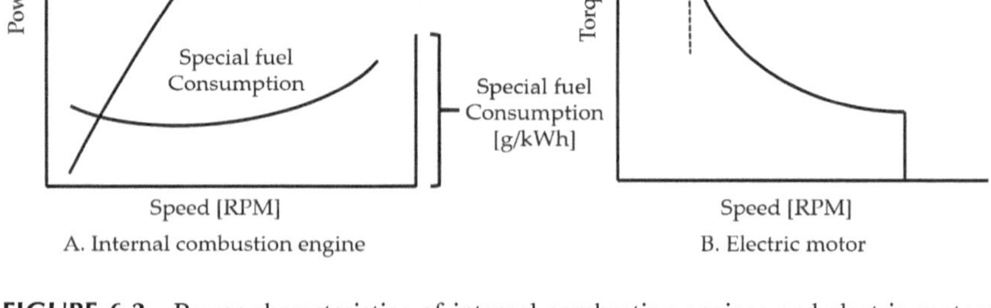

FIGURE 6.3 Power characteristics of internal combustion engines and electric motors (Han et al., 2021).

Figure 6.3(b) shows the power characteristic of the electric motor. The electric motor has the characteristic of generating a maximum torque at a low rotational speed and generating a low torque at a high rotational speed, thereby generating a constant maximum output regardless of the rotational speed. The characteristics of the electric motor show a constant power output with high torque at low speed and low torque at high speed (Lee and Hong, 2021).

Nevertheless, the reasons for using transmissions in electric vehicles are as follows.

Figure 6.4 is an energy efficiency graph of an electric vehicle with a single reducer and a two-speed transmission. The gray shaded area is where the electric motor operates at high efficiency. A wider high-efficiency range can be seen when using a two-speed transmission than when using a single reducer. Due to this, it can improve adaptability and responsiveness to various driving environments.

Many countries around the world have large and small hills and various types of roads that are hard on vehicles. To overcome this, a transmission is needed. When torque optimization is possible according to the driving speed, it can adapt more quickly to various driving environments and respond faster.

However, when one adapts to this driving environment without a transmission, the size of the motor increases to obtain greater torque, which increases the vehicle body weight and, so, battery consumption rate, which impairs practicality.

With a transmission, it is possible to improve both fuel economy and increase mileage. Figure 6.5 shows high characteristics in the operating range of the specified range. The range shaded in dark in Figure 6.5 shows high-energy efficiency. Equipped with a transmission, it is possible to drive with high efficiency in various driving environments. It uses energy from a limited energy source, the battery, more efficiently, resulting in improved fuel economy and increased mileage.

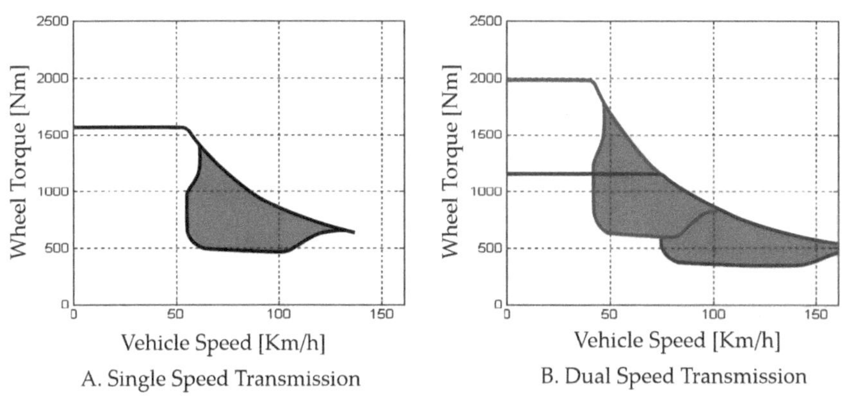

FIGURE 6.4 Increase in efficiency range of the two-speed electric vehicle (Han et al., 2021).

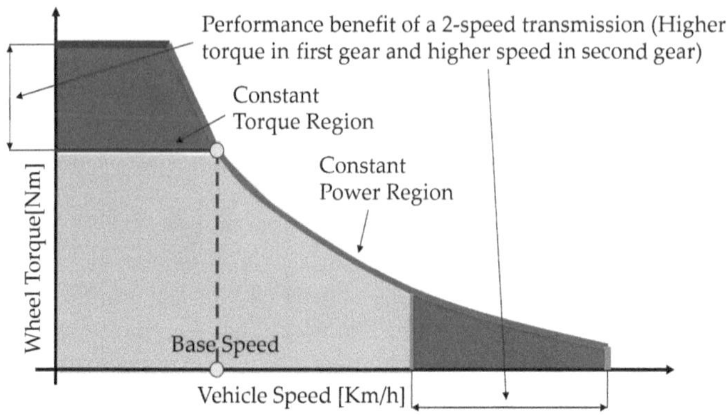

FIGURE 6.5 Torque and speed changes of a two-speed electric vehicle (Han et al., 2021).

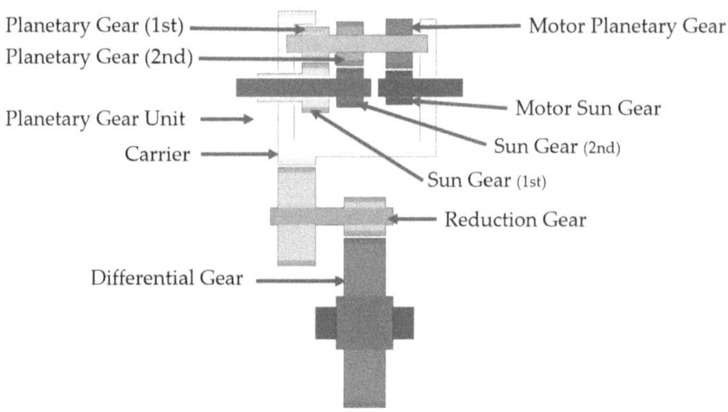

FIGURE 6.6 Two-speed transmission with epicyclic gearing (Han et al., 2021).

In the two-speed transmission with epicyclic gearing, as shown in Figure 6.6, the input to the planetary gearbox is given to the motor sun gear from the electric motor. In the first stage, the motor sun gear rotation is transferred to the motor planet gear. In the second stage, for the first speed selection, the first speed sun gear is held stationary, and the output is taken from the carrier (spider arm). In this case, the first speed planet gear transmits the power to the carrier (spider arm). For the second speed selection, the second speed sun gear is held stationary, and the output is taken from the carrier (spider arm). In this case, the second speed planet gear transmits the power to the carrier (spider arm). Hence, this arrangement comes under the first type of solar epicyclic gearing (where a sun gear is the fixed member).[2] Also in this arrangement, there is no ring gear.

Referring to Table 1.1, it is mentioned that, for the solar type of epicyclic gearing, the input rotation and output rotation are in the same sense. However, in the case of a 2 speed transmission, the motor sun gear sense of rotation and the carrier sense of rotation are opposite in direction. This can be explained by

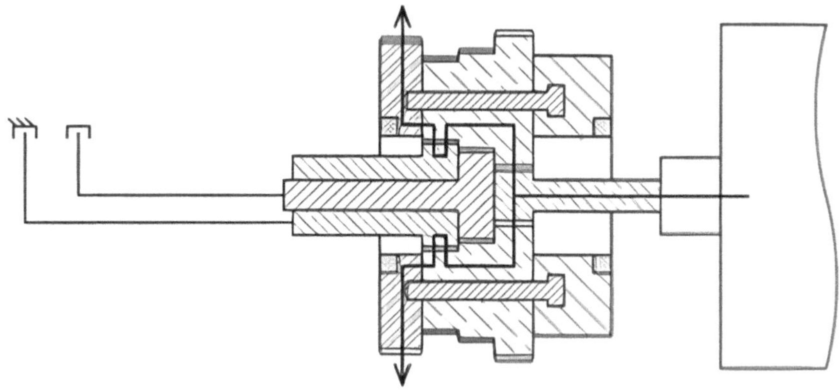

FIGURE 6.7 First gear shift mechanism (Han et al., 2021).

the fact that the motor sun–planet gearing combination and the first and second speed planets and their respective sun-gearing combinations form two separate stages of planetary gearing. Assuming the motor sun gear rotates in the clockwise direction, the mating motor planet gear will be rotating in the anticlockwise direction, so also are the first and second speed planet gears, since all planets are mounted rigidly to the same shaft. Hence, for the first and second speed planets and their respective sun gearing arrangement, if the planets rotate in the anticlockwise direction, then an imaginary ring gear, in their engagement as input, will also be rotating in the anticlockwise direction, and correspondingly, the carrier (spider arm) will also be turning in the anticlockwise direction since the respective sun gear is held stationary. Hence, the sense of rotation of the input planet gear (as well as an imaginary input ring gear) and the output carrier (spider arm) will be the same.

In the motor stage planetary gearing, the sun gear is the driver gear and the planet gear is the driven gear. In the first-stage and second-stage planetary gearing, the planet gear is the driver gear, and the sun gear is the driven gear (but fixed, alternately). Also, it is to be noted that, during the first speed gear selection, the second sun gear and its mating planet gear freely rotate without being subjected to any kind of loading because the second sun gear is not connected to any output loading (Figure 6.7).

Similarly, during the second speed gear selection, the first sun gear and its mating planet gear freely rotate without being subjected to any kind of loading because the first sun gear is not connected to any output loading (Figure 6.8).

The power transmitted from the carrier gear goes to the final reduction gear stage with a differential gearing arrangement that delivers power to the road wheels.

6.6.2 Case Study of an Electric Vehicle Gearbox

The design optimization is attempted for the electric vehicle planetary gear train with the specifications listed in Table 6.1 (Han et al., 2021).

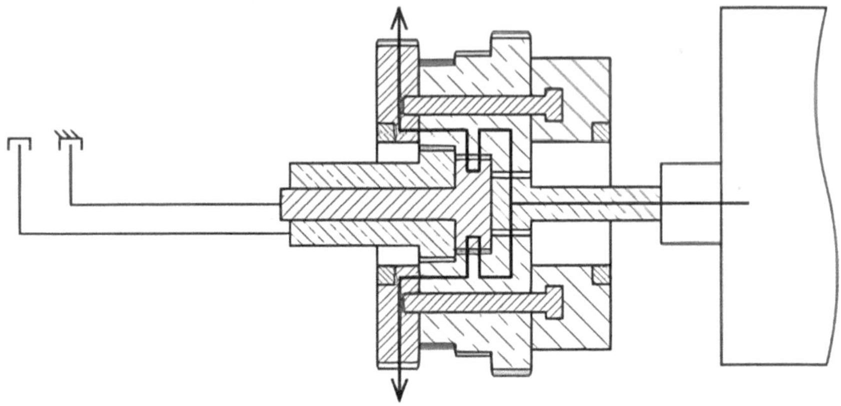

FIGURE 6.8 Second gear shift mechanism (Han et al., 2021).

TABLE 6.1
Electric Vehicle Gearing Specifications (Han et al., 2021)

	First-Stage Planetary Gearing (Motor Stage)		Second-Stage Solar Gearing: First Speed		Second-Stage Solar Gearing: Second Speed	
	S_M	P_M	S_1	P_1	S_2	P_2
Module (mm)	2.0		2.0			
Pressure angle (degree)	20		20			
Number of teeth (Z)	19	32	37	14	32	19
Helix angle (degree)	0		0			
Profile shift coefficient (X_n)	0.3	0.1	0.1	0.3	0.1	0.3
Face width (b mm)	30		30			
Number of planets	3 in each stage (motor, first speed, and second speed)					

To find the speed of individual elements of the existing planetary gear train, refer to Table 6.2.

The last row of Table 6.2 provides us with the equation for the speed of individual elements.

Input power $P = 100$ kW
Input speed of motor sun gear shaft = 1200 rpm (CW-assumed)

that is,

$$y + x = 1200 \tag{6.1}$$

For the first speed selection, the first speed sun gear is locked.

TABLE 6.2
Speed Calculation for Planetary Elements

Operation	Carrier (Spider Arm L)	Motor Sun Wheel Z_s	Planets Compound Z_p	First Speed Sun Gear	Second Speed Sun Gear
1. Arm L is locked and sun wheel S is given +1 revolution.	0	+1	$-\frac{Z_s}{Z_p}=-\frac{19}{32}$	$\frac{19}{32}\times\frac{14}{37}$	$\frac{19}{32}\times\frac{19}{32}$
2. Multiply by x	0	x	$-\frac{Z_s}{Z_p}x=-\frac{19}{32}x$	$\frac{19}{32}\times\frac{14}{37}x$	$\frac{19}{32}\times\frac{19}{32}x$
3. Add y to all columns	y	$y+x$	$y-\frac{Z_s}{Z_p}x=y-\frac{19}{32}x$	$y+\frac{19}{32}\times\frac{14}{37}x$	$y+\frac{19}{32}\times\frac{19}{32}x$

that is,

$$y+\frac{19}{32}\times\frac{14}{37}x=0$$

or

$$y+0.2247x=0$$

or

$$y=-0.2247x \tag{6.2}$$

Solving equations (6.1, 6.2),

$$x=1547.8$$
$$y=-347.8$$

Hence, for the first gear selection, the carrier (spider arm) rotates at 347.8 rpm (CCW) with a first gear ratio of 3.451:1.

Speed of the compound planets is given by,

$$y-\frac{Z_s}{Z_p}x=y-\frac{19}{32}x=1266.8 \text{ rpm}$$

that is, N_1 = 1266.8 rpm

Similarly for the second gear selection, the second speed sun gear is locked. that is,

$$y+\frac{19}{32}\times\frac{19}{32}x=0$$

$$y + 0.35254x = 0$$

$$y = -0.35254x \tag{6.3}$$

Solving equations (6.1, 6.3),

$$x = 1853.4$$

$$y = -653.4$$

Hence, for the second gear selection, the carrier (spider arm) rotates 653.4 rpm (CCW) with a second gear ratio of 1.837:1.

The first gear ratio and the second gear ratio are given in the following form of equations (Han et al., 2021):

$$i_1' = \frac{\left(\dfrac{Z_{P_1} Z_{SM}}{Z_{PM} Z_{S_1}}\right)}{\left(1 - \dfrac{Z_{P_1} Z_{SM}}{Z_{PM} Z_{S_1}}\right)}$$

$$i_2' = \frac{\left(\dfrac{Z_{P_2} Z_{SM}}{Z_{PM} Z_{S_2}}\right)}{\left(1 - \dfrac{Z_{P_2} Z_{SM}}{Z_{PM} Z_{S_2}}\right)}$$

As indicated already, the input motor shaft and output carrier (spider arm) shaft turn in opposite directions.

6.6.3 Calculation of Gear Parameters

Gear parameters are calculated using the procedure explained in Chapter 3 for Power Take-Off (PTO) gearbox.[3]

6.6.4 Design Space

The existing gear design problem for gears made of a chosen material can be reduced to a six-dimensional design problem where the six free parameters are the number of teeth on the planet pinion z_1, the number of teeth on the sun gear z_2, the module m_n, the two profile shift coefficients x_{n1} and x_{n2} of the planet gear and the sun gear, respectively, and the number of planets n_p. The inequality constraints which bound this design space are the bending stress, Hertz stress, contact ratio, top land thickness, and operating interference of gears.

6.6.5 Objective Functions and Constraints

Eight objective functions for minimization/maximization with an array of inequality and equality constraints have been considered, and the problem can be stated as follows:

The following objective functions are considered for the design optimization of the planetary gear train:

$$f_{obj} = \begin{cases} min.\ center\ distance\ between\ planetary\ gear\ pair \\ min.\ weight\ of\ the\ planetary\ gear\ train \\ max.\ efficiency\ of\ planetary\ gear\ train \\ Balancing\ the\ specific\ sliding \\ Balancing\ the\ Almen\ product \\ Balancing\ the\ flash\ temperature \\ Balancing\ the\ lambda\ ratio \end{cases}$$

The independent design variables of the planetary gear train include a module, the number of teeth of the pinion and the profile shift coefficients x_{n1} and x_{n2} of the planet gear and the sun gear, respectively. The vector of design variables is given below:

$$x(i) = \begin{cases} planet\ gear\ no.of teeth, Z_1 \\ sun\ gear\ no.of teeth, Z_2 \\ planet\ gear\ profile\ shift\ coefficient, x_{n_1} \\ sun\ gear\ profile\ shift\ coefficient, x_{n_3} \\ normal\ module, m_n \\ number\ of\ planets, n_p \end{cases}$$

where $i = 1, 2, \ldots,$ the number of variables.

A number of constraints, which act as sub functions to restrict the objective function, are introduced into the objective function to be optimized and are used to provide suitable design choices. The very often encountered failure types in a transmission gear are tooth fracture, surface fatigue, abrasion, and plastic deformation. Thus, the following constraints are considered for the design optimization of the planetary gear train:

$$g(j) = \begin{cases} Bending\ stress\ constraint \\ Surface\ contact\ compressive\ stress\ constraint \\ Contact\ ratio\ constraint \\ Top land\ thickness\ constraint \\ Operating\ interference\ of\ planet\ gear\ constraint \\ Operating\ interference\ of\ sun\ gear\ constraint \end{cases}$$

where $j = 1, 2, \ldots,$ the number of constraints.

6.6.6 Establishing a Mathematical Model

To establish a mathematical model for the planetary gear train design optimization, mathematical equations for the design variables, objective functions, and necessary constraints are to be developed. These sections are covered in more exhaustive detail in Chapter 3 for the PTO gearbox.[4]

6.6.7 Objective Functions for the Epicyclic Gearing

The following objective functions are considered for the design optimization of epicyclic gearing.

6.6.7.1 Minimization of Center Distance

The objective function for minimization of center distance is given by,

$$f_1 = \min a = \frac{m_n\left(z_2 + z_1\right)}{2}$$

6.6.7.2 Minimization of Weight

The second objective function may be written as,

Minimize f_2 = min total weight

$$\begin{aligned} f_2 &= W = W_1 + W_2 \\ &= n_w \rho \pi \bar{d}_1 A_1 + \rho \pi \bar{d}_2 A_2 \end{aligned}$$

where

W_1 and W_2 are the weights of the planet gear and the sun gear, respectively, in kg; A is the swept area, mm^2; $\bar{d}$ is the diameter at C.G. mm; ρ is the material density, kg/mm^2.

Referring to Figure 3.6, assuming a rim thickness of $1m_n$ for a gear, for the sun gear and the planet gears,

$$\bar{d} = d_w - 2m_n$$

The swept area is given by,

$$A = b \times 2m_n$$

6.6.7.3 Maximization of Efficiency

In the electric vehicle gearbox, the planet gear is the driver gear, and the sun gear is the driven gear.

The objective function for efficiency increase is given by,

Maximize

$$f_3 = \eta = \left[1 - \left(\mu_{m_1} H v_1\right)\right]100$$

where Hv_1 is the power loss factor.

The power loss factor Hv_1 for the planet gear and the sun gear pair is given by,

$$Hv_1 = \frac{\pi}{z_1 \cos\beta_b} \cdot \frac{1+i_1}{i_1}\left(1-\varepsilon_{\alpha_1}+\varepsilon_1^2+\varepsilon_2^2\right)$$

where ε_{α_1} the base contact ratio for the planet gear and sun gear pair is given by,

$$\varepsilon_{\alpha_1} = \varepsilon_1 + \varepsilon_2$$

The base pitch and the base transverse pitch of the planet gear are given by,

$$p_b = \frac{\pi d_{b_1}}{z_1}$$

$$p_{b_t} = \frac{p_b}{\cos\beta}$$

Referring to Figure 4.4, the addendum contact ratio for planet gear ε_1 and sun gear ε_2, in the planet sun gear engagement, is given by,

$$\varepsilon_1 = \frac{AP}{p_{b_t}}$$

$$\varepsilon_2 = \frac{BP}{p_{b_t}}$$

The base helix angle is given by,

$$\beta_b = \tan^{-1}\left(\frac{d_{b_1}\tan\beta}{\frac{m_n z_1}{\cos\beta}}\right)$$

Also,

$$\beta_b = \sin^{-1}\left(\sin\beta\cos\alpha_n\right)$$

The mean coefficient of friction is assumed to be constant for every contact point of the engagement region, even though it varies slightly along the path of contact. The approximate value of the mean coefficient of friction is given by,

$$\mu_{m_1} = 0.048\left(\frac{F_{t_{b_1}}/b}{\gamma_{\varepsilon c_1}\rho_{redc_1}}\right)^{0.2}\eta_{oil_1}^{-0.05}R_{a_1}^{+0.25}X_{L_1}$$

where $F_{t_{b_1}}$ is the circumferential force at the base circle (N); b is the face width for spur gears (mm); $\gamma_{\varepsilon c_1}$ is the sum velocity at the operating pitch circle (m/s); ρ_{redc_1} is the reduced radius of curvature at pitch point (mm); η_{oil} is the dynamic oil viscosity at oil temperature (mPa.s); R_a is the arithmetic mean roughness (μm); X_L is the lubricant correction factor.

The load-dependent losses are influenced by both gear geometry and lubricant properties.

The tangential force at the base circle for the planet gear and the sun gear engagement is,

$$F_{t_{b1}} = \frac{1000T_1}{r_{b1}}$$

where T_1 is the torque for the planet gear and is calculated as,

$$T_1 = \frac{60000P}{2\pi N_1 n_p}$$

The sum velocity at operating pitch circles for the planet gear and the sun gear pair is,

$$\gamma_{\varepsilon c_1} = 2v_{t_1} \sin \alpha_{wt_1}$$

Here, the tangential velocity at the pitch circle is calculated as,

$$v_{t_1} = r_{w1} \omega_1$$

The angular velocity of planet gear is,

$$\omega_1 = \frac{2\pi N_1}{60}$$

The curvature radius, or transverse profile radius, for planet gear (in the sun gear, planet gear pair) is,

$$\rho_1 = \frac{r_{w1} \sin \alpha_{wt_1}}{\cos \beta_b}$$

The curvature radius for sun gear (in the sun gear, planet gear pair) is,

$$\rho_2 = \frac{r_{w2} \sin \alpha_{wt_1}}{\cos \beta_b}$$

Then the reduced radius of curvature (in the sun gear, planet gear pair) is,

$$\rho_{redc_1} = \frac{\rho_1 \rho_2}{(\rho_1 + \rho_2) \cos \beta_b}$$

For the present case, the following values are assumed:

$$\eta_{oil} = 50\,\text{mPa.s}$$

$$R_{a1} = R_{a2} = 0.8\,\text{microns}$$

The lubricant correction factor X_{L_1} for the planet gear sun gear pair is given by,

$$X_{L_1} = \frac{1}{\left(\frac{F_{t_{b1}}}{b}\right)^d}$$

Here, it is assumed that lubricating oil is mineral oil for which the factor $d = 0.0651$ (Martinsa, R, 2006).

6.6.7.4 Sliding and Rolling Velocities

Linear velocities of any contact point B (refer to Figure 3.10) are denoted by v_{t_1} and v_{t_2} for the planet gear 1 and the sun gear 2, respectively, and are given by the following equations:

$$v_{t_1} = r_{B_1}\omega_1$$

$$v_{t_2} = r_{B_2}\omega_2$$

where r_{B_1} and r_{B_2} are the distance of the contact point from the gear centers O_1 and O_2, respectively.

The direction of linear velocity is perpendicular to the line joining the gear center and the contact point, in the direction of rotation. The components of velocity normal to the line of action, and therefore tangential to the involute tooth profiles, are called rolling velocities and are denoted by v_{r2} and v_{r2}. Rolling velocity is the velocity of the contact point in the direction of sliding.

The rolling velocities of the gear pair are calculated by the following equations:

$$v_{r_1} = v_{t_1}\sin\beta_1' = \rho_1\omega_1$$

$$v_{r_2} = v_{t_2}\sin\beta_2' = \rho_2\omega_2$$

Hence, the sliding velocity is calculated as:

$$v_{s_1} = v_{r_1} - v_{r_2}$$

where v_{s_1} is interpreted as the sliding velocity of the driving gear relative to the driven gear.

Similarly, v_{s_2} is the sliding velocity of the driven gear relative to the driving gear and is given by,

$$v_{s_2} = v_{r_2} - v_{r_1}$$

The above equations show that v_{s_1} and v_{s_2} have equal and opposite relationship, that is, $v_{s_1} = -v_{s_2}$

Since the sun gear is fixed,

$$v_{r_2} = v_{t_2} = 0$$

$$\therefore v_{s_1} = v_{r_1}$$

$$v_{s_2} = -v_{r_1}$$

6.6.7.5 Entraining Velocity

The entraining velocity is given by

$$v_e = v_{r_2} + v_{r_1}$$

And mean entraining velocity is given by

$$v_{e_{mean}} = \frac{v_{r_2} + v_{r_1}}{2}$$

Since the sun gear is fixed,

$$v_{r_2} = v_{t_2} = 0$$

$$\therefore v_e = v_{r_1}$$

6.6.7.6 Slide Roll Ratio

The slide roll ratio (SRR) is given by,

$$SRR = \frac{v_s}{v_e}$$

Since the sun gear is fixed,

$$v_s = v_e$$

that is, $SRR = 1$

Hence, the minimization of SRR is omitted from the problem statement.

6.6.7.7 Specific Sliding

For balancing or equalization of specific sliding, the following equations are used, which form the next set of objective functions:

$$\min\ f_4 = \left|\gamma_{1\max} - \gamma_{2\max}\right|$$

where $\gamma_{1\max}$ and $\gamma_{2\max}$ are the specific sliding ratios at SAP and EAP, respectively, for the planet gear and sun gear pair meshing.

To find out the values of $\gamma_{1\max}$ and $\gamma_{2\max}$, the following equations are used :

$$\gamma_{1\max} = \frac{\left(\tan\alpha_{at_1} - \tan\alpha_{wt_1}\right)}{\left(\left(\frac{i_1+1}{i_1}\right)\tan\alpha_{wt_1} - \tan\alpha_{at_1}\right)}\left(\frac{i_1+1}{i_1}\right)$$

$$\gamma_{2\max} = \frac{\left(\tan\alpha_{at_2} - \tan\alpha_{wt_1}\right)}{\left(\left(i_1+1\right)\tan\alpha_{wt_1} - \tan\alpha_{at_2}\right)}\left(i_1+1\right)$$

Here, α_{wt_1} is the working transverse pressure angle between the planet gear and sun gear pair meshing.

Where i_1 is the gear ratio between the planet gear-sun gear pair, and is defined as below,

$$i_1 = \frac{z_2}{z_1}$$

The addendum circle pressure angle of the planet gear is given by,

$$\alpha_{at_1} = \cos^{-1}\left(\frac{d_1\cos\alpha_t}{d_{a_1}'}\right)$$

The addendum circle pressure angle of the sun gear is given by,

$$\alpha_{at_2} = \cos^{-1}\left(\frac{d_2\cos\alpha_t}{d_{a_2}'}\right)$$

The transverse pressure angle α_t is given by,

$$\alpha_t = \tan^{-1}\left(\frac{\tan\alpha_n}{\cos\beta}\right)$$

6.6.7.8 Almen Product

For balancing or equalization of the Almen product, the following equations are used, which form the next set of objective functions:

Minimize $f_5 = \left|\sigma_{H_{a1}} v_{s_{1a}} - \sigma_{H_{r1}} v_{s_{1r}}\right|$ for the planet gear and the sun gear pair meshing.

For the planet gear and sun gear pair meshing, the contact stress at SAP is given by,

$$\sigma_{H_{a1}} = Z_{E_1}\left[F_{n_a}'\left(\frac{1}{\rho_{1_a}} + \frac{1}{\rho_{2_a}}\right)\right]^{0.5}$$

where Z_{E_1} is the elasticity factor, MPa; F_{n_a}'. is the normal load per unit length at SAP, N/mm; ρ_{1_a} is the radius of curvature of the planet gear at SAP, mm; ρ_{2_a} is the radius of curvature of the sun gear at SAP, mm.

Similarly, for the planet gear and sun gear pair meshing, the contact stress at EAP is given by,

$$\sigma_{H_r1} = Z_{E_1}\left[F_{n_r}'\left(\frac{1}{\rho_{1_r}}+\frac{1}{\rho_{2_r}}\right)\right]^{0.5}$$

where Z_{E_1} is the elasticity factor, MPa; F_{n_r}' is the normal load per unit length at EAP, N/mm; ρ_{1_r} is the radius of curvature of the planet gear at EAP, mm; ρ_{2_r} is the radius of curvature of the sun gear at EAP, mm.

The elasticity factor, or reduced modulus of elasticity, Z_E, depends on Young's moduli E_1, E_2 and Poisson's ratios ν_1,ν_2 of the planet gear and sun gear material according to the equation:

$$Z_{E_1} = \left[\frac{1}{\pi\left\{\frac{1-\nu_1^{2}}{E_1}+\frac{1-\nu_2^{2}}{E_2}\right\}}\right]^{0.5}$$

where E_1, E_2 are Young's moduli of the planet gear and sun gear material, MPa; ν_1, ν_2 are Poisson's ratios of the planet gear and sun gear material.

Referring to Figure 4.4, the normal load per unit length at SAP is,

$$F_{n_a}' = \frac{1000T_1}{\left(r_{b_1}^{2}+AM^{2}\right)^{0.5}\cos\alpha_t b}$$

The normal load per unit length at EAP is,

$$F_{n_r}' = \frac{1000T_1}{r_{a_1}'\cos\alpha_t b}$$

For the planet gear and the sun gear engagement, the radius of curvature of the planet gear at SAP is,

$$\rho_{1_a} = \frac{MP-AP}{\cos\beta_b}$$

The radius of curvature of the planet gear at EAP is,

$$\rho_{1_r} = \frac{MP + PB}{\cos \beta_b}$$

The radius of curvature of the sun gear at SAP is,

$$\rho_{2_a} = \frac{NP + AP}{\cos \beta_b}$$

The radius of curvature of the sun gear at EAP is,

$$\rho_{2_r} = \frac{NP - PB}{\cos \beta_b}$$

Referring to Figure 4.4,

$$MP = r_{w_1} \sin \alpha_{wt_1}$$
$$NP = r_{w_2} \sin \alpha_{wt_1}$$
$$AP = \left(r_{a_2}'^{\,2} - r_{b_2}^{\,2} \right)^{0.5} - r_{w_2} \sin \alpha_{wt_1}$$
$$PB = \left(r_{a_1}'^{\,2} - r_{b_1}^{\,2} \right)^{0.5} - r_{w_1} \sin \alpha_{wt_1}$$
$$AM = MP - AP$$

For the present case, the following material properties are taken from the standards:
Poisson's ratio and Young's modulus for the gear pair,

$$\nu_1 = \nu_2 = 0.3$$
$$E_1 = E_2 = 210GPa$$

The sliding velocity for the planet gear and the sun gear pair meshing at SAP is,

$$v_{s_{1a}} = AP\left(\omega_1 + \omega_2\right)$$

The sliding velocity for the planet gear and the sun gear pair meshing at EAP is,

$$v_{s_{1r}} = PB\left(\omega_1 + \omega_2\right)$$

The angular velocity of the planet gear and the sun gear are given by,

$$\omega_1 = \frac{2\pi N_1}{60}$$
$$\omega_2 = \frac{2\pi N_2}{60}$$

6.6.7.9 Flash Temperature

The flash temperature for a gear pair in engagement can be calculated for any local point on the involute flank along the path of contact, using the following equation (AGMA 925-A03):

$$\theta_{fl} = 31.62 K \mu_m \frac{X_r F_n}{\left(b_H\right)^{0.5}} \frac{\left|v_{r_1} - v_{r_2}\right|}{B_{m1}\left(v_{r_1}\right)^{0.5} + B_{m2}\left(v_{r_2}\right)^{0.5}}$$

where θ_{fl} is the flash temperature, °C; K is 0.80, a numerical factor valid for a semi-elliptic (Hertzian) distribution of frictional heat over the instantaneous width, 2 b_H, of the rectangular contact band; μ_m is the coefficient of friction; X_r is the load-sharing factor; F_n is the normal unit load, N/mm; b_H is the semi-width of the Hertzian contact band, mm; v_{r_1} is the rolling tangential velocity of the pinion, m/s; v_{r_2} is the rolling tangential velocity of the wheel, m/s; B_{m1} is thermal contact coefficient, pinion, N/(mm $s^{0.5}$K); B_{m2} is the thermal contact coefficient, wheel, N/(mm $s^{0.5}$K).

The semi-width of the Hertzian contact band is given by,

$$b_H = \left[\frac{4 F_n \rho_{redc}}{\pi E'}\right]^{0.5}$$

where F_n is the normal force at the contact point per unit width, N/mm; ρ_{redc} is the reduced radius of curvature at the contact point, mm; E' is the equivalent elastic modulus of gear pair materials, N/mm^2.

For balancing or equalization of flash temperature, the following equations are used, which form the next set of objective functions:

Minimize

$f_6 = \theta_{fla_1}$ for the planet gear and the sun gear pair meshing at SAP.
$f_7 = \theta_{fln_1}$ for the planet gear and the sun gear pair meshing at EAP.

The flash temperature for the planet gear and the sun gear pair engagement at SAP is,

$$\theta_{fla_1} = 31.62 K \mu_{m_1} \frac{X_{r_1} F_{n_a}'}{\left(b_{H_{a1}}\right)^{0.5}} \frac{\left|v_{r_1 a}' - v_{r_2 a}'\right|}{B_{m1}\left(v_{r_1 a}'\right)^{0.5} + B_{m2}\left(v_{r_2 a}'\right)^{0.5}}$$

Here, the rolling tangential velocity of the planet gear at SAP is,

$$v_{r_1 a}' = \rho_{1_a} \omega_1$$

The rolling tangential velocity of the sun gear at SAP is,

$$v_{r2_a}' = \rho_{2_a}\omega_2$$

The minimum Hertzian semi-contact width in one mesh cycle at SAP,

$$b_{H_{a1}} = \left[\frac{4F_{n_a}'\rho_{redc_{1a}}}{\pi E_1'}\right]^{0.5}$$

The reduced radius of curvature at SAP (in the sun gear, planet gear pair) is,

$$\rho_{redc_{1a}} = \frac{\rho_{1_a}\rho_{2_a}}{\left(\rho_{1_a} + \rho_{2_a}\right)\cos\beta_b}$$

Here, E_1' is the equivalent or effective modulus of elasticity, which depends on the modulus of elasticity and Poisson ratio of the sun gear material and planet gear material.

$$E_1' = \frac{1}{2}\left[\frac{1-\nu_1^2}{E_1} + \frac{1-\nu_2^2}{E_2}\right]$$

Here, E_1 and E_2 are in N/mm^2.

The value of the thermal contact coefficient B_{m1} for the planet gear is given by the following equation:

$$B_{m1} = \frac{\left(\lambda_m\rho_m c_m\right)^{0.5}}{10^3}$$

where λ_m is the heat conductivity of the planet gear material, W/(m.K); ρ_m is the planet gear material density, kg/m^3; c_m is the specific heat of planet gear material per unit of mass, J/(kg.K).

For the present case, the following material properties are taken from the standards:

$$\lambda_m = 40 \text{ W/(m.K)}$$
$$\rho_m = 7850 \text{ kg/m}^3$$
$$c_m = 465 \text{ J/(kg.K)}$$

The value thermal contact coefficient B_{m2} for the sun gear is given by the following equation:

$$B_{m2} = \frac{\left(\lambda_m\rho_m c_m\right)^{0.5}}{10^3}$$

where λ_m is the heat conductivity of the sun gear material, W/(m.K); ρ_m is the sun gear material density, kg/m^3; c_m is the specific heat of sun gear material per unit of mass, J/(kg.K).

The flash temperature for the planet gear and the sun gear pair engagement at EAP is,

$$\theta_{fl\eta} = 31.62K\mu_{m_1}\frac{X_{\eta}F_{n_r}'}{\left(b_{H_{r1}}\right)^{0.5}}\frac{\left|v_{\eta r}'-v_{r_2 r}'\right|}{B_{m1}\left(v_{\eta r}'\right)^{0.5}+B_{m2}\left(v_{r_2 r}'\right)^{0.5}}$$

Here, the rolling tangential velocity of the planet gear at EAP is,

$$v_{\eta_r}' = \rho_{1_r}\omega_1$$

The rolling tangential velocity of the sun gear at EAP is,

$$v_{r_{2r}}' = \rho_{2_r}\omega_2$$

The minimum Hertzian semi-contact width in one mesh cycle at EAP,

$$b_{H_{r1}} = \left[\frac{4F_{n_r}'\rho_{redc_{1r}}}{\pi E_1'}\right]^{0.5}$$

The reduced radius of curvature at EAP (in the sun gear, planet gear pair) is,

$$\rho_{redc_{1r}} = \frac{\rho_{1_r}\rho_{2_r}}{\left(\rho_{1_r}+\rho_{2_r}\right)\cos\beta_b}$$

6.6.7.10 Minimum Oil Film Thickness

Taking into account the unsteady speed and load conditions, the minimum oil film thickness for the external gear pair engagement is given by,

$$h_{\min} = 1.714R_i\left(A\right)^{-0.128}\left(B\right)^{0.694}\left(C\right)^{0.568}$$

where $h_{\min}$ is the minimum oil film thickness, m; R_i is the radius of curvature of the interface between the mating gears, m.

Here, R_i is given by,

$$R_i = \frac{r_{w1}r_{w2}\sin\alpha_{wt_1}}{1000\left(r_{w_1}+r_{w_2}\right)}$$

$$A = \frac{10^6 F_n}{bE''R_i}$$

where F_n is the normal load at pitch point, *N*; *b* is the face width, *m*; *E''* is the equivalent modulus of elasticity, N/m².

The equivalent modulus of elasticity in this case is given by,

$$E'' = \left[\frac{2}{\frac{1-\nu_1^2}{E_1} + \frac{1-\nu_2^2}{E_2}} \right]$$

Here, E_1 and E_2 are the modulus of elasticity mating gears, in N/m^2.

$$B = \frac{\eta_o \left(v_1 + v_2 \right) \cos \alpha_{wt_1}}{2 \times 10^3 E'' R_i}$$

where η_o is the viscosity at working temperature, mPa.s; v_1, v_2 is the tangential velocity of mating gears at the pitch circle, m/s

$$C = \alpha' E''$$

where α' is the lubricant pressure viscosity coefficient, m^2/N.

The minimum oil film thickness for the planet gear and the sun gear pair engagement is given by,

$$h_{\min_1} = 1.714 R_{i_1} \left(A' \right)^{-0.128} \left(B' \right)^{0.694} \left(C' \right)^{0.568}$$

where

$$A' = \frac{10^6 F_{n_1}}{b E_1'' R_{i_1}}$$

$$F_{n_1} = \left(\frac{1000 T_1}{r_{w_1}} \right)$$

The equivalent modulus of elasticity in this case is given by,

$$E_1'' = \left[\frac{2}{\frac{1-\nu_1^2}{E_1} + \frac{1-\nu_2^2}{E_2}} \right]$$

R_{i_1} is the radius of curvature of the interface between the planet gear and the sun gear and is given by,

$$R_{i_1} = \frac{r_{w_1} r_{w_2} \sin \alpha_{wt_1}}{10^3 \left(r_{w_1} + r_{w_2} \right)}$$

$$B' = \frac{\eta_o \left(v_1 + v_2 \right) \cos \alpha_{wt_1}}{2 \times 10^3 E_1'' R_i}$$

For the planet gear,

$$v_1 = \frac{r_{w1}\omega_1}{10^3} = \frac{v_{t1}}{10^3}$$

For the sun gear,

$$v_2 = \frac{r_{w2}\omega_2}{10^3} = \frac{v_{t2}}{10^3}$$

Here, v_1 and v_2 are in m/s.

$$C' = \alpha_1' E_1''$$

For the present case, the following values are taken from the standards:

$$\eta_o = 90 \text{ mPa.s}$$
$$\alpha_1' = 2 \times 10^{-8} \text{m}^2 / \text{N}$$

6.6.7.11 Lambda Ratio

Improvement in the lambda ratio value ensures better performance of the gear pair in engagement, and its value must be maximized. Hence, the following equations form the next set of objectives:

To maximize the lambda ratio,

Maximize

$$f_8 = \lambda_1 = \frac{h_{\min_1}}{R_{a1}'}$$

where

$$R_{a1}' = R_{a1} + R_{a2}$$

where R_{a1} and R_{a2} are the arithmetic mean roughness of the planet and sun gears, respectively (μm).

For the present case,

$$R_{a1}' = 0.1$$

6.6.8 Design Variables and Constraints

The independent design parameters of a planetary gear drive include the module m_n, the number of teeth of planet gear z_1, the number of teeth of sun gear z_2, the profile shift coefficients x_{n1}, x_{n2}, and the number of planets n_p.

Therefore, the design variables are,

$$X = \left[z_1, z_2, x_{n1}, x_{n2}, m_n, n_p\right]^T = \left[x_1, x_2, x_3, x_4, x_5, x_6\right]^T$$

A number of constraints, which are introduced into the objective functions to be optimized, are used to provide suitable design choices. For an overview of types of gear failures, see Section 3.6.14.

6.6.8.1 Bending Strength

Maximum allowable bending stress is given by the bending strength of spur gears and helical gears in the ISO6336:2006-based JGMA401-01 (1974) standard.

$$\sigma_F = F_t \frac{y_F y_\varepsilon y_\beta}{9.81 m_n b}\left(\frac{K_V K_O}{K_L K_{FX}}\right) S_F \ \text{kgf} / \text{mm}^2$$

where F_t is the tangential force at the working pitch circle, kgf; y_F is the tooth profile factor; y_ε is the load-sharing factor; y_β is the helix angle factor; K_V is the dynamic load factor; K_O is the overload factor; K_L is the life factor; K_{FX} is the size factor of root stress; S_F is the safety factor for bending failure.

For the sun gear–planet gear engagement,

$$F_t = F_{n_2}$$

The tooth profile factor y_F is found by the equation (refer to Figure 3.13),

$$y_F = \frac{6\left(\frac{h_A}{m_n}\right)\cos\alpha_{wt}}{\left(\frac{S_A{}^2}{m_n{}^2}\right)\cos\alpha_n}$$

The load sharing factor y_ε is given by,

$$y_\varepsilon = \frac{1}{\varepsilon_{\alpha_1}{}'}$$

The helix angle factor y_β is given by,

$$y_\beta = 1 - \frac{\beta}{120}$$

The dynamic load factor, K_V, is based on the precision of the gear and the tangential speed of gears at the working pitch circle. The tangential speed of gears at the working pitch circle is,

$$v_{t_1} = \frac{\pi d_{w1} N_1}{60000}$$

The dynamic load factor K_V is taken from the standards.

For tangential speed less than 1.0 m/s,

$$K_V = 1.0$$

The overload factor *Ko* is defined as,

$$K_O = \frac{\text{Actual tangential force}}{\text{Nominal tangential force } F_t}$$

Since the *EV* gearbox is expected to be subjected to heavy input load,

K_O is taken as 1.75 from the standards.

The life factor K_L depends on the number of cyclic repetitions or the total loaded meshing cycles during its lifetime. Its value is taken from the standards. For the number of cyclic repetitions of $>10^7$ cycles, K_L is taken as,

$$K_L = 1.0$$

The size factor of root stress K_{FX} is generally taken as 1.0.

S_F is the safety factor for bending failure. Since S_F is too complicated to be determined precisely, it is usually taken as at least 1.2.

6.6.8.2 Contact Compressive (Crushing) Strength

Contact compressive stress should be within allowable limits to prevent failures due to wear like scuffing, scoring, welding, peening, rippling, ridging, and tooth breakage.

Contact compressive strength of spur gears and helical gears as per ISO 6336:2006-based JGMA 402-01 (1975) standard is given by,

$$\sigma_H = \sqrt{\frac{F_t\left(i+1\right)}{9.81 d_w b i}\frac{Z_H Z_M Z_\varepsilon Z_\beta}{K_{HL} Z_L Z_R Z_V Z_W K_{HX}}}\sqrt{K_{H\beta} K_V K_O} S_H \quad \text{kgf / mm}^2$$

The "+" symbol in the above equation applies to two external gears in the mesh, whereas the "-" symbol is used for an internal gear and an external gear mesh.

Zone factor, Z_H
Material factor, Z_M
Contact ratio factor, Z_ε
Helix angle factor, Z_β
Life factor, K_{HL}
Lubricant factor, Z_L
Surface roughness factor, Z_R
Lubrication speed factor, Z_V
Hardness ratio factor, Z_W
Size factor, K_{HX}

Longitudinal load distribution factor, $K_{H\beta}$
Dynamic load factor, K_V
Overload factor, K_O
Safety factor for pitting, S_H

For the sun gear–planet gear engagement, Hertz contact stress is given by,

$$\sigma_{H_1} = \sqrt{\frac{F_t(i+1)}{9.81 d_{w_2} b i} \frac{Z_H Z_M Z_\varepsilon Z_\beta}{K_{HL} Z_L Z_R Z_V Z_W K_{HX}}} \sqrt{K_{H\beta} K_V K_O} S_H \ \text{kgf / mm}^2$$

$$i = i_1$$

$$F_t = F_{n_1}$$

Zone factor, Z_H, is given by,

$$Z_H = \sqrt{\frac{2\cos\beta_b \cos\alpha_{wt_1}}{\cos^2\alpha_t \sin\alpha_{wt_1}}} = \frac{1}{\cos\alpha_t}\sqrt{\frac{2\cos\beta_b}{\tan\alpha_{wt_1}}}$$

Material factor, Z_M, is given by,

$$Z_M = \left[\frac{1}{\pi\left\{\frac{1-\nu_1^2}{E_1} + \frac{1-\nu_2^2}{E_2}\right\}}\right]^{0.5}$$

For steel, $E_1 = 21000$ kgf/mm^2

Contact ratio factor, Z_ε, is given by the following:

For spur gear,

$$Z_\varepsilon = 1.0$$

For helical gear,

When the overlap ratio $\varepsilon_\beta \leq 1$,

$$Z_\varepsilon = \sqrt{1 - \varepsilon_\beta + \frac{\varepsilon_\beta}{\varepsilon_\alpha}}$$

When $\varepsilon_\beta > 1$,

$$Z_\varepsilon = \sqrt{\frac{1}{\varepsilon_\alpha}}$$

The helix angle factor, Z_β, is difficult to calculate and can be taken as 1.0.

The life factor, K_{HL}, depends on the duty cycle.

For duty cycle > 10^7 cycles,

$$K_{HL} = 1.0$$

Lubricant factor, Z_L, is based upon the lubricant's kinematic viscosity at 50°C, *cSt*.

From the standards,

$$Z_L = 1.0$$

The surface roughness factor, Z_R, is obtained on the basis of the average roughness $R_{\max}$ (μm).

The average roughness, $R_{\max}$, is calculated by the below-given equation using the surface roughness values of the pinion and gear, $R_{\max 1}$ and $R_{\max 2}$, and the center distance a in mm.

$$R_{\max} = \frac{R_{\max 1} + R_{\max 2}}{2}\left(\frac{100}{a'}\right)^{\frac{1}{3}}$$

For surface-hardened gear, and $R_{\max} = 10$ microns,

$$Z_R = 0.92$$

Lubrication speed factor, Z_V, depends on the tangential speed at the pitch circle, v (m/s).

For surface-hardened gear, and $v = 2.5$ m/s,

$$Z_V = 0.98$$

The hardness ratio factor, Z_W, applies only to the gear that is in mesh with a pinion, which is quenched and ground. It is given by the equation,

$$Z_W = 1.2 - \frac{H_{B2} - 130}{1700}$$

Here, H_{B2} is in the range of the Brinell hardness of gear material.

$$130 \leq H_{B2} \leq 470$$

If a gear is out of this range, the Z_W is assumed to be 1.0.

The size factor, K_{HX}, is usually taken as 1.0 because the conditions affecting this parameter are often unknown.

The longitudinal load distribution factor, $K_{H\beta}$, depends on the method of gear shaft support and the ratio, b / d_{w2}, of the gear face width b to the pitch diameter, d_{w2}. When tooth contact under load is good, the longitudinal load distribution factor can be taken to be in the range of 1.0–1.2. For the present case, let $K_{H\beta} = 1.0$.

The dynamic load factor, K_V, depends on the gear's precision grade and pitch circle tangential speed, v.

For speeds less than 1 m/s the value of K_V is taken as 1.0.

The overload factor, K_O, depends on the impact load on the gears on the prime mover side and on the load side. For uniform load from the load side, this factor can be taken as 1.0.

The safety factor for pitting, S_H, is normally taken as 1.15 since the causes of pitting are difficult to precisely define because it involves many environmental factors.

6.6.8.3 Contact Ratio

The contact ratio is given by,

$$\varepsilon_\alpha' = \frac{\text{Path of contact}}{p_b}$$

where ε_α' is the contact ratio; p_b is the base pitch.

The base pitch, p_b, is defined as the distance from one tooth to the next tooth measured along the line of action. The base pitch is given by,

$$p_b = p_c \cos\alpha_t = \pi m_n \cos\alpha_t$$

where p_c is the circular pitch.

This contact ratio should be greater than 1.4 and is normally less than 2.0 for standard spur gears. For contact with an internal gear at low ratios and for non-standard addenda and dedenda, it can exceed 2.0.

The transverse contact ratio for the sun gear–planet gear pair engagement is given by,

$$\varepsilon_{\alpha_1}' = \frac{AB}{p_b} = \frac{AB}{\pi m_t \cos\alpha_t}$$

Overlap ratio

$$\varepsilon_\beta = \frac{b\sin\beta}{\pi m_n}$$

The total contact ratio for the sun gear–planet gear pair engagement is given by,

$$\varepsilon_{\alpha_1}' + \varepsilon_\beta \geq 1.4$$

6.6.8.4 Top Land Thickness

The top land thickness is the tooth thickness on the tip circle. The top land thickness is normally referred to in terms of the percentage of the module. The ratio of top land thickness and the module is suggested to be between 0.2 and 0.4, depending on the material and heat treatment of the gear.

The top land thickness of the planet gear is given by,

$$S_{a_1} = d_{a_1}'\left(\frac{S_1}{d_1} + inv\alpha_n - \Omega_1\right)$$

where

$$S_1 = \frac{\pi m_n}{2} + 2x_{n_1} m_n \tan\alpha_n$$

$$\Omega_1 = \tan\alpha_{at_1} - \alpha_{at_1}\left(\frac{\pi}{180}\right)$$

Similarly, the top land thickness of the sun gear is given by,

$$S_{a_2} = d_{a_2}'\left(\frac{S_2}{d_2} + inv\alpha_n - \Omega_2\right)$$

where

$$S_2 = \frac{\pi m_n}{2} + 2x_{n_2} m_n \tan\alpha_n$$

$$\Omega_2 = \tan\alpha_{at_2} - \alpha_{at_2}\left(\frac{\pi}{180}\right)$$

The top land thickness should not be below 0.2 times the normal module, m_n, and correspondingly the constraint equations for the top land thickness of the planet gear and the sun gear are as given below:

$$S_{a_1} \geq 0.2 m_n$$

$$S_{a_2} \geq 0.2 m_n$$

6.6.8.5 Operating Interference

To avoid operating interference on the two mating gears, the following conditions are to be met (refer to Figure 4.4):

1. The distance O_2M is greater than the tip radius of the sun gear.
2. The distance O_1N is greater than the tip radius of the planet gear.

Hence, the constraint equations are,

$$g_p\left(x_{n_1}, x_{n_2}\right) = \left(r_{b_1}{}^2 + \left(a' \sin\alpha_{wt_1}\right)^2\right)^{0.5} > r_{a_1}'$$

$$g_s\left(x_{n_1}, x_{n_2}\right) = \left(r_{b_2}{}^2 + \left(a' \sin\alpha_{wt_1}\right)^2\right)^{0.5} > r_{a_2}'$$

6.6.9 Design Data for Optimization

The following are the input parameters considered for the planetary gear design based on the existing gear design as in Table 6.3.

6.6.10 Methodology of Optimization of Planetary Gear Using MOACO

This section describes how the MOACO algorithm is configured for solving the optimization problem of planetary gearing. This book utilizes the OPTIMO software package (Gokul Krishnan, 2024), which provides an efficient MOACO implementation in Python along with the complete problem set corresponding to this case study. Chapter 2 provides detailed instructions on how to obtain and use the software.[5]

6.6.10.1 Variable Bounds

In order to determine the values of design variables, such as a module, the number of teeth, the face width, and the helix angle through the optimization procedure, the upper and lower bounds of the design variables are to be established first.

The following variables are considered, and their lower and upper bounds are fixed. Further, the type of variable – whether discrete (integer) or real-valued (float) – is also specified.

Data for the first speed solar gearing

Number of planet gear teeth, z_1: 10 to 20 (integer)
Number of sun gear teeth, z_2: 30 to 45 (integer)
Profile shift coefficient for planet gear, x_{n1}: 0.0 to 0.5 (float)
Profile shift coefficient for sun gear, x_{n2}: 0.0 to 0.5 (float)
Module, m_n: 1 to 3 mm (float)

TABLE 6.3
Electric Vehicle Planetary Gear Parameters

Data	Gear
Density of the material, ρ	7.86×10^6 kg/mm^3
Material of sun gear, planet gear, and ring gear	Steel 20Mn5Cr5 (IS:4432-1988)
Design bending stress, σ_b	$400 \mathrm{N} / \mathrm{mm}^2$
Design crushing stress, σ_c	$2000 \mathrm{N} / \mathrm{mm}^2$
Input speed, N	1200 rpm
Power transmitted, P	100 kW
Young's modulus of the material (E)	$2.0 \times 10^6 \mathrm{N} / \mathrm{mm}^2$
Pressure angle, α_n	20°
Helix angle, β	0°

Number of planets, n_p: 2 to 5 (integer)

Data for the second speed solar gearing

Number of planet gear teeth, z_1: 15 to 25 (integer)
Number of sun gear teeth, z_2: 25 to 40 (integer)
Profile shift coefficient for planet gear, x_{n1}: 0.0 to 0.5 (float)
Profile shift coefficient for sun gear, x_{n2}: 0.0 to 0.5 (float)
Module, m_n: 1 to 3 *mm* (float)
Number of planets, n_p: 2 to 5 (integer)

6.6.10.2 Input Parameters

For this model, the most important data inputs are the gear design parameters of the two mating gears. The decision maker is required to make this decision based on experience and design requirements.

The problem sets provided in this book contain Python code corresponding to the objective functions and design constraints formulated in Sections 6.6.7 and 6.6.8. The upper and lower bounds of design variables are also specified in the same file(s). The seed for the random number generator is fixed at 0.123 for reproducibility of results.

The problem definitions corresponding to Chapter 6 can be found in *problemsets/problem_ch6a.py* and *problemsets/problem_ch6b.py* files within the software repository. The optimization run is started by entering the command specified in Listing 6.18 from the repository root.

LISTING 6.18 COMMAND TO EXECUTE THE ALGORITHM RUN

```
# macOS / Linux
python3 run.py -a aco -p ch6a
python3 run.py -a aco -p ch6b

# Windows
python.exe run.py -a aco -p ch6a
python.exe run.py -a aco -p ch6b
```

6.6.10.3 Decision-Making

The goal of decision-making is to select the most optimal solution from the Pareto solution set obtained from the optimization results. Selection of the single best solution from many conflicting non-dominated solutions from the Pareto-set is based on expertise along with any problem-specific requirements. Apart from the above, there are decision-making techniques such as the linear programming technique for multidimensional analysis of preference (LINMAP), Shannon Entropy (SE), and the technique of order preference by similarity to an ideal solution (TOPSIS), which can be used as a post-processing step to obtain a relative gradation of results.

The OPTIMO software makes use of a Common Objective Function, or COF, which assigns a scalar value to each solution in the Pareto-optimal set. This scalar value is essentially the normalized average of the set of objective function values. For a given solution i,

$$P_i = \left[p_1, p_2, \ldots p_N\right]$$
$$S_i = COF\left(P_i\right)$$

where P_i is the vector of objective function values; N is the number of objectives; S_i is the scalar output assigned to the solution.

The solution set is then sorted on its COF value, such that the solution with the lowest COF value is considered to be, on average, the overall best solution (assuming minimization of objectives).

6.6.11 Interpretation of MOACO Results

In this section, the results obtained from MOACO are discussed and compared with the design of existing electric vehicle epicyclic gearing. For detailed information on interpreting the tables and plots in the subsequent sections, refer to Section 3.6.17.

6.6.11.1 Results for the First Speed Epicyclic Gearing

Table 6.4 shows a sample of the MOACO output for the first speed epicyclic gearing. Since constraints are specified in the form of ≥ 0 only, all the columns corresponding to the constraints shown in Table 6.4 indicate results obtained by modifying the constraint equations accordingly. Table 6.4 lists the values of the objective functions, constraints, and design variables from the generated solutions. The output represents a non-dominated Pareto-optimal solution set.

6.6.11.1.1 Comparison of Results

In Table 6.5, a comparison has been made between the results that are obtained with MOACO and the design of the existing epicyclic gearing. The solutions corresponding to number of teeth of the planet gear, namely, 20, 20, and 19, were selected from the results. The values of design variables and objective functions corresponding to each solution are indicated in Table 6.5. The total weight of epicyclic gearing in kg, the efficiency in percentage, and the center distance in mm were taken for comparison.

The comparison shows that the results obtained from MOACO are an improvement over the existing design by optimizing the eight objective functions within the given constraints. The results were obtained with minimal computational

TABLE 6.4
Sample of MOACO Output for First Speed Gearing

Objectives								Constraints							Variables					
Center Distance	Weight	Efficiency	Specific Sliding	Almen Product	Flash Temperature		Lambda Ratio	Interference		Top Land Thickness		Contact Ratio	Bending Stress	Contact Stress	z_1	z_2	x_{n1}	x_{n2}	m_n	n_p
77.55	1.47	99.35	1.03	454.74	34.79	51.39	5.56	8.42	2.23	1.29	1.43	0.30	326.97	747.83	20	35	0.17	0.27	2.82	5
82.32	1.62	99.36	1.65	679.15	32.13	48.17	5.90	9.56	2.09	1.45	1.41	0.31	332.95	808.77	20	36	0.09	0.38	2.94	5
78.65	1.51	99.34	1.24	516.44	33.27	50.43	5.60	8.19	2.05	1.33	1.46	0.33	329.87	749.58	20	35	0.13	0.23	2.86	5
75.99	1.59	99.32	0.07	101.73	34.74	51.89	5.78	6.50	2.69	1.20	1.61	0.30	334.44	773.20	20	31	0.27	0.06	2.98	5
76.50	1.57	99.36	0.49	267.66	37.06	53.05	5.73	8.40	2.74	1.25	1.52	0.24	329.51	784.85	19	32	0.29	0.29	3.00	5
65.52	1.04	99.18	3.10	1121.40	41.35	66.39	4.09	8.71	1.12	1.23	1.16	0.27	290.12	424.11	17	35	0.07	0.46	2.52	5
70.74	1.02	99.28	1.30	669.19	51.77	74.59	4.55	9.68	1.91	1.21	1.26	0.23	277.42	423.97	18	36	0.21	0.49	2.62	4
49.02	**0.56**	**99.25**	**0.57**	**313.93**	**63.56**	**86.22**	**3.13**	**6.49**	**1.80**	**0.77**	**0.86**	**0.22**	**194.00**	**57.28**	**20**	**37**	**0.32**	**0.49**	**1.72**	**5**
79.20	1.23	99.24	2.99	1183.55	40.30	65.35	4.95	9.81	1.29	1.40	1.41	0.32	303.51	507.63	18	37	0.04	0.35	2.88	4
79.20	1.09	99.31	0.48	328.39	55.10	80.16	5.33	8.24	2.48	1.22	1.53	0.30	283.11	411.31	20	35	0.25	0.17	2.88	3
66.78	1.04	99.16	3.04	1035.24	39.64	66.19	3.99	7.93	0.97	1.14	1.29	0.32	293.18	376.94	17	36	0.09	0.23	2.52	5
77.38	1.22	99.31	0.62	389.46	52.23	72.18	4.96	11.19	2.37	1.17	1.49	0.18	292.41	539.97	17	36	0.36	0.45	2.92	4
69.96	0.91	99.37	0.14	160.92	70.05	92.19	5.05	8.89	3.45	1.11	1.34	0.15	248.45	375.44	20	33	0.46	0.50	2.64	3
44.72	**0.54**	**99.21**	**0.64**	**332.43**	**64.65**	**89.69**	**3.07**	**5.24**	**1.85**	**0.79**	**0.82**	**0.21**	**192.83**	**10.00**	**20**	**32**	**0.29**	**0.47**	**1.72**	**5**
61.25	0.89	99.24	0.71	399.62	57.96	82.93	4.21	7.23	2.16	1.06	1.22	0.21	263.36	313.95	18	31	0.29	0.38	2.50	4
59.40	0.83	99.16	0.28	26.52	54.44	76.38	3.66	6.72	1.91	0.70	1.25	0.25	262.07	268.55	18	36	0.43	0.05	2.20	5
67.20	0.81	99.25	0.85	461.35	55.59	80.44	3.90	8.54	1.60	0.94	1.20	0.31	247.86	203.75	19	41	0.23	0.22	2.24	4
91.00	1.14	99.35	1.61	904.96	51.01	74.87	5.34	12.51	1.83	1.35	1.44	0.33	278.79	432.44	20	45	0.11	0.38	2.80	3
57.00	0.77	99.24	0.19	12.15	65.57	89.26	4.01	6.32	2.67	0.82	1.22	0.18	242.14	244.29	19	31	0.47	0.27	2.28	4
48.36	**0.60**	**99.10**	**2.26**	**839.86**	**52.84**	**83.07**	**3.09**	**5.34**	**1.10**	**0.92**	**0.87**	**0.30**	**222.90**	**0.03**	**19**	**33**	**0.05**	**0.37**	**1.86**	**5**

TABLE 6.5
Optimized Design Parameters for First Speed Gearing Using MOACO

	Variables			Objective Functions		
	Module m (mm)	Number of Teeth (Planet Gear) Z_1	Number of Teeth (Sun Gear) Z_2	Total Weight W (kg)	Efficiency η	Center Distance (mm)
Existing first speed epicyclic gearing design	2.0	14	37	0.43	98.81	51.0
MOACO design	1.72	20	37	0.56	99.25	49.02
	1.72	20	32	0.54	99.21	44.72
	1.86	19	33	0.60	99.10	48.36

effort and time. The three best results chosen from the solution set of MOACO output are analyzed for selecting a single best solution. The total weight of the existing first speed gearing was 0.43 kg, the center distance was 51.0 mm, and the efficiency was 98.81%.

Referring to Table 6.5, in the case of the first result, by selecting the number of teeth in the planet gear as 20, the weight has increased to 0.56 kg, the efficiency has increased to 99.25%, and the center distance has reduced to 49.02 mm. There is an increase of 30.23% in weight, an increase of 0.44% in efficiency, and a space reduction of 3.88%. Hence, there is an improvement in efficiency and center distance, along with a marginal weight increase, in the solution obtained with MOACO.

In the case of the second result, by selecting the number of teeth in the planet gear as 20, the weight has increased to 0.54 kg, the efficiency has increased to 99.21%, and the center distance has reduced to 44.72 mm. There is an increase of 25.58% in weight, an increase of 0.40% in efficiency, and a space reduction of 12.31%. Hence, there is an improvement in efficiency and center distance, along with a marginal weight increase, in the solution obtained with MOACO.

In the case of the third result, by selecting the number of teeth in the planet gear as 19, the weight has increased to 0.60 kg, the efficiency has increased to 99.10%, and the center distance has reduced to 48.36 mm. There is an increase of 39.53% in weight, an increase of 0.29% in efficiency, and a space reduction of 5.18%. Hence, there is an improvement in efficiency and center distance, along with a marginal weight increase, in the solution obtained with MOACO.

6.6.11.1.2 Visualization of Results

The formation of the Pareto front from the non-dominated solutions containing eight objective functions is depicted in the subsequent graphs in combinations of three objective functions for every graph. The graphs show the 3D scatter plot of the solution set generated by MOACO in various combinations of three objective functions.

Even though the graphs help in better visualization of the non-dominated Pareto-optimal solution set, one should be cautious of the fact that all the eight objective functions put together form a single set of Pareto-optimal solutions that cannot be visualized in totality in graphical format. Hence, for the sake of decision-making, the solutions better suited for the individual requirements of the user/decision maker must be given priority in the choice of objective function combinations taken for visualization.

As a result of optimization, a Pareto-optimal front is formed in the objective space of the solutions. In the scatter plot, the circular points represent non-dominated solutions with respect to all eight objectives taken together, whereas the triangular points represent a non-dominated front formed with respect to the three objectives that are taken for visualization. It can be observed that the distribution of solutions along the front is uniform, which means the MOACO algorithm offers a high-quality solution set.

Referring to Figure 6.9(a), the solutions close to the upper right corner have higher total weight, higher center distance, and higher efficiency, which means they have better efficiency but are heavier and less compact. On the contrary, the solutions near the lower left corner present better compactness and better total weight but lesser efficiency. The solutions lying in the lower right corner have the best value in terms of higher efficiency, lesser weight, and better compactness.

Referring to Figure 6.9(b), the solutions close to the upper right corner offer higher efficiency, better compactness, and a higher lambda ratio for the sun gear/planet gear engagement.

Referring to Figure 6.10(a), the solutions close to the lower right corner offer higher efficiency, lower value of specific sliding, and lower value of Almen product at SAP and at EAP of the gear tooth profile engagement during running.

Referring to Figure 6.10(b), the solutions close to the lower right corner offer higher efficiency and lower flash temperatures that are balanced at SAP and EAP of the gear tooth profile engagement during running.

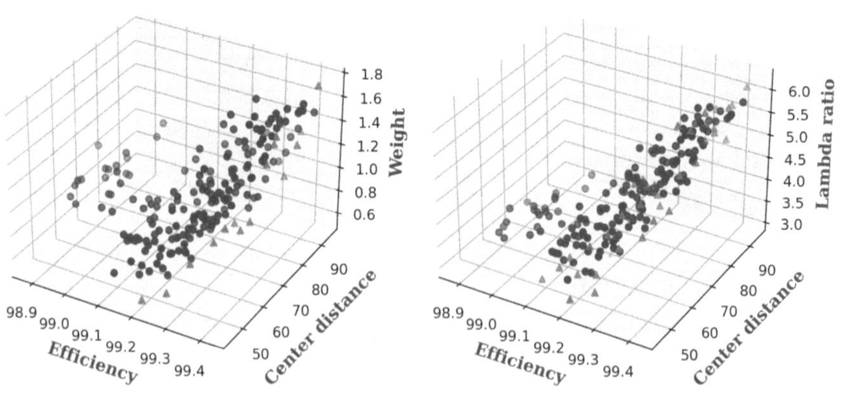

FIGURE 6.9 (a) and (b) 3D scatter plot of MOACO results for first speed gearing.

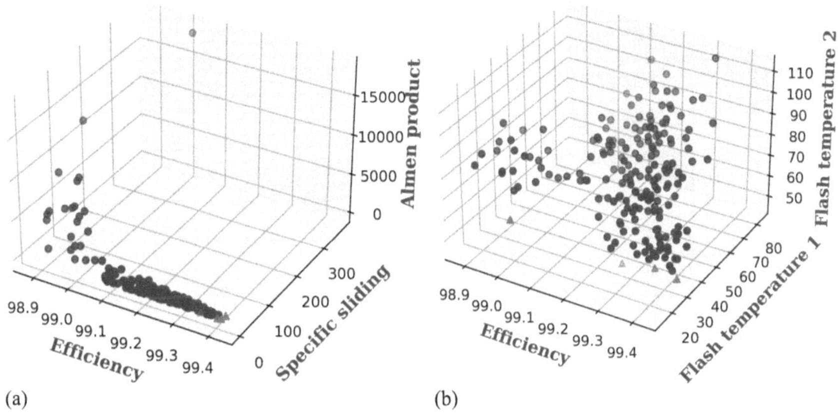

FIGURE 6.10 (a) and (b) 3D scatter plot of MOACO results for first speed gearing.

6.6.11.2 Results for the Second Speed Epicyclic Gearing

Table 6.6 shows a sample of the MOACO output for the second speed epicyclic gearing. Since constraints are specified in the form of ≥ 0 only, all the columns corresponding to the constraints shown in Table 6.6 indicate results obtained by modifying the constraint equations accordingly. Table 6.6 lists the values of the objective functions, constraints, and design variables from the generated solutions. The output represents a non-dominated Pareto-optimal solution set.

6.6.11.2.1 Comparison of Results

In Table 6.7, a comparison has been made between the results that are obtained with MOACO and the design of the existing epicyclic gearing. The solutions corresponding to the number of teeth of the planet gear, viz., 24, 24, and 25, were selected from the results. The values of design variables and objective functions corresponding to each solution are indicated in Table 6.7. The total weight of epicyclic gearing in kg, the efficiency in percentage, and the center distance in mm were taken for comparison.

The comparison shows that the results obtained from MOACO are an improvement over the existing design by optimizing the eight objective functions within the given constraints. The results were obtained with minimal computational effort and time. The three best results chosen from the solution set of MOACO output are analyzed for selecting a single best solution. The total weight of the existing second speed gearing was 0.49 kg, the center distance was 51.0 mm, and the efficiency was 99.09%.

Referring to Table 6.7, in the case of the first result, by selecting the number of teeth in the planet gear as 24, the weight has increased to 0.58 kg, the efficiency has increased to 99.25%, and the center distance has reduced to 49.20 mm. There is an increase of 18.37% in weight, an increase of 0.16% in efficiency, and a space reduction of 3.53%. Hence, there is an improvement in efficiency and center

TABLE 6.6
Sample of MOACO Output for Second Speed Gearing

Objectives								Constraints							Variables					
Center Distance	Weight	Efficiency	Specific Sliding	Almen Product	Flash Temperature		Lambda Ratio	Interference		Top Land Thickness		Contact Ratio	Bending Stress	Contact Stress	z_1	z_2	x_{n1}	x_{n2}	m_n	n_p
79.20	1.76	99.50	0.01	43.02	34.97	46.02	7.18	8.03	5.06	1.28	1.45	0.19	337.72	978.96	24	31	0.47	0.44	2.88	5
76.20	1.42	99.49	0.12	11.61	38.57	50.04	6.23	8.47	4.22	1.07	1.36	0.22	321.27	862.68	24	36	0.49	0.38	2.54	5
74.73	1.69	99.43	0.28	52.85	32.75	46.62	6.86	5.22	4.52	1.24	1.45	0.28	340.97	912.87	25	28	0.33	0.17	2.82	5
74.40	1.37	99.45	0.19	126.18	34.93	48.22	6.05	6.85	3.46	1.15	1.28	0.31	325.07	815.08	25	35	0.26	0.25	2.48	5
76.50	1.84	99.44	0.09	85.96	32.06	45.50	7.14	5.85	4.57	1.39	1.45	0.25	344.68	952.71	24	27	0.29	0.29	3.00	5
68.44	0.85	99.44	0.29	224.73	62.26	82.35	5.48	7.12	3.97	1.14	1.14	0.22	254.63	444.63	25	33	0.35	0.50	2.36	3
42.35	0.51	99.27	0.03	35.35	61.34	84.43	3.29	3.52	2.52	0.70	0.78	0.26	199.66	70.33	25	30	0.35	0.28	1.54	5
49.20	**0.58**	**99.25**	**0.06**	**80.11**	**53.51**	**76.35**	**3.38**	**4.25**	**1.91**	**0.73**	**0.90**	**0.36**	**225.48**	**107.92**	**24**	**36**	**0.24**	**0.07**	**1.64**	**5**
64.90	0.95	99.30	0.34	200.54	43.78	62.79	4.30	6.81	2.00	0.93	1.20	0.33	287.25	458.85	21	38	0.25	0.12	2.20	5
62.92	1.06	99.24	0.36	55.17	44.41	64.61	4.49	5.49	2.38	0.89	1.35	0.29	299.88	503.05	20	32	0.36	0.00	2.42	5
75.26	1.25	99.32	0.09	121.29	45.34	65.03	5.43	7.43	2.76	1.13	1.52	0.28	308.56	602.18	20	33	0.32	0.14	2.84	4
65.55	1.06	99.31	1.63	665.40	38.35	56.80	4.78	7.02	1.95	1.21	1.06	0.32	301.16	579.99	22	35	0.03	0.43	2.30	5
56.10	0.78	99.28	0.11	22.95	57.22	81.29	4.32	4.71	2.85	0.92	1.14	0.26	260.01	308.32	22	29	0.34	0.20	2.20	4
46.72	**0.48**	**99.31**	**0.20**	**141.60**	**63.23**	**83.85**	**3.09**	**5.41**	**2.01**	**0.66**	**0.77**	**0.28**	**170.36**	**11.95**	**24**	**40**	**0.34**	**0.36**	**1.46**	**5**
59.95	0.70	99.39	0.00	59.26	70.24	93.49	4.89	5.56	3.98	0.99	1.08	0.21	228.03	287.39	25	30	0.43	0.41	2.18	3
65.52	0.89	99.37	0.50	322.46	60.29	83.59	5.49	5.84	3.84	1.25	1.13	0.23	267.31	429.62	24	28	0.25	0.46	2.52	3
60.42	0.86	99.30	0.63	334.82	51.64	74.57	4.54	5.62	2.49	1.08	1.11	0.28	271.74	376.44	22	31	0.19	0.31	2.28	4
54.29	0.64	99.32	0.25	173.69	59.81	78.61	3.50	7.37	2.15	0.77	0.93	0.22	216.79	221.97	21	40	0.40	0.47	1.78	5
47.85	0.63	99.22	0.65	185.75	54.79	77.94	3.59	3.47	2.49	0.69	0.97	0.31	240.71	189.72	24	31	0.39	0.00	1.74	5
47.30	**0.54**	**99.29**	**0.05**	**30.65**	**66.23**	**91.71**	**3.66**	**3.85**	**2.77**	**0.78**	**0.87**	**0.27**	**200.23**	**59.27**	**25**	**30**	**0.34**	**0.26**	**1.72**	**4**

TABLE 6.7
Optimized Design Parameters for Second Speed Gearing Using MOACO

	Variables			Objective Functions		
	Module, m (mm)	Number of Teeth (Planet Gear) Z_1	Number of Teeth (Sun Gear) Z_2	Total Weight, W (kg)	Efficiency, η	Center Distance (mm)
Existing second speed epicyclic gearing design	2.0	19	32	0.49	99.09	51.0
MOACO design	1.64	24	36	0.58	99.25	49.20
	1.46	24	40	0.48	99.31	46.72
	1.72	25	30	0.54	99.29	47.30

distance, along with a marginal weight increase, in the solution obtained with MOACO.

In the case of the second result, by selecting the number of teeth in the planet gear as 24, the weight has reduced to 0.48 kg, the efficiency has increased to 99.31%, and the center distance has reduced to 46.72 mm. There is a reduction of 2.04% in weight, an increase of 0.22% in efficiency, and a space reduction of 8.39%. Hence, there is an improvement in all three objective functions in the solution obtained with MOACO.

In the case of the third result, by selecting the number of teeth in the planet gear as 25, the weight has increased to 0.54 kg, the efficiency has increased to 99.29%, and the center distance has reduced to 47.30 mm. There is an increase of 10.20% in weight, an increase of 0.20% in efficiency, and a space reduction of 7.25%. Hence, there is an improvement in efficiency and center distance, along with a marginal weight increase, in the solution obtained with MOACO.

6.6.11.2.2 *Visualization of Results*

The formation of the Pareto front from the non-dominated solutions containing eight objective functions is depicted in the subsequent graphs in combinations of three objective functions for every graph. The graphs show the 3D scatter plot of the solution set generated by MOACO in various combinations of three objective functions.

Even though the graphs help in better visualization of the non-dominated Pareto-optimal solution set, one should be cautious of the fact that all the eight objective functions put together form a single set of Pareto-optimal solutions that cannot be visualized in totality in graphical format. Hence, for the sake of decision-making, the solutions better suited for the individual requirements of the user/decision maker must be given priority in the choice of objective function combinations taken for visualization.

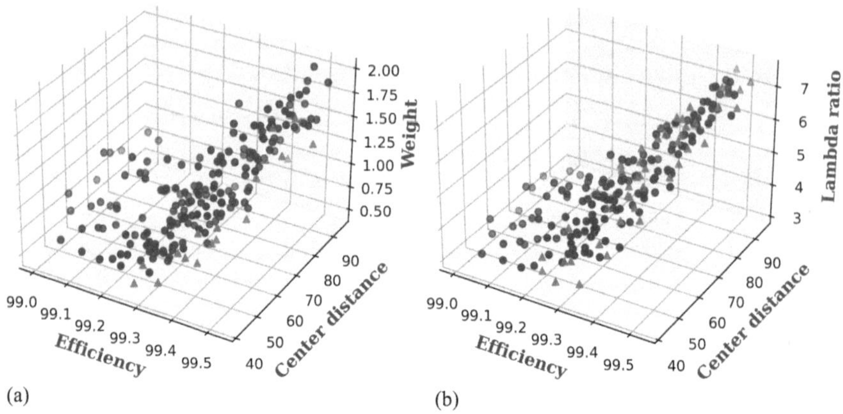

FIGURE 6.11 (a) and (b) 3D scatter plot of MOACO results for second speed gearing.

As a result of optimization, a Pareto-optimal front is formed in the objective space of the solutions. In the scatter plot, the circular points represent non-dominated solutions with respect to all eight objectives taken together, whereas the triangular points represent a non-dominated front formed with respect to the three objectives that are taken for visualization. It can be observed that the distribution of solutions along the front is uniform, which means the MOACO algorithm offers a high-quality solution set.

Referring to Figure 6.11(a), the solutions close to the upper right corner have higher total weight, higher center distance, and higher efficiency, which means they have better efficiency but are heavier and less compact. On the contrary, the solutions near the lower left corner present better compactness and better total weight but lesser efficiency. The solutions lying in the lower right corner have the best value in terms of higher efficiency, lesser weight, and better compactness.

Referring to Figure 6.11(b), the solutions close to the upper right corner offer higher efficiency, better compactness, and a higher lambda ratio for the sun gear/planet gear engagement.

Referring to Figure 6.12(a), the solutions close to the lower right corner offer higher efficiency, lower value of specific sliding, and lower value of Almen product at SAP and at EAP of the gear tooth profile engagement during running.

Referring to Figure 6.12(b), the solutions close to the lower right corner offer higher efficiency and lower flash temperatures that are balanced at SAP and EAP of the gear tooth profile engagement during running.

6.6.11.3 Selection of Best Solution

Comparing the results of Tables 6.5 and 6.7, since the center distance of first speed and second speed gearing must be the same, the selection of a single best acceptable solution is based on satisfying the above criterion. Hence, the combination of first speed and second speed gearing listed in Table 6.8 is selected as an acceptable solution.

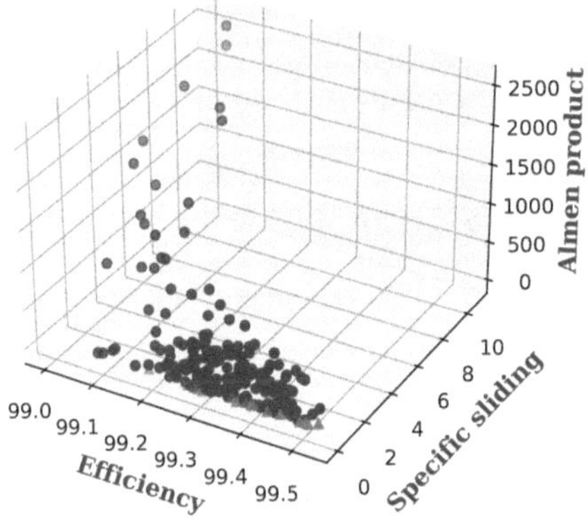

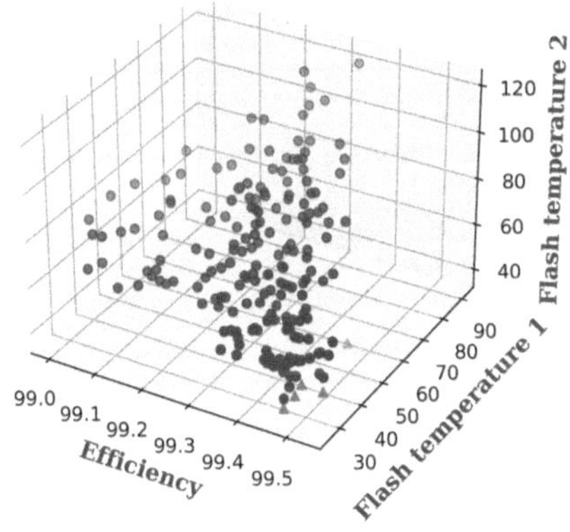

FIGURE 6.12 (a) and (b) 3D scatter plot of MOACO results for second speed gearing.

TABLE 6.8
Acceptable Optimized Design Parameters for Electric Vehicle Epicyclic Gearing

	Variables			Objective Functions		
	Module, m (mm)	Number of Teeth (Planet Gear) Z_1	Number of Teeth (Sun Gear) Z_2	Total Weight, W (kg)	Efficiency, η	Center Distance (mm)
MOACO design for first speed gearing	1.72	20	37	0.56	99.25	49.02
MOACO design for second speed gearing	1.64	24	36	0.58	99.25	49.20

From the above analysis, we can conclude that the solution in Table 6.8, obtained through the design optimization of the electric vehicle 2-speed epicyclic gearing using MOACO, can be considered the best solution.

NOTES

1 For more information on Pareto-optimality, refer to Section 3.4.
2 Refer Table 1.1.
3 Refer to Section 3.6.5.
4 Refer to Section 3.6.8 and onward.
5 Refer to Section 2.9.

REFERENCES

Alaya, I., et al., "Ant Colony Optimization for Multi-objective Optimization Problems," *19th IEEE International Conference on Tools with Artificial Intelligence (ICTAI)*, October 2007, Patras, Greece. pp. 450–457.

Deb, K., et al., 2002, "A Fast and Elitist Multiobjective Genetic Algorithm: NSGA-II," *IEEE Transactions on Evolutionary Computation*, 6, 2, 182–197. https://doi.org/10.1109/4235.996017

Dorigo, M., 1992, "Optimization, Learning and Natural Algorithms," Ph.D. Thesis, Politecnico di Milano, Italy.

Dorigo, M. and Di Caro, G., 1999, "Ant Colony Optimization: A New Meta-Heuristic," *IEEE*, 2, 1477. https://doi.org/10.1109/CEC.1999.782657

Han, J.-O., Jeong, W.-H., Lee, J.-S., and Oh, S.-H., 2021, "The Structure and Optimal Gear Tooth Profile Design of Two-Speed Transmission for Electric Vehicles," *Energies*, 14, 3736. https://doi.org/10.3390/en14133736. Used under the terms and conditions of the Creative Commons Attribution (CC BY) license (https://creativecommons.org/licenses/by/4.0/)

Jin, Y., et al., 2001, "Dynamic Weighted Aggregation for evolutionary multi-objective optimization: why does it work and how," *Proceedings of the 3rd Annual Conference on Genetic and Evolutionary Computation (GECCO'01)*. Morgan Kaufmann Publishers Inc., San Francisco, CA, USA, 1042–1049.

Krishnan, G., 2024, OPTIMO software – Algorithms for Multi-Objective optimization. https://github.com/epicyclic-book/optimo/. Used under the terms and conditions of the MIT license.

Lee, T.-W. and Hong, D.-K., 2021, "Electrical and Mechanical Characteristics of a High-Speed Motor for Electric Turbochargers in Relation to Eccentricity," *Energies*, 14, 3340. https://doi.org/10.3390/en14113340. Used under the terms and conditions of the Creative Commons Attribution (CC BY) license (https://creativecommons.org/licenses/by/4.0/)

Liu, L., et al., 2014, "Ant Colony Optimization Algorithm for Continuous Domains Based on Position Distribution Model of Ant Colony Foraging," *The Scientific World Journal*, 428539, 9. https://doi.org/10.1155/2014/428539

7 Design Optimization of Wolfrom Epicyclic Gearing for Robotic Actuator Using Multi-Objective Optimization Techniques

Chandrasekaran G

7.1 INTRODUCTION

Human–robot interaction (HRI) devices are becoming increasingly customary to assist humans, both in industrial and clinical environments. Traditional robotic gearbox technologies, like cycloid reducers or Harmonic Drives® (HD), perform well to solve the need for extreme precision and high speed of industrial robots, but when used in HRI devices, they are usually forced above their limits, ultimately restraining the performance and the widespread use of such devices. A closer analysis of the specific needs behind the actuation for human limb assistance and replication highlights the differences with traditional robotic applications. Matching those needs with the main characteristics of different robotic gearboxes displays the limitations of traditional robotic transmissions, but it also unveils the potential of a virtually unexploited kind of planetary gear train (Garcia et al., 2019).

Industrial robots have an extreme need for high precision, required to be able to perform their tasks. This need has been central to the current usage ranking of robotic gearbox technologies. Industrial robots have an extreme need for high precision, required to be able to perform their tasks. This need has been central to the current usage ranking of robotic gearbox technologies (Garcia et al., 2019).

One of the most used technologies in robotics is the HD, a system that takes advantage of the ability of strain wave gears with multiple teeth contact to provide high gear ratios in a very compact shape and with extreme positioning accuracy. Cycloidal reducers are based on the same fundamental principle as the strain wave gears, the small-tooth-difference, internal contact type gear train. They use rolling

DOI: 10.1201/9781003144465-7

contacts instead of teeth to transfer torque and speed. Planetary gear trains can achieve very high efficiencies, which are not strongly dependent on the operating conditions, and their ability to split power between the different planets makes them very compact. Wolfrom Planetary Gear Train (PGTs) could make high gear ratios compatible with maintaining good efficiencies over a very broad range of operating conditions, resulting in a transmission extremely well suited for HRI actuation (Garcia et al., 2019).

7.2 OVERVIEW OF WOLFROM GEARBOX

In Wolfrom PGTs (refer to Figure 7.1), the two meshing contacts of the (compound) planet gearwheels with both the grounded and the output annular rings are subjected simultaneously to high torque levels and high relative speeds (Garcia et al., 2019).

7.3 CASE STUDY: DESIGN OPTIMIZATION OF A WOLFROM GEARBOX HAVING HIGH BACK DRIVABILITY USED IN ROBOTIC ACTUATORS

For safety purposes, cooperative robots are installed with an actuator composed of a low-power servo motor, a reduction gearbox, and a torque sensor. When cooperative robots make contact with humans or the environment, they must detect the contact force with a force sensor, a contact sensor, or a joint torque sensor. Equipping these sensors increases the cost and size of the application but can be avoided under sufficient back drivability of the actuator (Hiroshi Matsuki et al., 2019).

The design optimization is attempted for the Wolfrom epicyclic gearing used in the robotic actuator with high back drivability, as shown in Figure 7.1 with specifications as given in Table 7.1.

The Wolfrom gearing (Figure 7.2) has an arrangement of two stage gear reduction. In the first-stage gearing, input from the electric motor is given to the sun gear S, which in turn rotates a set of three planet gears P_1. The ring gear R_1

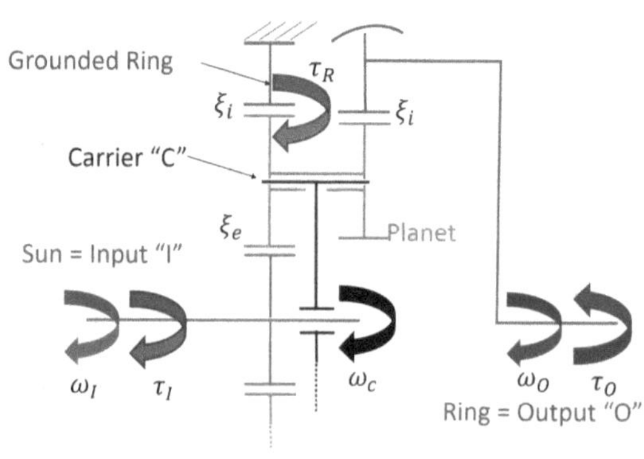

FIGURE 7.1 Wolfrom planetary gearing (Garcia et al., 2019).

TABLE 7.1
Wolfrom Gearbox Specifications

	First-Stage Gearing			Second-Stage Gearing	
	Sun Gear	**Planet Gear**	**Ring Gear**	**Planet Gear**	**Ring Gear**
Module (mm)		0.8		0.8467	
Pressure angle (degree)		20		20	
Number of teeth (Z)	12	39	90	32	81
Helix angle (degree)		0		0	
Profile shift coefficient (X_n)	0.476	0.762	2.0	0.536	1.21

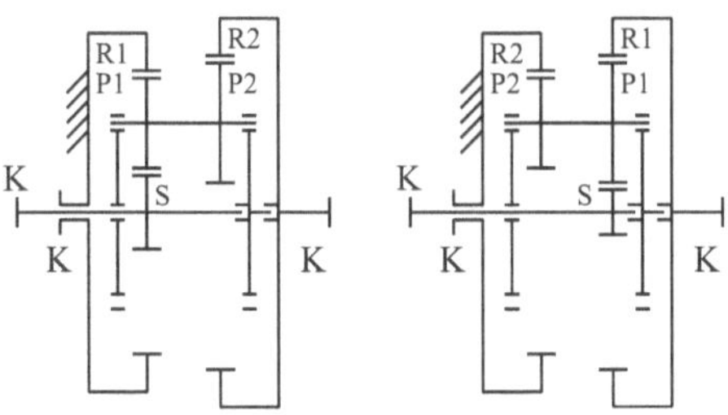

FIGURE 7.2 Wolfrom epicyclic gearing stages (Matsuki et al., 2019).

is fixed, and thereby the planet carrier rotates. Hence, this stage of gearing comes under the first type of planetary gearing.[1] In the second-stage gearing, the other end of the planet carrier holds another set of three planet gears, P_2, which engage with the second-stage ring gear, R_2. This ring gear shaft is the output shaft, which supplies the required output torque. Hence, the second stage gearing comes under the second type of solar gearing.[1] It should be noted that for the second-stage gearing, the sun gear is not present, and for the purpose of visualization, it should be taken as an imaginary fixed sun gear (Figure 7.3).

The Wolfrom gearbox has the following specifications:

To find the speed of individual elements of the existing planetary gear train (Figure 7.1),

For the first-stage reduction, let,

Number of teeth on planet gears, $z_1 = 39$
Number of teeth on sun wheel, $z_2 = 12$
Number of teeth on annular ring gear, $z_3 = 90$

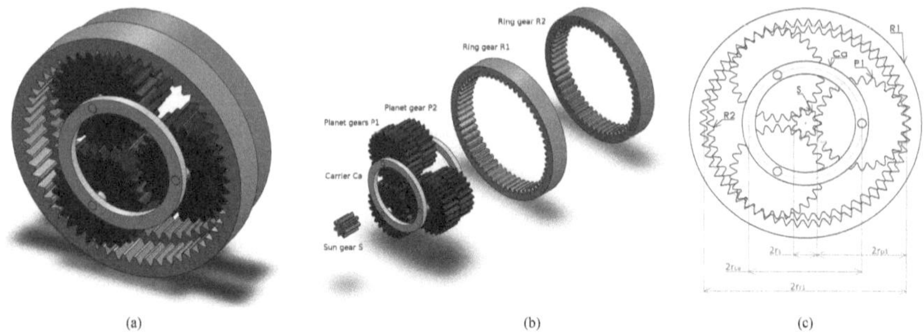

FIGURE 7.3 Illustrative model of the type-3K compound planetary gearbox: (a) assembled model, (b) disassembled model, and (c) kinematics of gears (Matsuki et al., 2019).

For the second-stage reduction, let,

Number of teeth on planet gears, $z_1 = 32$
Number of teeth on the sun wheel, $z_2 = 0$ (since the sun gear is not present in the second stage)
Number of teeth on the annular ring gear, $z_3 = 81$

The last row of Table 7.2 provides us with the equation for the speed of individual elements.

For the present case of Wolfrom gearing,

Input speed of motor sun gear = 3000 rpm (CW-assumed)
Motor power input = 400 W

that is, $y + x = 3000$

Since the annulus of stage 1 is locked,

$$y - \frac{12}{90}x = 0$$

Hence,

$$x = 2647.84$$

$$y = 352.16$$

$$\text{Speed of planets} = y - \frac{12}{39}x = 352.16 - \frac{12}{39} \times 2647.84 = -462.56 \text{ rpm}(\text{ACW})$$

Output speed or speed of the second-stage ring gear or annular wheel,

$$y - \frac{12}{39} \times \frac{32}{81} x = 31.17 \text{ rpm}(\text{CW})$$

Hence, the gear ratio effectively achieved is 96.2:1.

TABLE 7.2
Speed Calculation for Wolfrom Epicyclic Gearing Elements

Operation	Spider Arm L	Sun Wheel, Z_s	First-Stage Planet Wheel, Z_{p1}	First-Stage Annular Wheel, Z_{a1}	Second-Stage Planet Wheel, Z_{p2}	Second-Stage Annular Wheel, Z_{a2}
1. Arm L is locked and sun wheel S is given +1 revolution.	0	+1	$-\frac{Z_{s_1}}{Z_{p_1}}=-\frac{12}{39}$	$-\frac{Z_{s_1}}{Z_{a_1}}=-\frac{12}{90}$	$-\frac{Z_{s_1}}{Z_{p_1}}=-\frac{12}{39}$	$-\frac{12}{39}\times\frac{32}{81}$
2. Multiply by x	0	x	$-\frac{Z_{s_1}}{Z_{p_1}}x=-\frac{12}{39}x$	$-\frac{Z_{s_1}}{Z_{a_1}}x=-\frac{12}{90}x$	$-\frac{Z_{s_1}}{Z_{p_1}}x=-\frac{12}{39}x$	$-\frac{12}{39}\times\frac{32}{81}x$
3. Add y to all columns	y	$y+x$	$y-\frac{Z_{s_1}}{Z_{p_1}}x=y-\frac{12}{39}x$	$y-\frac{Z_{s_1}}{Z_{a_1}}x=y-\frac{12}{90}x$	$y-\frac{Z_{s_1}}{Z_{p_1}}x=y-\frac{12}{39}x$	$y-\frac{12}{39}\times\frac{32}{81}x$

7.4 MULTI-OBJECTIVE DESIGN OPTIMIZATION OF FIRST-STAGE EPICYCLIC GEARING

The first-stage planetary gearing receives input power from the drive motor, which runs at 3000 rpm. The sun gear is connected to the drive motor and hence runs at the same speed as the motor. The ring gear is fixed, and hence the spider arm rotates and transfers the torque to the second-stage gearing.

7.4.1 Calculation of Gear Parameters

Gear parameters are calculated using the procedure explained in Chapter 3 for Power Take-Off (PTO) gearbox.[2]

7.4.2 Design Space

The existing gear design problem for the first-stage planetary gearing can be reduced to an eight-dimensional design problem where the eight free parameters are the number of teeth on the planet pinion z_1, number of teeth on the sun gear z_2, number of teeth on the ring gear z_3, the module m_n, the three profile shift coefficients x_{n1}, x_{n2}, and x_{n3} of the planet gear, the sun gear and the ring gear, respectively, and the number of planets n_p. The inequality constraints, which bound this design space, are the bending stress, Hertz stress, contact ratio, top land thickness, operating interference of gears, and involute and tip interference of internal gears. The gear pair ratios and conditions for proper assembly of planetary gearing form equality constraints.

7.4.3 Objective Functions and Constraints

Fourteen objective functions for minimization/maximization with an array of inequality and equality constraints have been considered, and the problem can be stated as follows.

The following objective functions are considered for the design optimization of the planetary gear train:

$$f_{obj} = \begin{cases} Min.center\ distance\ between\ planetary\ gear\ pair \\ Min.weight\ of\ the\ planetary\ gear\ train \\ Max.efficiency\ of\ planetary\ gear\ train \\ Min.slide\ roll\ ratio \\ Balancing\ the\ specific\ sliding \\ Balancing\ the\ Almen\ product \\ Balancing\ the\ flash\ temperature \\ Balancing\ the\ lambda\ ratio \end{cases}$$

The independent design variables of the planetary gear train include a module, the number of teeth of the pinion and the three profile shift coefficients x_{n1}, x_{n2}, and x_{n3} of the planet gear, the sun gear, and the ring gear, respectively. The vector of design variables is given below:

$$x(i) = \begin{cases} planet\ gear\ no.of teeth, Z_1 \\ sun\ gear\ no.of teeth, Z_2 \\ ring\ gear\ no.of teeth, Z_3 \\ planet\ gear\ profile\ shift\ coefficient, x_{n_1} \\ sun\ gear\ profile\ shift\ coefficient, x_{n_2} \\ ring\ gear\ profile\ shift\ coefficient, x_{n_3} \\ normal\ module, m_n \\ number\ of\ planets, n_p \end{cases}$$

where $i = 1, 2, \ldots,$ the number of variables.

A number of constraints, which act as sub functions to restrict the objective function, are introduced into the objective function to be optimized and are used to provide suitable design choices. The very often encountered failure types in a transmission gear are tooth fracture, surface fatigue, abrasion, and plastic deformation. Thus, the following constraints are considered for the design optimization of the planetary gear train:

$$g(j) = \begin{cases} Bending\ stress\ constraint \\ Surface\ contact\ compressive\ stress\ constraint \\ Contact\ ratio\ constraint \\ Top land\ thickness\ constraint \\ Operating\ interference\ of\ planet\ gear\ constraint \\ Operating\ interference\ of\ sun gear\ constraint \\ Operating\ interference\ of\ ring\ gear\ constraint \\ Involute\ interference\ of\ internal\ gear\ constraint \\ Tip interference\ of\ internal\ gear\ constraint \\ Proper\ assembly\ of\ planetary\ gearing\ constraint \end{cases}$$

where $j = 1, 2, \ldots,$ the number of constraints.

7.4.4 Establishing a Mathematical Model

To establish a mathematical model for the planetary gear train design optimization, mathematical equations for the design variables, objective functions, and necessary constraints are to be developed.

7.4.5 Objective Functions for the Epicyclic Gearing

Since the problem formulation is identical to that of the PTO gearbox in Chapter 3, the objective functions are formulated based on the principles and procedures explained in Section 3.6.9 for the PTO gearbox.

7.4.6 Design Variables and Constraints

The independent design parameters of planetary gear drive include a module m_n, the number of teeth of planet gear z_1, the number of teeth of sun gear z_2, the number of teeth of ring gear z_3, the three profile shift coefficients x_{n1}, x_{n2}, and x_{n3}, and the number of planets n_p.

Therefore, the design variables are,

$$X = \left[z_1, z_2, z_3, x_{n1}, x_{n2}, x_{n3}, m_n, n_p \right]^T = \left[x_1, x_2, x_3, x_4, x_5, x_6, x_{7,} x_8 \right]^T$$

A number of constraints, which are introduced into the objective function to be optimized, are used to provide suitable design choices. The constraints are formulated based on the principles and procedures explained in Section 3.6.14 for the PTO gearbox.

7.4.7 Design Data for Optimization

Following are the input parameters considered for the planetary gear design based on the existing gear design as in Table 7.3.

7.4.8 Methodology of Optimization of Wolfrom First-Stage Epicyclic Gearing

This section describes how the three algorithms, namely, multi-objective particle swarm optimization (MOPSO), multi-objective simulated annealing (MOSA), and multi-objective ant colony optimization (MOACO), are configured for

TABLE 7.3
Wolfrom Epicyclic Gearing Parameters

Data	Gear
Density of the material, ρ	7.86×10^6 kg/mm^3
Material of sun gear, planet gear, and ring gear	Steel SCM 440
Design bending stress, σ_b	$420 \mathrm{N} / \mathrm{mm}^2$
Design crushing stress, σ_c	$2000 \mathrm{N} / \mathrm{mm}^2$
Input speed, N	3000 rpm
Power transmitted, P	400 W
Young's modulus of the material (E)	$2.0 \times 10^5 \mathrm{kg} / \mathrm{mm}^2$
Pressure angle, α_n	20°

solving the optimization problem of Wolfrom's first-stage epicyclic gearing. The results obtained from the three multi-objective optimization techniques are analyzed, and comparisons are made on the performance of these three techniques.

This book utilizes the OPTIMO software package (Gokul Krishnan, 2024), which provides efficient implementations of the algorithms in Python along with the complete problem set corresponding to this case study. Chapter 2 provides detailed instructions on how to obtain and use the software.[3]

7.4.8.1 Variable Bounds

In order to determine the values of design variables, such as module, number of teeth, face width, and helix angle, through the optimization procedure, the upper and lower bounds of the design variables are to be established first.

The following variables are considered, and their lower and upper bounds are fixed. Further, the type of variable – whether discrete (integer) or real-valued (float) – is also specified.

Data for the first-stage epicyclic gearing

Number of planet gear teeth, z_1: 30 to 55 (integer)
Number of sun gear teeth, z_2: 10 to 25 (integer)
Number of sun gear teeth, z_3: 80 to 100 (integer)
Profile shift coefficient for planet gear, x_{n1}: 0.0 to 2.0 (float)
Profile shift coefficient for sun gear, x_{n2}: 0.0 to 2.0 (float)
Profile shift coefficient for sun gear, x_{n3}: 1.0 to 3.0 (float)
Module, m_n: 0.5 to 2.0 mm (float)
Number of planets, n_p: 2 to 8 (integer)

7.4.8.2 Input Parameters

For this model, the most important data inputs are the gear design parameters of the two mating gears. The decision maker is required to make this decision based on experience and design requirements.

The problem sets provided in this book contain Python code corresponding to the objective functions and design constraints formulated in Sections 7.4.4 and 7.4.5. The upper and lower bounds of design variables are also specified in the same file(s). The seed for the random number generator is fixed at 0.123 for reproducibility of results.

The problem definition corresponding to Chapter 7, first-stage planetary gearing, can be found in *problemsets/problem_ch7a.py* file within the software repository. The optimization run is started by entering the command specified in Listing 7.1 from the repository root.

LISTING 7.1 COMMAND TO EXECUTE THE ALGORITHM RUN

```
# macOS / Linux
python3 run.py -a pso -p ch7a
```

```
python3 run.py -a sa  -p ch7a
python3 run.py -a aco -p ch7a
# Windows
python.exe run.py -a pso -p ch7a
python.exe run.py -a sa  -p ch7a
python.exe run.py -a aco -p ch7a
```

7.4.9 Interpretation of MOPSO Results

In this section, the results obtained from MOPSO are discussed and compared with the design of existing Wolfrom first-stage planetary gearing. For detailed information on the interpretation of the tables and plots in the subsequent sections, refer to Section 3.6.17.

Table 7.4 shows a sample of the MOPSO output. Since constraints are specified in the form of ≥ 0 only, all the columns corresponding to the constraints shown in Table 7.4 indicate results obtained by modifying the constraint equations accordingly. Table 7.4 lists the values of the objective functions, constraints, and design variables from the generated solutions. The output represents a non-dominated Pareto-optimal solution set.

7.4.9.1 Comparison of Results

In Table 7.5, a comparison has been made between the results that are obtained with MOPSO and the design of the existing planetary gearing. The solutions correspond to the number of teeth of the planet gear, namely, 51 and 51, which were selected from the results. The values of design variables and objective functions corresponding to each solution are indicated in Table 7.5. The total weight of planetary gearing in kg, the efficiency in percentage, and the center distance in mm were taken for comparison.

The two best results chosen from the solution set of MOPSO output are analyzed for selecting a single best solution. The total weight of the existing planetary gearing was 0.112 kg, the center distance was 20.4 mm, and the efficiency was 99.51%.

Referring to Table 7.5, in the case of the first result, by selecting the number of teeth in the planet gear as 51, the weight has reduced to 0.05 kg, the efficiency has increased to 99.56%, and the center distance has reduced to 17.75 mm. There is a reduction of 55.36% in weight, an increase of 0.05% in efficiency, and a space reduction of 12.99%. Hence, there is an improvement in all three objective functions in the solution obtained with MOPSO.

In the case of the second result, by selecting the number of teeth in the planet gear as 51, the weight has reduced to 0.06 kg, the efficiency has increased to 99.63%, and the center distance has reduced to 20.35 mm. There is a reduction of 46.43% in weight, an increase of 0.12% in efficiency, and a space reduction of 0.25%. Hence, there is an improvement in all three objective functions in the solution obtained with MOPSO.

TABLE 7.4
Sample of MOPSO Output for First-Stage Gearing

				Objectives									
Center distance	Weight	Efficiency	Slide roll ratio	Specific Sliding		Almen Product			Flash Temperature			Lambda Ratio	
66.00	0.67	99.53	0.13	0.32	0.33	5.03	2653.51	0.60	1.09	0.25	0.53	10.23	19.65
68.24	0.82	99.80	0.46	1.68	0.54	164.21	205.13	0.02	0.31	0.23	0.31	18.91	17.43
51.00	0.57	99.24	0.28	1.10	0.13	91.48	563.00	0.11	0.72	0.36	0.44	9.81	13.90
70.00	1.00	99.83	0.28	4.26	0.77	295.87	216.17	0.46	0.36	0.22	0.32	14.78	21.43
56.33	0.60	99.59	0.36	0.12	0.39	4.13	495.80	0.01	0.71	0.44	0.55	12.08	13.70
70.00	0.70	99.50	0.28	0.85	0.25	114.63	882.91	0.38	0.81	0.45	0.60	12.99	18.57
52.06	0.51	99.66	0.43	0.53	0.49	53.92	492.22	0.00	0.50	0.28	0.40	12.09	11.57
68.05	0.80	99.67	0.42	0.17	0.41	30.43	393.51	0.04	0.52	0.38	0.47	16.37	16.19
80.00	1.28	99.62	0.49	0.15	0.21	18.53	436.84	0.06	0.25	0.16	0.21	23.44	19.92
55.19	0.55	99.59	0.41	0.10	0.40	23.94	546.24	0.04	0.67	0.40	0.53	12.49	12.65
41.81	0.28	99.19	0.57	1.12	0.35	128.49	546.19	0.61	1.31	0.93	1.15	10.71	6.70
20.00	0.06	98.62	0.49	0.33	0.11	21.58	1331.00	0.44	1.74	0.70	1.18	3.71	3.29
17.75	**0.05**	**99.56**	**0.44**	**2.54**	**0.98**	**220.20**	**274.16**	**0.66**	**1.76**	**1.07**	**1.67**	**3.06**	**2.93**
47.02	0.46	99.21	0.69	1.37	0.33	114.23	216.30	0.40	0.78	1.03	1.13	15.63	6.41
38.40	0.21	99.67	0.57	0.06	1.04	0.20	29.12	0.54	1.22	1.58	1.67	9.94	5.44
30.18	0.13	98.91	0.57	0.35	0.14	46.87	1082.84	0.22	1.57	1.01	1.45	6.43	4.66
25.86	0.11	99.10	0.48	1.18	0.60	138.73	1282.19	0.94	2.38	0.90	1.52	5.19	4.09
52.00	0.52	99.13	0.65	1.62	0.61	170.49	413.53	0.54	1.08	1.06	1.22	16.15	7.10
78.00	0.67	99.44	0.51	1.64	3.08	213.52	330.65	0.01	0.49	0.74	0.74	21.31	15.76
20.35	**0.06**	**99.63**	**0.28**	**1.99**	**0.79**	**243.42**	**181.61**	**1.97**	**2.66**	**2.10**	**2.54**	**2.93**	**3.98**

(Continued)

TABLE 7.4 (CONTINUED)
Sample of MOPSO Output for First-Stage Gearing

Constraints														
Interference		Top Land Thickness			Contact Ratio		Bending Stress		Contact Stress		Involute interference		Tip interference	Proper assembly
1.07	15.96	1.20	0.23	1.19	0.03	0.38	396.73	397.49	1732.20	1947.89	0.27	0.40	0.34	0.00
1.97	13.90	0.64	1.22	1.10	0.12	0.53	398.03	399.27	1840.72	1968.32	1.63	0.53	0.40	0.00
1.57	10.21	0.93	0.10	1.16	0.05	0.29	396.97	398.29	1767.76	1956.56	2.36	0.35	0.23	0.00
0.65	16.21	0.88	1.15	1.14	0.17	0.20	398.53	399.17	1833.24	1966.41	1.11	0.58	0.47	0.00
1.57	10.78	0.99	0.53	1.08	0.12	0.32	396.97	398.57	1765.63	1951.63	2.24	0.36	0.23	0.00
1.04	13.62	1.12	0.67	1.42	0.22	0.24	397.13	398.38	1752.66	1953.41	2.11	0.37	0.26	0.00
1.32	8.33	0.75	0.61	0.86	0.29	0.68	397.11	398.86	1773.62	1953.39	1.15	0.38	0.26	0.00
2.00	11.95	1.07	0.77	1.23	0.19	0.36	397.79	399.09	1805.86	1959.76	3.14	0.37	0.22	0.00
2.84	11.84	1.10	0.90	1.61	0.29	0.78	398.91	399.63	1870.12	1975.79	2.93	0.36	0.23	0.00
1.60	9.66	0.91	0.60	1.04	0.18	0.48	396.77	398.65	1760.55	1951.09	1.79	0.36	0.24	0.00
2.60	5.08	0.87	0.28	0.88	0.12	0.74	390.86	397.51	1622.70	1921.35	2.56	0.25	0.10	0.00
0.66	2.51	0.30	0.19	0.43	0.35	1.15	371.84	390.47	1299.78	1879.80	0.27	0.33	0.24	0.00
0.30	**3.05**	**0.23**	**0.28**	**0.30**	**0.29**	**0.55**	**368.61**	**387.69**	**1242.11**	**1839.72**	**0.22**	**0.53**	**0.41**	**0.00**
4.13	3.89	1.07	0.42	1.31	0.13	0.36	394.96	399.07	1716.14	1942.25	6.23	0.20	0.01	0.00
2.31	5.87	0.61	0.52	0.57	0.10	0.30	387.61	396.64	1602.35	1897.74	3.60	0.37	0.11	0.00
0.87	2.93	0.51	0.42	0.83	0.41	0.79	383.86	395.55	1416.08	1902.69	0.84	0.30	0.20	0.00
1.34	3.90	0.56	0.12	0.49	0.06	0.95	377.76	392.21	1397.87	1864.10	0.37	0.27	0.18	0.00
4.36	4.80	1.30	0.32	1.14	0.08	0.60	394.27	398.80	1695.33	1930.13	4.91	0.15	0.01	0.00
2.34	13.64	0.77	1.29	1.11	0.21	0.11	397.04	399.04	1797.65	1951.38	5.06	0.85	0.51	0.00
0.30	**4.81**	**0.30**	**0.32**	**0.34**	**0.09**	**0.02**	**366.00**	**380.94**	**1207.60**	**1827.99**	**0.75**	**0.53**	**0.37**	**0.00**

Variables							
Z_1	Z_2	Z_3	x_{n1}	x_{n2}	x_{n3}	m_n	n_p
55	11	85	0.52	0.66	1.79	2.00	2
55	23	89	1.61	0.07	2.54	1.75	4
55	15	93	0.24	0.87	3.00	1.46	4
55	15	81	1.18	0.09	1.54	2.00	4
49	16	86	0.53	0.58	1.92	1.73	3
55	15	91	0.40	0.38	2.26	2.00	2
54	21	95	0.45	0.30	1.36	1.39	4
51	19	92	0.56	0.47	1.92	1.94	3
55	25	100	0.47	0.39	2.90	2.00	5
49	18	87	0.51	0.49	1.84	1.65	3
39	22	84	0.11	0.92	2.04	1.37	2
55	25	98	0.00	0.43	3.00	0.50	3
51	**20**	**82**	**0.82**	**0.00**	**1.00**	**0.50**	**3**
30	25	80	0.04	0.95	2.70	1.71	3
44	25	97	0.83	0.70	1.00	1.11	2
43	24	82	0.00	0.17	3.00	0.90	2
41	18	80	0.18	0.94	1.02	0.88	2
30	22	80	0.00	1.03	1.31	2.00	2
53	25	83	1.28	0.00	1.00	2.00	2
51	**14**	**80**	**1.13**	**0.30**	**1.53**	**0.63**	**2**

TABLE 7.5
Optimized Design Parameters for First-Stage Gearing Using MOPSO

	Variables				Objective Functions		
	Module, m (mm)	Number of Teeth (Planet Gear) Z_1	Number of Teeth (Sun Gear) Z_2	Number of Teeth (Ring Gear) Z_3	Total Weight, W (kg)	Efficiency, η	Center Distance (mm)
Existing first-stage planetary gearing design	0.8	39	12	90	0.112	99.51	20.4
MOPSO design	0.5	51	20	82	0.05	99.56	17.75
	0.63	51	14	80	0.06	99.63	20.35

7.4.9.2 Visualization of Results

The formation of the Pareto front from the non-dominated solutions containing 14 objective functions is depicted in the subsequent graphs in combinations of three objective functions for every graph. The graphs show the 3D scatter plot of the solution set generated by MOPSO in various combinations of three objective functions.

Referring to Figure 7.4(a), the solutions close to the upper right corner have higher total weight, higher center distance, and higher efficiency, which means they have better efficiency but are heavier and less compact. On the contrary, the solutions near the lower left corner present better compactness and better total weight but lesser efficiency. The solutions lying in the lower right corner have the best value in terms of higher efficiency, lesser weight, and better compactness.

Referring to Figure 7.4(b), the solutions close to the lower right corner offer higher efficiency, better compactness, and lesser slide roll ratio values.

Referring to Figure 7.5(a), the solutions close to the lower right corner offer higher efficiency and lesser specific sliding values. A lower value of specific sliding indicates a balanced specific sliding at start of active profile (SAP) and end of active profile (EAP) of the gear tooth profile engagement during running.

Referring to Figure 7.5(b), the solutions close to the lower right corner offer higher efficiency and lower Almen product values at SAP and EAP of the gear tooth profile engagement during running.

Referring to Figure 7.6(a) and (b), the solutions close to the lower right corner offer higher efficiency and lower flash temperatures that are balanced at SAP and EAP of the gear tooth profile engagement during running.

Referring to Figure 7.7, the solutions close to the upper right corner offer higher efficiency and higher lambda ratio for the sun gear/planet gear engagement, and the planet gear/ring gear engagement, respectively.

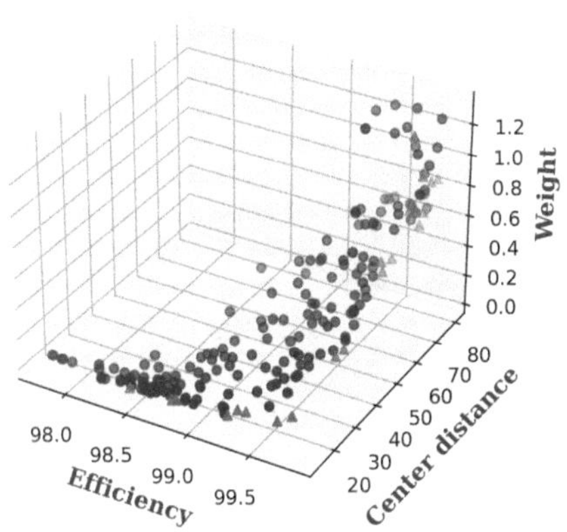

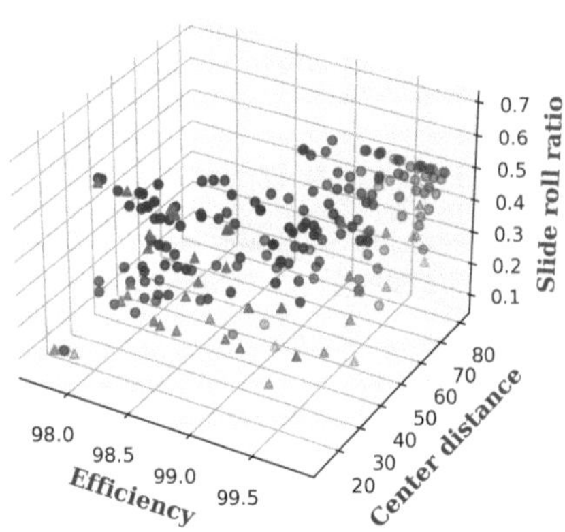

FIGURE 7.4 (a) and (b) 3D scatter plot of MOPSO results for first-stage gearing.

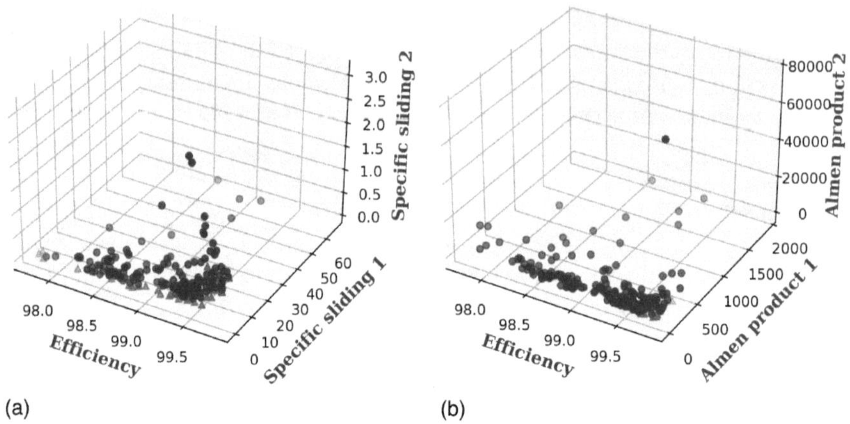

FIGURE 7.5 (a) and (b) 3D scatter plot of MOPSO results for first-stage gearing.

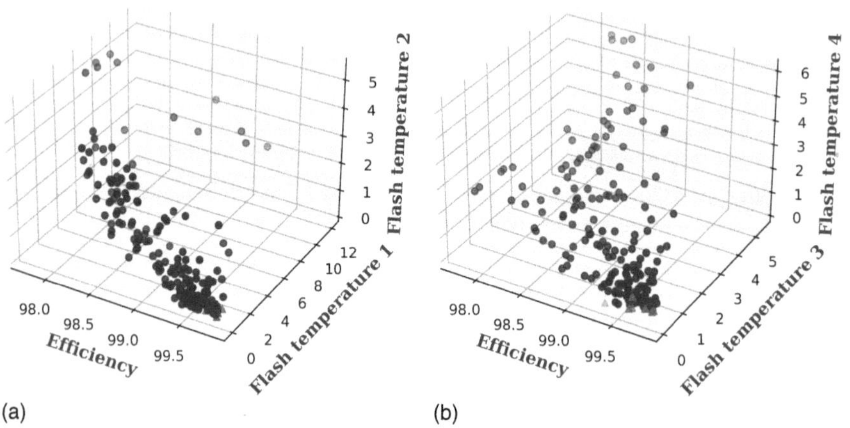

FIGURE 7.6 (a) and (b) 3D scatter plot of MOPSO results for first-stage gearing.

7.4.10 Interpretation of MOSA Results

In this section, the results obtained from MOSA are discussed and compared with the design of existing Wolfrom first-stage planetary gearing. For detailed information on the interpretation of the tables and plots in the subsequent sections, refer to Section 3.6.17.

Table 7.6 shows a sample of the MOSA output. Since constraints are specified in the form of ≥ 0 only, all the columns corresponding to the constraints shown in Table 7.6 indicate results obtained by modifying the constraint equations accordingly. Table 7.6 lists the values of the objective functions, constraints, and design variables from the generated solutions. The output represents a non-dominated Pareto-optimal solution set.

TABLE 7.6
Sample of MOSA Output for first-Stage Gearing

				Objectives										
Center Distance	Weight	Efficiency	Slide roll Ratio	Specific Sliding		Almen Product			Flash Temperature			Lambda Ratio		
41.30	0.38	99.45	0.26	0.07	0.20	21.80	667.84	0.19	0.80	0.38	0.52	7.49	11.32
34.65	0.47	99.41	0.18	0.94	0.19	68.12	499.71	0.43	0.53	0.27	0.36	5.88	10.44
36.03	0.41	99.40	0.37	0.04	0.27	7.95	484.89	0.01	0.52	0.27	0.37	7.59	8.76
57.58	0.45	99.38	0.46	0.13	0.36	1.81	1133.31	0.10	0.89	0.44	0.67	12.71	11.42
53.93	0.40	99.59	0.44	0.29	0.52	23.10	543.94	0.20	0.97	0.58	0.77	11.88	10.86
46.45	0.39	99.60	0.44	0.01	0.62	15.27	250.67	0.07	0.79	0.77	0.88	10.27	8.93
49.01	0.62	99.74	0.56	1.17	0.72	96.37	60.85	0.07	0.32	0.40	0.44	15.36	9.85
50.75	0.84	99.35	0.60	1.25	0.21	73.89	143.10	0.16	0.31	0.36	0.39	16.80	9.74
17.50	0.07	99.38	0.57	0.03	0.70	4.52	209.54	0.38	1.20	1.09	1.31	3.87	2.36
23.54	0.13	98.93	0.48	1.35	0.22	107.87	249.00	0.54	1.15	0.95	1.06	5.07	4.15
19.75	**0.05**	**99.61**	**0.50**	**0.79**	**0.83**	**103.51**	**66.86**	**0.40**	**1.55**	**0.80**	**1.23**	**4.14**	**3.37**
19.42	0.08	99.06	0.63	0.90	0.35	75.77	159.16	0.77	1.47	1.98	2.12	4.96	2.49
18.75	0.04	99.47	0.53	1.64	0.97	210.11	28.59	0.12	2.14	2.19	2.61	3.67	2.62
18.00	0.06	98.36	0.40	1.40	0.13	146.55	1420.24	1.05	2.56	0.73	1.25	3.01	3.27
14.50	0.05	99.21	0.66	0.24	0.80	12.21	129.20	0.69	1.49	2.08	2.25	3.71	1.56
21.07	0.09	99.05	0.56	2.11	0.58	142.68	350.04	0.22	1.64	2.22	2.52	4.15	2.75
16.25	0.05	99.26	0.56	1.67	0.53	142.74	246.18	0.01	1.90	2.83	3.11	3.20	2.09
33.40	0.19	99.48	0.67	0.18	0.93	30.98	127.63	0.57	1.36	2.51	2.61	9.52	3.81
14.75	0.06	98.93	0.65	0.91	0.31	66.42	143.98	0.85	1.60	2.18	2.34	3.77	1.75
17.50	**0.04**	**99.53**	**0.44**	**0.32**	**0.83**	**34.69**	**202.93**	**1.03**	**3.16**	**2.01**	**2.49**	**3.08**	**2.63**

(Continued)

TABLE 7.6 (CONTINUED)
Sample of MOSA Output for first-Stage Gearing

Constraints															
Interference		Top land Thickness			Contact Ratio		Bending Stress		Contact Stress		Involute Interference		Tip interference	Proper Assembly
0.94	9.60	0.69	0.40	0.92	0.05	0.28	395.59	397.38	1719.72	1947.93	1.12	0.41	0.31	0.00
0.47	8.43	0.57	0.34	0.82	0.07	0.08	397.09	397.99	1753.52	1955.34	0.96	0.41	0.32	0.00
0.91	6.37	0.61	0.33	0.73	0.19	0.48	396.80	398.53	1753.86	1952.69	1.02	0.35	0.24	0.00
1.54	7.45	0.95	0.59	1.05	0.33	0.81	395.52	398.36	1700.99	1941.15	1.09	0.30	0.20	0.00
1.86	8.79	0.88	0.50	0.90	0.19	0.63	394.58	397.87	1698.87	1934.68	1.88	0.35	0.21	0.00
1.33	7.13	0.78	0.49	0.84	0.24	0.15	395.52	398.26	1709.42	1933.65	3.14	0.32	0.13	0.00
2.20	8.67	0.54	0.96	0.84	0.08	0.26	397.51	399.30	1822.30	1960.90	3.16	0.45	0.24	0.00
3.46	5.63	0.99	0.31	1.24	0.14	0.24	398.30	399.57	1842.28	1967.97	6.30	0.25	0.04	0.00
0.77	2.18	0.27	0.21	0.30	0.27	0.66	378.45	394.03	1407.77	1866.20	0.95	0.31	0.13	0.00
1.25	3.39	0.45	0.07	0.47	0.09	0.29	387.71	395.68	1571.50	1910.68	2.19	0.27	0.09	0.00
0.86	**3.92**	**0.22**	**0.34**	**0.25**	**0.06**	**0.96**	**365.69**	**388.60**	**1380.04**	**1869.63**	**0.26**	**0.60**	**0.47**	**0.00**
1.51	2.25	0.40	0.17	0.44	0.10	0.20	376.60	394.67	1431.15	1877.85	2.58	0.24	0.03	0.00
0.60	3.15	0.19	0.32	0.29	0.21	0.48	349.81	384.56	1015.45	1595.63	0.73	0.54	0.35	0.00
0.80	2.95	0.33	0.02	0.38	0.07	1.06	364.91	384.94	1262.09	1864.21	0.25	0.32	0.24	0.00
1.07	1.62	0.27	0.21	0.28	0.13	0.39	367.94	393.49	1308.34	1837.93	1.50	0.24	0.04	0.00
0.35	1.92	0.39	0.35	0.45	0.44	0.05	380.70	394.49	1300.45	1855.35	1.56	0.23	0.07	0.00
0.42	1.99	0.24	0.27	0.34	0.34	0.06	363.03	389.61	1153.45	1824.89	1.25	0.31	0.12	0.00
2.32	3.78	0.57	0.57	0.66	0.14	0.13	385.07	397.06	1522.10	1885.84	4.07	0.25	0.03	0.00
1.24	1.61	0.31	0.14	0.38	0.08	0.24	367.80	393.26	1328.54	1860.03	1.99	0.24	0.03	0.00
0.80	**3.32**	**0.29**	**0.19**	**0.25**	**0.05**	**0.62**	**343.93**	**378.44**	**1126.63**	**1792.36**	**0.72**	**0.43**	**0.26**	**0.00**

Variables							
z_1	z_2	z_3	x_{n1}	x_{n2}	x_{n3}	m_n	n_p
54	14	86	0.84	0.60	3.00	1.21	4
54	12	84	0.81	0.48	3.00	1.05	8
51	17	88	0.37	0.51	2.22	1.06	7
50	21	93	0.00	0.38	1.40	1.62	2
51	20	94	0.37	0.59	1.28	1.52	2
48	19	95	0.29	0.45	1.00	1.39	3
44	24	84	1.42	0.22	2.28	1.44	6
41	25	95	0.00	1.01	3.00	1.54	8
45	25	95	0.35	0.43	1.00	0.50	5
48	21	99	0.00	1.05	2.46	0.68	5
54	**25**	**86**	**1.61**	**0.38**	**1.99**	**0.50**	**3**
37	25	91	0.28	0.99	2.68	0.63	4
50	25	87	1.23	0.00	1.50	0.50	2
53	19	92	0.00	1.04	2.73	0.50	3
33	25	85	0.52	0.67	1.44	0.50	5
39	21	87	0.00	0.00	1.03	0.70	3
42	23	88	0.49	0.00	1.76	0.50	3
32	25	85	0.65	0.50	1.40	1.17	2
34	25	85	0.29	1.00	3.00	0.50	5
50	**20**	**90**	**0.82**	**0.79**	**1.15**	**0.50**	**2**

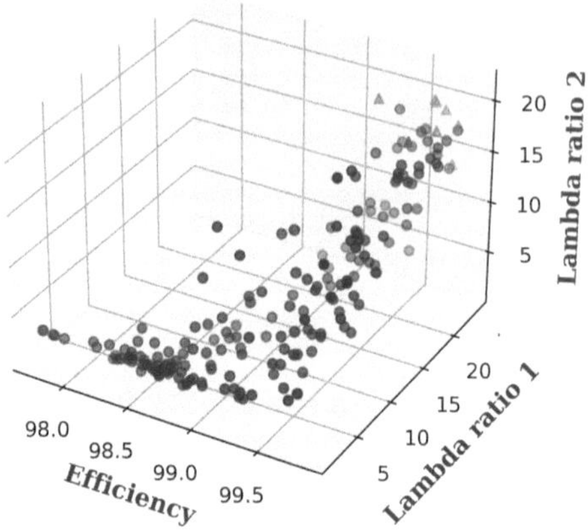

FIGURE 7.7 3D scatter plot of MOPSO results for first-stage gearing.

7.4.10.1 Comparison of Results

In Table 7.7, a comparison has been made between the results that are obtained with MOSA and the design of the existing planetary gearing. The solutions correspond to the number of teeth of planet gear, namely, 54 and 50, which were selected from the results. The values of design variables and objective functions corresponding to each solution are indicated in Table 7.7. The total weight of planetary gearing in kg, the efficiency in percentage, and the center distance in mm were taken for comparison.

The two best results chosen from the solution set of MOSA output are analyzed for selecting a single best solution. The total weight of the existing planetary gearing was 0.112 kg, the center distance was 20.4 mm, and the efficiency was 99.51%.

Referring to Table 7.7, in the case of the first result, by selecting the number of teeth in the planet gear as 54, the weight has reduced to 0.05 kg, the efficiency has increased to 99.61%, and the center distance has reduced to 19.75 mm. There is a reduction of 55.36% in weight, an increase of 0.10% in efficiency, and a space reduction of 3.19%. Hence, there is an improvement in all three objective functions in the solution obtained with MOSA.

In the case of the second result, by selecting the number of teeth in the planet gear as 50, the weight has reduced to 0.04 kg, the efficiency has increased to 99.53%, and the center distance has reduced to 17.50 mm. There is a reduction of 64.29% in weight, an increase of 0.02% in efficiency, and a space reduction of 14.22%. Hence, there is an improvement in all three objective functions in the solution obtained with MOSA.

7.4.10.2 Visualization of Results

The formation of the Pareto front from the non-dominated solutions containing 14 objective functions is depicted in the subsequent graphs in combinations of three objective functions for every graph. The graphs show the 3D scatter plot of the

TABLE 7.7
Optimized Design Parameters for First-Stage Gearing Using MOSA

	Variables				Objective Functions		
	Module, m (mm)	Number of Teeth (Planet Gear) Z_1	Number of Teeth (Sun Gear) Z_2	Number of Teeth (Ring Gear) Z_3	Total Weight, W (kg)	Efficiency, η	Center Distance (mm)
Existing first-stage planetary gearing design	0.8	39	12	90	0.112	99.51	20.4
MOSA design	0.50	54	25	86	0.05	99.61	19.75
	0.50	50	20	90	0.04	99.53	17.50

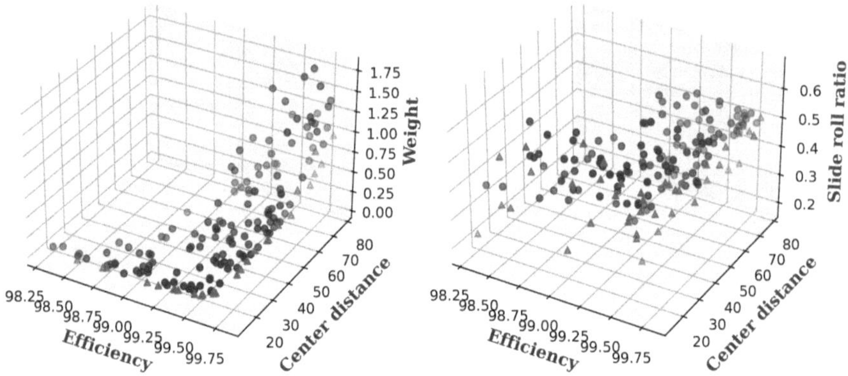

(a) (b)

FIGURE 7.8 (a) and (b) 3D scatter plot of MOSA results for first-stage gearing.

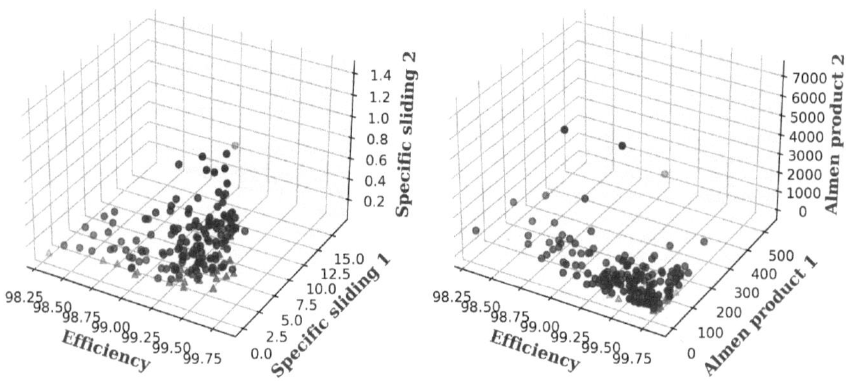

(a) (b)

FIGURE 7.9 (a) and (b) 3D scatter plot of MOSA results for first-stage gearing.

solution set generated by MOSA in various combinations of three objective functions.

Referring to Figure 7.8(a), the solutions close to the upper right corner have higher total weight, higher center distance, and higher efficiency, which means they have better efficiency but are heavier and less compact. On the contrary, the solutions near the lower left corner present better compactness and better total weight but lesser efficiency. The solutions lying in the lower right corner have the best value in terms of higher efficiency, lesser weight, and better compactness.

Referring to Figure 7.8(b), the solutions close to the lower right corner offer higher efficiency, better compactness, and lesser slide roll ratio values.

Referring to Figure 7.9(a), the solutions close to the lower right corner offer higher efficiency and lesser specific sliding values. A lower value of specific sliding indicates a balanced specific sliding at SAP and EAP of the gear tooth profile engagement during running.

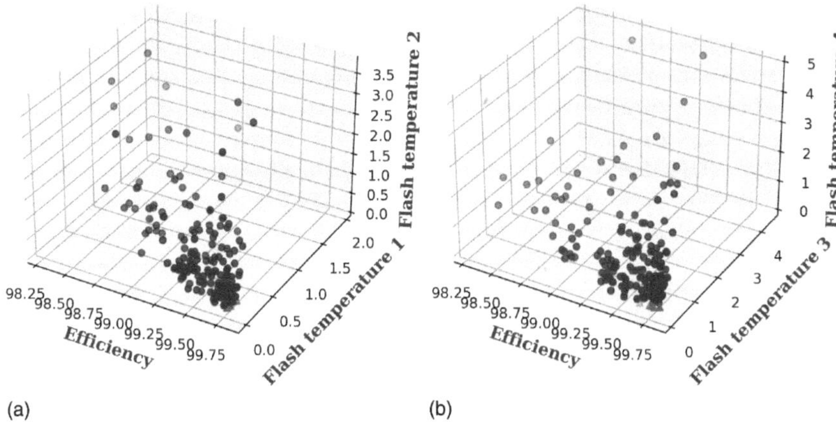

FIGURE 7.10 (a) and (b) 3D scatter plot of MOSA results for first-stage gearing.

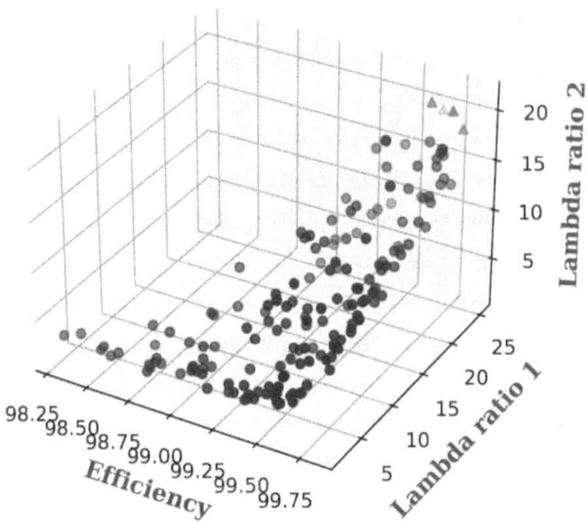

FIGURE 7.11 3D scatter plot of MOSA results for first-stage gearing.

Referring to Figure 7.9(b), the solutions close to the lower right corner offer higher efficiency and lower Almen product values at SAP and EAP of the gear tooth profile engagement during running.

Referring to Figure 7.10(a) and (b), the solutions close to the lower right corner offer higher efficiency and lower flash temperatures that are balanced at SAP and EAP of the gear tooth profile engagement during running.

Referring to Figure 7.11, the solutions close to the upper right corner offer higher efficiency and higher lambda ratios for the sun gear/planet gear engagement and the planet gear/ring gear engagement, respectively.

7.4.11 Interpretation of MOACO Results

In this section, the results obtained from MOACO are discussed and compared with the design of existing Wolfrom first-stage planetary gearing. For detailed information on the interpretation of the tables and plots in the subsequent sections, refer to Section 3.6.17.

Table 7.8 shows a sample of the MOACO output. Since constraints are specified in the form of ≥ 0 only, all the columns corresponding to the constraints shown in Table 7.8 indicate results obtained by modifying the constraint equations accordingly. Table 7.8 lists the values of the objective functions, constraints, and design variables from the generated solutions. The output represents a non-dominated Pareto-optimal solution set.

7.4.11.1 Comparison of Results

In Table 7.9, a comparison has been made between the results that are obtained with MOACO and the design of the existing planetary gearing. The solutions correspond to number of teeth of the planet gear, namely, 46 and 51, which were selected from the results. The values of design variables and objective functions corresponding to each solution are indicated in Table 7.9. The total weight of planetary gearing in kg, the efficiency in percentage, and the center distance in mm were taken for comparison.

The two best results chosen from the solution set of MOACO output are analyzed for selecting a single best solution. The total weight of the existing planetary gearing was 0.112 kg, the center distance was 20.4 mm, and the efficiency was 99.51%.

Referring to Table 7.9, in the case of the first result, by selecting the number of teeth in the planet gear as 46, the weight has reduced to 0.09 kg, the efficiency has increased to 99.56%, and the center distance has increased to 20.65 mm. There is a reduction of 19.64% in weight, an increase of 0.05% in efficiency, and a space increase of 1.23%. Hence, there is an improvement in weight and efficiency, along with a marginal space increase in the solution obtained with MOACO.

In the case of the second result, by selecting the number of teeth in the planet gear as 51, the weight has reduced to 0.05 kg, the efficiency has increased to 99.58%, and the center distance has reduced to 19.62 mm. There is a reduction of 55.36% in weight, an increase of 0.07% in efficiency, and a space reduction of 3.82%. Hence, there is an improvement in all three objective functions in the solution obtained with MOACO.

7.4.11.2 Visualization of Results

The formation of the Pareto front from the non-dominated solutions containing 14 objective functions is depicted in the subsequent graphs in combinations of three objective functions for every graph. The graphs show the 3D scatter plot of the solution set generated by MOACO in various combinations of three objective functions.

Referring to Figure 7.12(a), the solutions close to the upper right corner have higher total weight, higher center distance, and higher efficiency, which means they have better efficiency but are heavier and less compact. On the contrary, the solutions near the lower left corner present better compactness and better total weight but lesser efficiency. The solutions lying in the lower right corner have the best value in terms of higher efficiency, lesser weight, and better compactness.

TABLE 7.8
Sample of MOACO Output for First-Stage Gearing

					Objectives								
Center Distance	**Weight**	**Efficiency**	**Slide Roll Ratio**	**Specific Sliding**		**Almen Product**			**Flash Temperature**			**Lambda Ratio**	
67.27	1.41	99.72	0.31	0.65	0.32	54.05	366.52	0.07	0.26	0.15	0.20	15.87	21.10
59.50	1.10	99.41	0.35	0.30	0.17	8.73	1018.49	0.00	0.32	0.11	0.19	13.82	17.18
58.13	0.95	99.65	0.32	0.66	0.31	57.66	566.86	0.08	0.35	0.17	0.25	13.12	16.93
66.21	1.47	99.77	0.38	0.07	0.63	3.62	169.58	0.02	0.24	0.16	0.20	17.06	17.80
59.07	0.86	99.66	0.48	0.12	0.32	18.23	287.32	0.09	0.37	0.30	0.36	16.52	14.09
39.04	0.42	99.48	0.29	0.25	0.33	4.44	394.87	0.08	0.72	0.44	0.54	7.39	9.84
41.06	0.57	99.74	0.45	0.13	0.53	4.71	105.35	0.12	0.38	0.34	0.37	11.02	9.41
69.94	1.36	99.68	0.54	0.11	0.36	2.61	406.79	0.08	0.22	0.12	0.17	22.92	16.61
49.70	0.80	99.71	0.49	0.17	0.40	7.62	108.63	0.11	0.30	0.35	0.37	14.49	11.28
49.46	0.47	99.67	0.47	0.17	0.34	24.64	427.68	0.14	0.52	0.31	0.41	12.78	11.14
51.42	0.70	99.71	0.55	0.01	0.46	5.79	226.04	0.14	0.35	0.26	0.32	15.99	10.91
66.98	1.06	99.62	0.51	0.23	0.31	12.19	336.27	0.11	0.33	0.23	0.29	20.04	15.45
20.65	**0.09**	**99.56**	**0.54**	**0.97**	**0.52**	**83.60**	**223.01**	**0.18**	**0.93**	**0.83**	**1.02**	**4.80**	**3.35**
19.08	0.08	99.42	0.46	0.56	0.45	46.98	286.24	0.54	1.31	0.74	0.94	3.92	3.39
20.95	0.14	99.38	0.44	2.47	0.41	119.77	302.19	0.25	0.66	0.56	0.70	4.11	3.98
16.96	0.10	99.01	0.55	0.64	0.35	32.05	576.54	0.33	0.92	0.44	0.69	3.88	2.74
35.38	0.29	99.29	0.49	0.83	0.54	53.61	340.21	0.22	0.87	0.71	0.85	8.12	6.19
35.12	0.35	99.12	0.55	1.44	0.26	108.67	345.31	0.36	0.65	0.37	0.46	9.90	6.55
32.81	0.22	99.28	0.58	0.63	0.28	46.94	227.89	0.34	0.82	0.94	1.04	8.55	5.26
19.62	**0.05**	**99.58**	**0.46**	**1.47**	**0.52**	**215.98**	**200.18**	**0.06**	**2.22**	**2.14**	**2.50**	**3.68**	**3.37**

(Continued)

TABLE 7.8 (CONTINUED)
Sample of MOACO Output for First-Stage Gearing

Constraints															
Interference		Top land Thickness			Contact Ratio		Bending Stress		Contact Stress		Involute Interference		Tip interference	Proper Assembly
1.48	14.90	0.99	0.85	1.27	0.11	0.30	399.03	399.49	1871.46	1975.01	1.95	0.43	0.31	0.00
1.51	10.46	1.00	0.46	1.30	0.19	0.72	398.80	399.43	1849.00	1972.91	0.62	0.35	0.28	0.00
1.25	12.16	0.89	0.71	1.13	0.14	0.44	398.54	399.26	1838.24	1969.02	1.08	0.41	0.32	0.00
1.94	12.38	1.06	0.65	1.06	0.16	0.37	399.11	399.60	1877.46	1972.42	2.85	0.40	0.24	0.00
2.41	10.43	0.96	0.79	1.30	0.11	0.36	398.18	399.36	1834.86	1966.92	3.32	0.37	0.21	0.00
0.93	8.02	0.71	0.28	0.78	0.09	0.15	396.32	397.98	1732.06	1945.39	1.73	0.36	0.23	0.00
1.96	8.18	0.68	0.54	0.63	0.01	0.24	397.41	399.02	1817.53	1958.64	2.91	0.40	0.20	0.00
3.47	10.61	1.09	0.84	1.31	0.16	1.01	398.98	399.69	1880.41	1976.17	1.73	0.35	0.23	0.00
2.44	8.81	0.80	0.60	0.88	0.08	0.06	398.21	399.39	1846.52	1966.24	4.76	0.35	0.13	0.00
2.06	8.98	0.72	0.66	0.91	0.12	0.74	396.44	398.73	1779.33	1956.15	1.51	0.40	0.27	0.00
2.91	8.62	0.83	0.74	0.93	0.06	0.71	397.77	399.34	1831.12	1964.24	2.43	0.37	0.21	0.00
3.10	10.33	1.12	0.75	1.39	0.16	0.62	398.54	399.53	1851.85	1970.37	3.33	0.33	0.19	0.00
0.81	**3.31**	**0.26**	**0.34**	**0.38**	**0.19**	**0.56**	**384.09**	**395.31**	**1520.58**	**1901.50**	**0.86**	**0.40**	**0.24**	**0.00**
0.98	3.52	0.33	0.17	0.29	0.03	0.75	380.16	392.59	1494.48	1889.46	0.79	0.37	0.22	0.00
0.30	3.08	0.31	0.30	0.39	0.37	0.29	391.93	396.83	1588.03	1919.24	0.78	0.35	0.22	0.00
0.81	2.14	0.31	0.16	0.35	0.20	1.03	386.54	396.05	1524.24	1906.04	0.33	0.29	0.19	0.00
1.33	4.52	0.70	0.28	0.71	0.21	0.37	394.40	398.10	1670.04	1926.30	2.24	0.24	0.09	0.00
2.58	4.71	0.77	0.09	0.72	0.01	0.91	395.25	398.64	1748.52	1947.83	2.05	0.26	0.12	0.00
1.78	3.85	0.57	0.32	0.73	0.23	0.30	392.19	397.93	1649.81	1929.03	3.29	0.27	0.08	0.00
0.64	**4.31**	**0.20**	**0.40**	**0.34**	**0.03**	**0.23**	**353.23**	**382.53**	**1240.23**	**1844.29**	**0.85**	**0.52**	**0.35**	**0.00**

Variables							
z_1	z_2	z_3	x_{n1}	x_{n2}	x_{n3}	m_n	n_p
55	16	89	0.96	0.44	2.54	1.90	7
53	17	88	0.26	0.56	2.60	1.70	7
53	16	86	0.80	0.42	2.38	1.69	6
53	18	94	0.56	0.56	1.16	1.87	8
46	20	85	0.80	0.56	2.80	1.79	5
50	14	86	0.48	0.62	2.04	1.22	5
49	20	92	1.04	0.78	2.14	1.19	8
47	24	88	0.64	0.62	2.36	1.97	7
49	22	98	0.84	0.74	2.54	1.40	8
51	22	90	0.94	0.56	2.66	1.36	4
44	23	85	0.94	0.70	2.42	1.54	6
46	22	88	0.54	0.64	2.56	1.97	5
46	**24**	**86**	**1.02**	**0.22**	**2.26**	**0.59**	**5**
51	21	94	0.72	0.94	1.94	0.53	5
51	20	92	0.38	0.02	1.68	0.59	8
42	22	82	0.18	0.66	1.92	0.53	8
42	19	89	0.00	0.62	1.02	1.16	4
41	22	86	0.10	1.26	2.42	1.12	6
43	25	95	0.18	0.68	2.64	0.97	4
51	**21**	**85**	**1.76**	**0.20**	**2.86**	**0.55**	**2**

TABLE 7.9
Optimized Design Parameters for First-Stage Gearing Using MOACO

	Variables				Objective Functions		
	Module, m (mm)	Number of Teeth (Planet Gear), Z_1	Number of Teeth (Sun Gear), Z_2	Number of Teeth (Ring Gear), Z_3	Total Weight, W (kg)	Efficiency, η	Center Distance (mm)
Existing first-stage planetary gearing design	0.8	39	12	90	0.112	99.51	20.4
MOACO design	0.59	46	24	86	0.09	99.56	20.65
	0.55	51	21	85	0.05	99.58	19.62

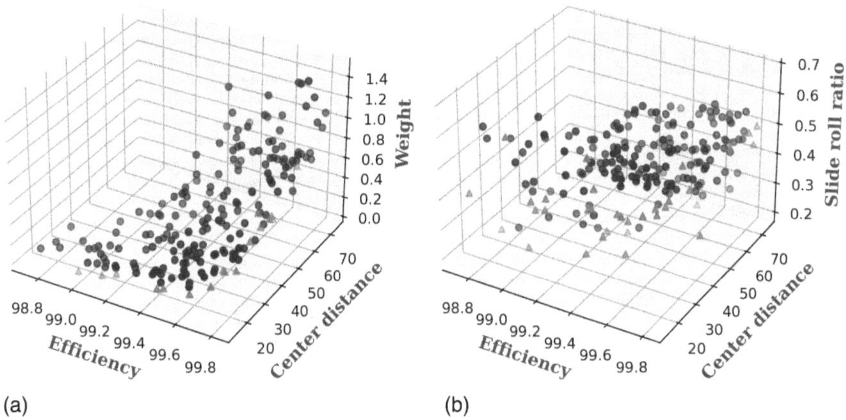

FIGURE 7.12 (a) and (b) 3D scatter plot of MOACO results for first-stage gearing.

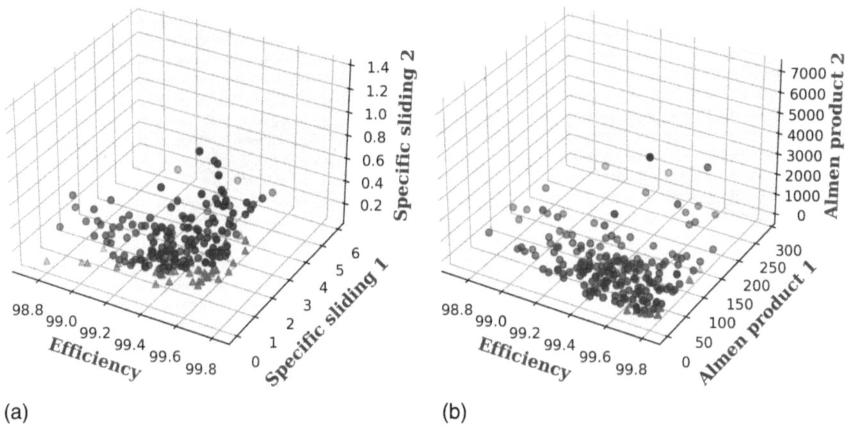

FIGURE 7.13 (a) and (b) 3D scatter plot of MOACO results for first-stage gearing.

Referring to Figure 7.12(b), the solutions close to the lower right corner offer higher efficiency, better compactness, and lesser slide roll ratio values.

Referring to Figure 7.13(a), the solutions close to the lower right corner offer higher efficiency and lesser specific sliding values. A lower value of specific sliding indicates a balanced specific sliding at SAP and EAP of the gear tooth profile engagement during running.

Referring to Figure 7.13(b), the solutions close to the lower right corner offer higher efficiency and lower Almen product values at SAP and EAP of the gear tooth profile engagement during running.

Referring to Figure 7.14(a) and (b), the solutions close to the lower right corner offer higher efficiency and lower flash temperatures that are balanced at SAP and EAP of the gear tooth profile engagement during running.

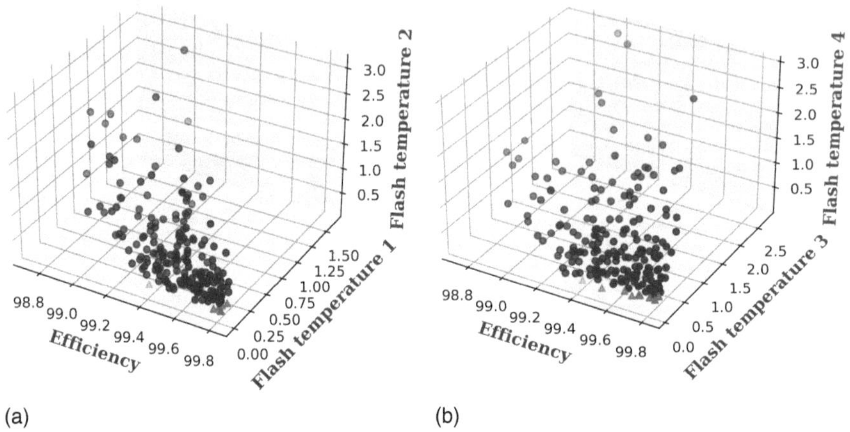

FIGURE 7.14 (a) and (b) 3D scatter plot of MOACO results for first-stage gearing.

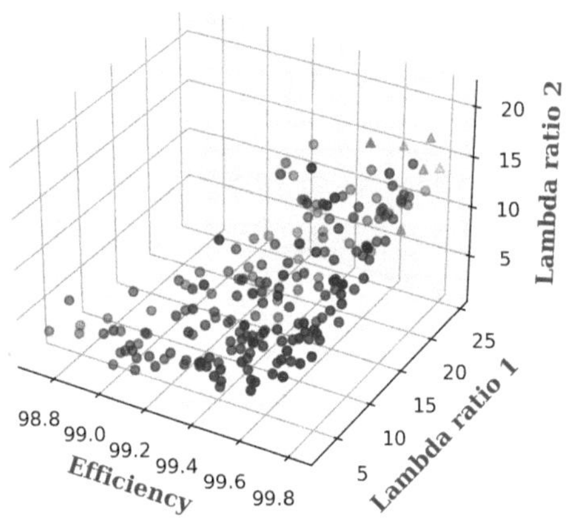

FIGURE 7.15 3D scatter plot of MOACO results for first-stage gearing.

Referring to Figure 7.15, the solutions close to the upper right corner offer higher efficiency and higher lambda ratios for the sun gear/planet gear engagement and the planet gear/ring gear engagement, respectively.

7.4.12 Comparison of Wolfrom First-Stage Planetary Gearing Optimization Using MOPSO, MOSA, and MOACO

In this section, a comparison of the results obtained by applying the three optimization techniques, namely, MOPSO, MOSA, and MOACO, for the first-stage gearing is carried out to find out the relative performance of each technique.

From Table 7.10, we can conclude that the three techniques offer a similar range of good quality results, with MOSA offering better weight and space

TABLE 7.10
Comparison of Optimized Design Parameters for First-Stage Gearing

	Variables				Objective Functions		
	Module, m (mm)	Number of Teeth (Planet Gear), Z_1	Number of Teeth (Sun Gear), Z_2	Number of Teeth (Ring Gear), Z_3	Total Weight, W (kg)	Efficiency, η	Center Distance (mm)
Existing first-stage planetary gearing design	0.8	39	12	90	0.112	99.51	20.4
MOPSO design	0.50	51	20	82	0.05	99.56	17.75
MOSA design	0.50	50	20	90	0.04	99.53	17.50
MOACO design	0.55	51	21	85	0.05	99.58	19.62

reduction, whereas MOACO offers higher efficiency. Also, since the planet gears of the first stage and second stage are mounted on the same Spider arm, the center distance for both stages must be the same.

7.5 MULTI-OBJECTIVE DESIGN OPTIMIZATION OF SECOND-STAGE SOLAR GEARING

The second stage of solar gearing receives input power from the first stage of gearing, and the compound planets rotate at 462.56 rpm. The second-stage planet gears drive the second-stage ring gear, which is connected to the output shaft. As already mentioned, since there is no sun gear, it should be assumed to be present for the purpose of visualization as an imaginary fixed member.

7.5.1 Calculation of Gear Parameters

Gear parameters are calculated using the procedure explained in Chapter 3 for Power Take-Off (PTO) gearbox.[4]

7.5.2 Design Space

The existing gear design problem for the second-stage solar gearing can be reduced to a six-dimensional design problem where the six free parameters are the number of teeth on the planet pinion z_1, the number of teeth on the ring gear z_3, the module m_n, the two profile shift coefficients x_{n1} and x_{n3} of the planet gear and the ring gear, respectively, and the number of planets n_p. The inequality constraints which bound this design space are the bending stress, Hertz stress, contact ratio, top land thickness, operating interference of gears, and involute and tip interference of internal gears. The gear pair ratios form equality constraints.

7.5.3 Objective Functions and Constraints

Nine objective functions for minimization/maximization with an array of inequality and equality constraints have been considered, and the problem can be stated as follows.

The following objective functions are considered for the design optimization of the solar gear train:

$$f_{obj} = \begin{cases} Min.center\ distance\ between\ solar\ gear\ pair \\ Min.weight\ of\ the\ solar\ gear\ train \\ Max.efficiency\ of\ solar\ gear\ train \\ Min.slide\ roll\ ratio \\ Balancing\ the\ specific\ sliding \\ Balancing\ the\ Almen\ product \\ Balancing\ the\ flash\ temperature \\ Balancing\ the\ lambda\ ratio \end{cases}$$

The independent design variables of the solar gear train include a module, the number of teeth of the pinion, and the two profile shift coefficients x_{n1} and x_{n3} of the planet gear and the ring gear, respectively. The vector of design variables is given below:

$$x(i) = \begin{cases} planet\ gear\ no.of teeth, Z_1 \\ ring\ gear\ no.og\ teeth, Z_3 \\ planet\ gear\ profile\ shift\ coefficient, x_{n_1} \\ ring\ gear\ profile\ shift\ coefficient, x_{n_3} \\ normal\ module\ m_n \\ no.of\ planets, n_p \end{cases}$$

where $i = 1, 2, \ldots,$ the number of variables.

A number of constraints, which act as sub functions to restrict the objective function, are introduced into the objective function to be optimized and are used to provide suitable design choices. The very often encountered failure types in a transmission gear are tooth fracture, surface fatigue, abrasion, and plastic deformation. Thus, the following constraints are considered for the design optimization of the solar gear train:

$$g(j) = \begin{cases} Bending\ stress\ constraint \\ Surface\ contact\ compressive\ stress\ constraint \\ Contact\ ratio\ constraint \\ Top\ land\ thickness\ constraint \\ Involute\ interference\ of\ internal\ gear\ constraint \\ Tip\ interference\ of\ internal\ gear\ constraint \end{cases}$$

where $j = 1, 2, \ldots,$ the number of constraints.

7.5.4 Establishing a Mathematical Model

To establish a mathematical model for the solar gear train design optimization, mathematical equations for the design variables, objective functions, and necessary constraints have to be developed. These sections are covered in more exhaustive detail in Chapter 3 for the PTO gearbox.[5]

7.5.5 Objective Functions for the Solar Gearing

The following objective functions are considered for the design optimization of solar gearing.

7.5.5.1 Minimization of Center Distance

The objective function for minimization of center distance is given by,

$$f_1 = \min a = \frac{m_n(z_3 - z_1)}{2}$$

7.5.5.2 Minimization of Weight

The second objective function may be written as,

$$\begin{aligned} &Minimize\ f_2 = \min \text{total weight} \\ &f_2 = W = W_1 + W_3 \\ &\quad = n_w \rho \pi \bar{d}_1 A_1 + \rho \pi \bar{d}_3 A_3 \end{aligned}$$

where W_1 and W_3 are the weights of the planet gear and the ring gear, respectively, in kg; A is the swept area, mm^2; $\bar{d}$ is the diameter at C.G. mm; ρ is the material density, kg/mm^2.

Referring to Figure 3.6, assuming a rim thickness of $1m_n$ for a gear, for the planet gears,

$$\bar{d} = d_w - 2m_n$$

For the ring gear,

$$\bar{d} = d_m + 2m_n$$

The swept area is given by,

$$A = b \times 2m_n$$

7.5.5.3 Maximization of Efficiency

In the second stage of solar gearing, the planet gear is the driver gear and the ring gear is the driven gear.

The objective function for efficiency increase is given by,

Maximize

$$f_3 = \eta = \left[1 - \left(\mu_{m_2} Hv_2\right)\right]100$$

where Hv_2 is the power loss factor.

The power loss factor Hv_2 for the planet gear and the ring gear pair is given by,

$$Hv_2 = \frac{\pi}{z_1 \cos \beta_b} \cdot \frac{1 + i_2}{i_2}\left(1 - \varepsilon_{\alpha_2} + \varepsilon_1'^2 + \varepsilon_3^2\right)$$

where ε_{α_2}, the base contact ratio for the planet gear and ring gear pair, is given by,

$$\varepsilon_{\alpha_2} = \varepsilon_1' + \varepsilon_3$$

The base pitch and the base transverse pitch of the planet gear are given by,

$$p_b = \frac{\pi d_{b_1}}{z_1}$$

$$p_{b_t} = \frac{p_b}{\cos\beta}$$

Referring to Figure 4.6, for the planet gear and the ring gear pair engagement, the contact path is along the common tangent to the respective base circles. The contact takes place along the line KF, and point E denotes the start of gear pair contact, and point F denotes the end of gear pair contact. As before, the pitch point P is where the pitch circles of the mating gear pair meet and also at the line connecting their centers of rotation.

Similarly, the addendum contact ratio for planet gear ε'_1 and ring gear ε_3, in the planet ring gear engagement, is given by,

$$\varepsilon'_1 = \frac{FP}{p_{b_t}}$$

$$\varepsilon_3 = \frac{EP}{p_{b_t}}$$

The base helix angle is given by,

$$\beta_b = \tan^{-1}\left(\frac{\dfrac{d_{b_1}\tan\beta}{m_n z_1}}{\cos\beta}\right)$$

Also,

$$\beta_b = \sin^{-1}\left(\sin\beta\cos\alpha_n\right)$$

The mean coefficient of friction is assumed to be constant for every contact point of the engagement region, even though it varies slightly along the path of contact. The approximate value of the mean coefficient of friction is given by,

$$\mu_{m_1} = 0.048\left(\frac{F_{t_{b_1}}/b}{\gamma_{\varepsilon c_1}\rho_{redc_1}}\right)^{0.2}\eta_{oil_1}{}^{-0.05}R_{a_1}{}^{+0.25}X_{L_1}$$

where $F_{t_{b_1}}$ is the circumferential force at the base circle (N); b is the face width for spur gears (mm); $\gamma_{\varepsilon c_1}$ is the sum velocity at the operating pitch circle (m/s); ρ_{redc_1} is the reduced radius of curvature at pitch point (mm); η_{oil} is the dynamic oil viscosity at oil temperature (mPa.s); R_a is the arithmetic mean roughness (μm); X_L is the lubricant correction factor.

The load-dependent losses are influenced by both gear geometry and lubricant properties.

The tangential velocity at the pitch circle is calculated as,

$$v_{t_1} = r_{w_1}\omega_1$$

The angular velocity of planet gear is,

$$\omega_1 = \frac{2\pi N_1}{60}$$

The curvature radius for planet gear (in the ring gear, planet gear pair) is,

$$\rho_1' = \frac{r_{w_1}\sin\alpha_{wt_2}}{\cos\beta_b}$$

The curvature radius for the ring gear is (in the ring gear, planet gear pair),

$$\rho_3 = \frac{r_{w_3}\sin\alpha_{wt_2}}{\cos\beta_b}$$

The coefficient of friction for the planet gear and ring gear pair is given by,

$$\mu_{m_2} = 0.048\left(\frac{F_{tb_2}/b}{\gamma_{\varepsilon c_2}\rho_{redc_2}}\right)^{0.2}\eta_{oil_2}^{-0.05}R_{a_2}^{+0.25}X_{L_2}$$

The power transmitted is,

$$P = 0.4kW$$

The tangential force at the base circle for the planet gear and the ring gear engagement is,

$$F_{tb_2} = F_{tb_1}$$

The sum velocity at operating pitch circles for the planet gear and the ring gear pair is,

$$\gamma_{\varepsilon c_2} = 2v_{t_1}\sin\alpha_{wt_2}$$

The reduced radius of curvature (in the ring gear, planet gear pair) is,

$$\rho_{redc_2} = \frac{\rho_1'\rho_3}{\left(\rho_3 - \rho_1'\right)\cos\beta_b}$$

For the present case, the following values are taken from the standards:

$$\eta_{oil} = 50\,\text{mPa.s}$$
$$R_{a1} = R_{a2} = 0.8\,\text{microns}$$

The lubricant correction factor X_{L_2} for the planet gear ring gear pair is given by,

$$X_{L_2} = \frac{1}{\left(\frac{F_{tb2}}{b}\right)^d}$$

Here, it is assumed that lubricating oil is mineral oil for which the factor d = 0.0651 (Martinsa, R, 2006).

7.5.5.4 Slide Roll Ratio (SRR)

For the second stage, the rolling velocities of the gear pair are calculated by the following equations:

$$v_{r1} = \rho_1^{'} \omega_1$$
$$v_{r3} = \rho_3 \omega_3$$

Since these tangential components of velocity are unequal at all the contact points along the contact path, sliding action occurs everywhere except the pitch point, *P*. Sliding velocity is the relative tangential velocity of the tooth profiles and is defined as the difference between the rolling velocities of teeth in mesh.

Hence, the sliding velocity is calculated as:

$$v_{s3} = v_{r1} - v_{r3}$$
$$v_{s4} = v_{r3} - v_{r1}$$

For the second stage, the entraining velocity is given by,

$$v_{e2} = v_{r1} + v_{r3}$$

For the second stage, the slide roll ratio is given by,

$$SRR = \frac{v_s}{v_{e2}}$$

Hence, the objective function for SRR is given by,

$$\min f_4 = \min \text{SRR}$$

7.5.5.5 Specific Sliding

For balancing or equalization of specific sliding between the planet gear and the output ring gear, the following equations are used, which form the next set of objective functions:

$$\min \; f_5 = \left| \gamma_{3\max} - \gamma_{1\max}' \right|$$

where $\gamma_{1\max}'$ and $\gamma_{3\max}$ are the specific sliding ratios at SAP and EAP, respectively, for the planet gear and ring gear pair meshing.

To find out the values of $\gamma_{1\max}'$ and $\gamma_{3\max}$, the following equations are used:

$$\gamma_{1\max}' = \frac{\left(\tan\alpha_{at_1} - \tan\alpha_{wt_2}\right)}{\left(\left(1+i_2\right)\tan\alpha_{wt_2} - \tan\alpha_{at_1}\right)}\left(i_2+1\right)$$

$$\gamma_{3\max} = \frac{\left(\tan\alpha_{at_3} - \tan\alpha_{wt_2}\right)}{\left(\left(\frac{i_2+1}{i_2}\right)\right)\tan\alpha_{wt_2} - \tan\alpha_{at_3}}\left(\frac{i_2+1}{i_2}\right)$$

Here, α_{wt_2} is the working transverse pressure angle between the planet gear-ring gear pair meshing.

Where i_2 is the gear ratio between the planet gear-ring gear pair and is defined as below,

$$i_2 = \frac{z_3}{z_1}$$

The addendum circle pressure angle of the planet gear is given by,

$$\alpha_{at_1} = \cos^{-1}\left(\frac{d_1\cos\alpha_t}{d_{a_1}'}\right)$$

The addendum circle pressure angle of the ring gear is given by,

$$\alpha_{at_3} = \cos^{-1}\left(\frac{d_3\cos\alpha_t}{d_{a_3}'}\right)$$

The transverse pressure angle α_t is given by,

$$\alpha_t = \tan^{-1}\left(\frac{\tan\alpha_n}{\cos\beta}\right)$$

7.5.5.6 Almen Product

For balancing or equalization of the Almen product, the following equations are used, which form the next set of objective functions:

Minimize $f_6 = \left|\sigma_{H_{a2}} v_{s2_a} - \sigma_{H_{r2}} v_{s2_r}\right|$ for the planet gear and the ring gear pair meshing.

For the planet gear and the ring gear engagement, the contact stress at SAP is given by,

$$\sigma_{H_{a2}} = Z_{E_2}\left[F_{n_a}''\left(\frac{1}{\rho_{1_a}'} - \frac{1}{\rho_{3_a}}\right)\right]^{0.5}$$

where Z_{E_2} is the elasticity factor, MPa; F_{n_a}'' is the normal load per unit length at SAP, N/mm; ρ_{1_a}' is the radius of curvature of the planet gear at SAP, mm; ρ_{3_a} is the radius of curvature of the ring gear at SAP, mm.

Here, the elasticity factor is,

$$Z_{E_2} = \left[\frac{1}{\pi\left\{\frac{1-\nu_1^2}{E_1} + \frac{1-\nu_3^2}{E_3}\right\}}\right]^{0.5}$$

where E_1, E_3 are Young's moduli of the planet gear and ring gear material, MPa; ν_1, ν_3 are Poisson's ratios of the planet gear and ring gear material.

Referring to Figure 4.5, the normal load per unit length at SAP is,

$$F_{n_a}'' = \frac{1000T_1}{\left(r_{b_1}^2 + LE^2\right)^{0.5}\cos\alpha_t b}$$

The contact stress at EAP is given by,

$$\sigma_{H_{r2}} = Z_{E_2}\left[F_{n_r}''\left(\frac{1}{\rho_{1_r}'} - \frac{1}{\rho_{3_r}}\right)\right]^{0.5}$$

where F_{n_r}'' is the normal load per unit length at EAP, N/mm.

The normal load per unit length at EAP is,

$$F_{n_r}'' = \frac{1000T_1}{r_{a_1}'\cos\alpha_t b}$$

The radius of curvature of the planet gear at SAP is,

$$\rho_{1_a}' = \frac{LE}{\cos\beta_b}$$

The radius of curvature of the planet gear at EAP is,

$$\rho_{1_r}' = \frac{LF}{\cos\beta_b}$$

The radius of curvature of the ring gear at SAP is,

$$\rho_{3_a} = \frac{KE}{\cos\beta_b}$$

The radius of curvature of the ring gear at EAP is,

$$\rho_{3_r} = \frac{KF}{\cos\beta_b}$$

Referring to Figure 4.5,

$$\begin{aligned}
LF &= \left(r_{a_1}'^2 - r_{b_1}^2\right)^{0.5} \\
KE &= \left(r_{a_3}'^2 - r_{b_3}^2\right)^{0.5} \\
EF &= LF - LE \\
EP &= KP - KE \\
PF &= LF - LP \\
KP &= \left(r_{w_3}^2 - r_{b_3}^2\right)^{0.5} \\
LP &= \left(r_{w_1}^2 - r_{b_1}^2\right)^{0.5} \\
LE &= KE - KL \\
KL &= a' \sin\alpha_{wt_2} \\
KF &= KP + PF
\end{aligned}$$

For the present case, the following material properties are taken from the standards:

Poisson's ratio and Young's modulus for the gear pair,

$$\begin{aligned}
\nu_1 &= \nu_3 = 0.3 \\
E_1 &= E_3 = 210 \text{ GPa}
\end{aligned}$$

The sliding velocity for the planet gear and the ring gear pair meshing at SAP is,

$$v_{s2_a} = EP\left(\omega_2 - \omega_3\right)$$

The sliding velocity for the planet gear and the sun gear pair meshing at EAP is,

$$v_{s2_r} = PF\left(\omega_2 - \omega_3\right)$$

The angular velocity of the planet gear is given by,

$$\omega_1 = \frac{2\pi N_1}{60}$$

The angular velocity of the ring gear is given by,

$$\omega_3 = \frac{2\pi N_3}{60}$$

7.5.5.7 Flash Temperature

The flash temperature for a gear pair in engagement can be calculated for any local point on the involute flank along the path of contact using the following equation (AGMA 925-A03):

$$\theta_{fl} = 31.62 K \mu_m \frac{X_r F_n}{\left(b_H\right)^{0.5}} \frac{\left|v_{r1} - v_{r2}\right|}{B_{m1}\left(v_{r1}\right)^{0.5} + B_{m2}\left(v_{r2}\right)^{0.5}}$$

where θ_{fl} is the flash temperature, °C; K is 0.80, a numerical factor valid for a semi-elliptic (Hertzian) distribution of frictional heat over the instantaneous width, 2 b_H, of the rectangular contact band; μ_m is the coefficient of friction; X_r is the load-sharing factor; F_n is the normal unit load, N/mm; b_H is the semi-width of the Hertzian contact band, mm; v_{r1} is rolling tangential velocity of the pinion, m/s; v_{r2} is rolling tangential velocity of the wheel, m/s; B_{m1} is thermal contact coefficient, pinion, N/(mm $s^{0.5}$K); B_{m2} is the thermal contact coefficient, wheel, N/(mm $s^{0.5}$K).

The semi-width of the Hertzian contact band is given by,

$$b_H = \left[\frac{4 F_n \rho_{redc}}{\pi E'}\right]^{0.5}$$

where F_n is the normal force at the contact point per unit width, N/mm; ρ_{redc} is the reduced radius of curvature at the contact point, mm; E' is the equivalent elastic modulus of gear pair materials, N/mm^2.For balancing or equalization of flash temperature, the following equations are used, which form the next set of objective functions:

Minimize

$f_7 = \theta_{fla_2}$, for the planet gear and the ring gear pair meshing at SAP.
$f_8 = \theta_{flr_2}$, for the planet gear and the ring gear pair meshing at EAP.

The flash temperature for the planet gear and the ring gear pair engagement at SAP is,

$$\theta_{fla_2} = 31.62 K \mu_{m2} \frac{X_{r2} F_{na}''}{\left(b_{H_{a2}}\right)^{0.5}} \frac{\left|v_{r1a}'' - v_{r3a}''\right|}{B_{m1}\left(v_{r1a}''\right)^{0.5} + B_{m2}\left(v_{r3a}''\right)^{0.5}}$$

The minimum Hertzian semi-contact width in one mesh cycle at SAP is,

$$b_{H_{a2}} = \left[\frac{4F_{n_a}'' \rho_{redc2a}}{\pi E_2'} \right]^{0.5}$$

The flash temperature for the planet gear and the ring gear pair engagement at EAP is,

$$\theta_{flr_2} = 31.62K\mu_{m_2} \frac{X_{r_2} F_{n_r}''}{\left(b_{H_{r2}}\right)^{0.5}} \frac{\left| v_{r_1r}'' - v_{r_3r}'' \right|}{B_{m1}\left(v_{r_1r}''\right)^{0.5} + B_{m2}\left(v_{r_3r}''\right)^{0.5}}$$

The minimum Hertzian semi-contact width in one mesh cycle at EAP,

$$b_{H_{r2}} = \left[\frac{4F_{n_r}'' \rho_{redc2r}}{\pi E_2'} \right]^{0.5}$$

Here, E_2' is the equivalent or effective modulus of elasticity, which depends on the modulus of elasticity and Poisson ratio of the sun gear material and planet gear material.

$$E_2' = \frac{1}{2}\left[\frac{1-\nu_1^2}{E_1} + \frac{1-\nu_3^2}{E_3} \right]$$

Here, E_1 and E_3 are in N/mm^2.

The reduced radius of curvature at SAP (in the planet gear, ring gear pair meshing) is,

$$\rho_{redc2a} = \frac{\rho_{1_a}' \rho_{3_a}}{\left(\rho_{3_a} - \rho_{1_a}'\right)\cos\beta_b}$$

The reduced radius of curvature at EAP (in the planet gear, ring gear pair meshing) is,

$$\rho_{redc2r} = \frac{\rho_{1_r}' \rho_{3_r}}{\left(\rho_{3_r} - \rho_{1_r}'\right)\cos\beta_b}$$

The rolling tangential velocity of the planet gear at SAP (in the planet gear, ring gear pair meshing) is,

$$v_{r_1a}'' = \rho_{1_a}' \omega_2$$

The rolling tangential velocity of the ring gear at SAP (in the planet gear, ring gear pair meshing) is,

$$v_{r3a}'' = \rho_{3_a}\omega_3$$

The rolling tangential velocity of the planet gear at EAP (in the planet gear, ring gear pair meshing) is,

$$v_{r1r}'' = \rho_{1_r}'\omega_2$$

The rolling tangential velocity of the ring gear at EAP (in the planet gear, ring gear pair meshing) is,

$$v_{r3r}'' = \rho_{3_r}\omega_3$$

7.5.5.8 Minimum Oil Film Thickness

Taking into account the unsteady speed and load conditions, the minimum oil film thickness for a gear pair engagement is given by,

$$h_{\min} = 1.714R_i\left(A\right)^{-0.128}\left(B\right)^{0.694}\left(C\right)^{0.568}$$

where $h_{\min}$ is the minimum oil film thickness, *m*; R_i is the radius of curvature of the interface between the mating gears, *m*.

Here, R_i is given by,

$$R_i = \frac{r_{w1}r_{w3}\sin\alpha_{wt2}}{1000\left(r_{w3} - r_{w1}\right)}$$

$$A = \frac{10^6 F_n}{bE''R_i}$$

where F_n is the normal load at pitch point, *N*; *b* is the face width, *m*; E'' is the equivalent modulus of elasticity, N/m².

The equivalent modulus of elasticity in this case is given by,

$$E'' = \left[\frac{2}{\dfrac{1-\nu_1^2}{E_1} + \dfrac{1-\nu_2^2}{E_2}}\right]$$

Here, E_1 and E_2 are the modulus of elasticity mating gears, in N/m².

$$B = \frac{\eta_o\left|v_1 - v_3\right|\cos\alpha_{wt2}}{2\times10^3 E''R_i}$$

where η_o is the viscosity at working temperature; v_1, v_3 is the tangential velocity of mating gears at the pitch circle, m/s.

$$C = \alpha' E''$$

where α' is the lubricant pressure viscosity coefficient, m²/N.

The minimum oil film thickness for the planet gear and the ring gear pair engagement is given by,

$$h_{\min_2} = 1.714 R_{i_2} \left(A''\right)^{-0.128} \left(B''\right)^{0.694} \left(C''\right)^{0.568}$$

where

$$A'' = \frac{10^6 F_{n_2}}{b E_2'' R_{i_2}}$$

$$F_{n_2} = \left(\frac{10^3 T_1}{r_{w1}}\right)$$

The equivalent modulus of elasticity in this case is given by,

$$E_2'' = \left[\frac{2}{\frac{1-\nu_1^2}{E_1} + \frac{1-\nu_3^2}{E_3}}\right]$$

$$R_{i_2} = \frac{r_{w1} r_{w3} \sin \alpha_{wt_2}}{10^3 \left(r_{w3} - r_{w1}\right)}$$

$$B'' = \frac{\eta_{o_2} \left|v_1 - v_3\right|}{E_2'' R_{i_2}}$$

For the planet gear,

$$v_1 = \frac{r_{w1}\omega_1}{10^3} = \frac{v_{t_1}}{10^3}$$

For the ring gear,

$$v_3 = \frac{r_{w3}\omega_3}{10^3} = \frac{v_{t3}}{10^3}$$

Here, v_1 and v_3 are in m/s.

$$C'' = \alpha_2' E_2''$$

For the present case, the following values are taken from the standards:

$$\eta_o = 90 \text{ mPa.s}$$
$$\alpha_2' = 2\times10^{-8}\text{m}^2/\text{N}$$

7.5.5.9 Lambda Ratio

Improvement in the lambda ratio value ensures better performance of the gear pair in engagement, and its value must be maximized. Hence, the following equations form the next set of objectives:

To maximize the lambda ratio,

$$Maximize\ f_9 = \lambda_2 = \frac{h_{\min 2}}{R_{a2}'}$$

where

$$R_{a2}' = R_{a1} + R_{a3}$$

where R_{a1} and R_{a3} are the arithmetic mean roughness of the planet and ring gears, respectively (μm).

For the present case,

$$R_{a2}' = 0.1$$

7.5.6 Design Variables and Constraints

The independent design parameters of planetary gear drive include a module m_n, the number of teeth of planet gear z_1, the number of teeth of ring gear z_3, the profile shift coefficients x_{n1}, x_{n3}, and the number of planets n_p.

Therefore, the design variables are,

$$X = \left[z_1, z_3, x_{n1}, x_{n3}, m_n, n_p\right]^T = \left[x_1, x_2, x_3, x_4, x_5, x_6\right]^T$$

A number of constraints, which are introduced into the objective functions to be optimized, are used to provide suitable design choices. For an overview of types of gear failures, see Section 3.6.14.

7.5.6.1 Bending Strength

Maximum allowable bending stress is given by bending strength of spur gears and helical gears ISO6336:2006-based JGMA401-01 (1974) standard.

$$\sigma_F = F_t \frac{y_F y_\varepsilon y_\beta}{9.81 m_n b}\left(\frac{K_V K_O}{K_L K_{FX}}\right) S_F \ \text{kgf/mm}^2$$

where F_t is the tangential force at the working pitch circle, kgf; y_F is the tooth profile factor; y_ε is the load-sharing factor; y_β is the helix angle factor; K_V is the dynamic load factor; K_O is the overload factor; K_L is the life factor; K_{FX} is the size factor of root stress; S_F is the safety factor for bending failure.

For the sun gear–planet gear engagement,

$$F_t = F_{n_1}$$

The tooth profile factor y_F is found by the equation (refer to Figure 3.13),

$$y_F = \frac{6\left(\dfrac{h_A}{m_n}\right)\cos\alpha_{wt}}{\left(\dfrac{S_A{}^2}{m_n{}^2}\right)\cos\alpha_n}$$

The load sharing factor y_ε is given by,

$$y_\varepsilon = \frac{1}{\varepsilon_{\alpha_2}{}'}$$

The helix angle factor y_β is given by,

$$y_\beta = 1 - \frac{\beta}{120}$$

The dynamic load factor, K_V, is based on the precision of the gear and the tangential speed of gears at the working pitch circle. The tangential speed of gears at the working pitch circle is,

$$v_{t_1} = \frac{\pi d_{w_1} N_1}{60000}$$

The dynamic load factor K_V is taken from the standards.

For tangential speed less than 1.0 m/s,

$$K_V = 1.0$$

The overload factor K_O is defined as,

$$K_O = \frac{\text{Actual tangential force}}{\text{Nominal tangential force } F_t}$$

Since the Wolfrom gearbox is expected to be subjected to heavy input load, K_O is taken as 1.75 from the standards.

The life factor K_L depends on the number of cyclic repetitions or the total loaded meshing cycles during its lifetime. Its value is taken from the standards. For the number of cyclic repetitions of $>10^7$ cycles, K_L is taken as,

$$K_L = 1.0$$

The size factor of root stress K_{FX} is generally taken as 1.0.

S_F is the safety factor for bending failure. Since S_F is too complicated to be determined precisely, it is usually taken as at least 1.2.

7.5.6.2 Contact Compressive (Crushing) Strength

Contact compressive stress should be within allowable limits to prevent failures due to wear like scuffing, scoring, welding, peening, rippling, ridging, and tooth breakage.

Contact compressive strength of spur gears and helical gears as per ISO 6336:2006-based JGMA 402-01 (1975) standard is given by,

$$\sigma_H = \sqrt{\frac{F_t(i+1)}{9.81 d_w b i} \frac{Z_H Z_M Z_\varepsilon Z_\beta}{K_{HL} Z_L Z_R Z_V Z_W K_{HX}}} \sqrt{K_{H\beta} K_V K_O} S_H \ \text{kgf/mm}^2$$

The "+" symbol in the above equation applies to two external gears in the mesh, whereas the "-" symbol is used for an internal gear and an external gear mesh.

Zone factor, Z_H
Material factor, Z_M
Contact ratio factor, Z_ε
Helix angle factor, Z_β
Life factor, K_{HL}
Lubricant factor, Z_L
Surface roughness factor, Z_R
Lubrication speed factor, Z_V
Hardness ratio factor, Z_W
Size factor, K_{HX}
Longitudinal load distribution factor, $K_{H\beta}$
Dynamic load factor, K_V
Overload factor, K_O
Safety factor for pitting, S_H

For the planet gear–ring gear engagement, Hertz contact stress is given by,

$$\sigma_{H_2} = \sqrt{\frac{F_t(i-1)}{9.81 d_{w_1} b i} \frac{Z_H Z_M Z_\varepsilon Z_\beta}{K_{HL} Z_L Z_R Z_V Z_W K_{HX}}} \sqrt{K_{H\beta} K_V K_O} S_H$$

$$i = i_2$$
$$F_t = F_{n_1}$$

Zone Factor, Z_H, is given by,

$$Z_H = \sqrt{\frac{2\cos\beta_b \cos\alpha_{wt_2}}{\cos^2\alpha_t \sin\alpha_{wt_2}}} = \frac{1}{\cos\alpha_t}\sqrt{\frac{2\cos\beta_b}{\tan\alpha_{wt_2}}}$$

Material Factor, Z_M, is given by,

$$Z_M = \left[\frac{1}{\pi\left\{\dfrac{1-\nu_1^{\,2}}{E_1} + \dfrac{1-\nu_3^{\,2}}{E_2}\right\}}\right]^{0.5}$$

For steel, $E_1 = 21000$ kgf/mm^2

Contact ratio factor, Z_ε, is given by the following:

For spur gear,

$$Z_\varepsilon = 1.0$$

For helical gear,

When the overlap ratio $\varepsilon_\beta \le 1$,

$$Z_\varepsilon = \sqrt{1 - \varepsilon_\beta + \frac{\varepsilon_\beta}{\varepsilon_\alpha}}$$

When $\varepsilon_\beta > 1$,

$$Z_\varepsilon = \sqrt{\frac{1}{\varepsilon_\alpha}}$$

The helix angle factor, Z_β, is difficult to calculate and can be taken as 1.0.

The life factor, K_{HL}, depends on the duty cycle.

For duty cycle > 10^7 cycles,

$$K_{HL} = 1.0$$

Lubricant factor, Z_L, is based on the lubricant's kinematic viscosity at 50°C, *cSt*.

From the standards,

$$Z_L = 1.0$$

The surface roughness factor, Z_R, is obtained on the basis of the average roughness $R_{\max}$ (μm).

The average roughness, R_{max}, is calculated by the below-given equation using the surface roughness values of the pinion and gear, R_{max1} and R_{max2}, and the center distance a in mm.

$$R_{max} = \frac{R_{max1} + R_{max2}}{2}\left(\frac{100}{a'}\right)^{\frac{1}{3}}$$

For surface-hardened gear and $R_{max} = 10$ microns,

$$Z_R = 0.92$$

Lubrication speed factor, Z_V, depends on the tangential speed at the pitch circle, v (m/s).

For surface-hardened gear and v < 1 m/s,

$$Z_V = 0.97$$

The hardness ratio factor, Z_W, applies only to the gear that is in mesh with a pinion, which is quenched and ground. It is given by the equation,

$$Z_W = 1.2 - \frac{H_{B2} - 130}{1700}$$

Here, H_{B2} is in the range of the Brinell hardness of gear material.

$$130 \le H_{B2} \le 470$$

If a gear is out of this range, Z_W is assumed to be 1.0.

The size factor, K_{HX}, is usually taken as 1.0 because the conditions affecting this parameter are often unknown.

The longitudinal load distribution factor, $K_{H\beta}$, depends on the method of gear shaft support and the ratio, b / d_{w1}, of the gear face width b to the pitch diameter, d_{w1}. When tooth contact under load is good, the longitudinal load distribution factor can be taken to be in the range of 1.0–1.2. For the present case, let $K_{H\beta} = 1.0$.

The dynamic load factor, K_V, depends on the gear's precision grade and pitch circle tangential speed, v.

For speeds less than 1 m/s, the value of K_V is taken as 1.0.

The overload factor, K_O, depends on the impact load on the gears on the prime mover side and on the load side. For a uniform load from the load side, this factor can be taken as 1.0.

The safety factor for pitting, S_H, is normally taken as 1.15 since the causes of pitting are difficult to precisely define because it involves many environmental factors.

7.5.6.3 Contact Ratio

The contact ratio is given by,

$$\varepsilon'_\alpha = \frac{\text{Path of contact}}{p_b}$$

where ε'_α is the contact ratio; p_b is the base pitch.

The base pitch, p_b, is defined as the distance from one tooth to the next tooth measured along the line of action. The base pitch is given by,

$$p_b = p_c \cos\alpha_t = \pi m_n \cos\alpha_t$$

where p_c is the circular pitch.

This contact ratio should be greater than 1.4 and is normally less than 2.0 for standard spur gears. For contact with an internal gear at low ratios and for non-standard addenda and dedenda, it can exceed 2.0.

The transverse contact ratio for the planet gear-ring gear pair engagement is given by,

$$\varepsilon_{\alpha_2}' = \frac{EF}{p_b} = \frac{EF}{\pi m_t \cos\alpha_t}$$

$$\text{Overlap ratio } \varepsilon_\beta = \frac{b\sin\beta}{\pi m_n}$$

The total contact ratio for the planet gear–ring gear pair engagement is given by,

$$\varepsilon_{\alpha_2}' + \varepsilon_\beta \geq 1.4$$

7.5.6.4 Top Land Thickness

The top land thickness is the tooth thickness on the tip circle. The top land thickness is normally referred to in terms of the percentage of the module. The ratio of top land thickness and the module is suggested to be between 0.2 and 0.4 depending on the material and heat treatment of the gear.

The top land thickness of the planet gear is given by,

$$S_{a_1} = d_{a_1}'\left(\frac{s_1}{d_1} + inv\alpha_n - \Omega_1\right)$$

where

$$S_1 = \frac{\pi m_n}{2} + 2x_{n_1} m_n \tan\alpha_n$$

$$\Omega_1 = \tan\alpha_{at_1} - \alpha_{at_1}\left(\frac{\pi}{180}\right)$$

Also, the top land thickness of the ring gear is given by,

$$S_{a_3} = d_{a_3}{}'\left(\frac{s_3}{d_3} - inv\alpha_n + \Omega_3\right)$$

where

$$S_3 = \frac{\pi m_n}{2} - 2x_{n_3} m_n \tan\alpha_n$$

$$\Omega_3 = \tan\alpha_{at_3} - \alpha_{at_3}\left(\frac{\pi}{180}\right)$$

The top land thickness should not be below 0.2 times the normal module, m_n, and correspondingly the constraint equations for the top land thickness of the planet gear, the sun gear, and the ring gear are as given below:

$$S_{a_1} \geq 0.2m_n$$

$$S_{a_3} \geq 0.2m_n$$

7.5.6.5 Constraints of Internal Gear

For proper tooth engagement and disengagement, involute interference (contact below the base circle) and secondary interference between the pinion tooth tip and the internal gear tooth tip for internal gearing must be avoided.

7.5.6.5.1 Involute Interference

For contact with an internal gear, involute interference is avoided if,

$$KL \leq KE$$

that is,

$$a' \sin\alpha_{wt_2} \leq KE$$

Also, to avoid involute interference, the following condition must be satisfied (Buckingham Earle, 1988):

$$\frac{z_1}{z_3} \geq 1 - \frac{\tan\alpha_{at_3}}{\tan\alpha_{wt_2}}$$

7.5.6.5.2 Tip Interference

To avoid tip interference, the following conditions must be satisfied (Buckingham, 1988):

$$\cos\varepsilon_1 = \frac{\left(r_{a_3}{}^2 - r_{a_1}{}^2 - a'^2\right)}{2a'r_{a_1}} = A$$

$$\therefore \varepsilon_1 = \cos^{-1} A$$

$$\varepsilon_2 = \left(\frac{z_1}{z_3}\right)\varepsilon_1$$

$$\theta_{t1} = \sin^{-1}\left[\frac{r_{a_1}\sin\varepsilon_1}{r_{a_3}}\right] - \varepsilon_2 = B$$

$$\theta_{t1r} = \frac{B \times \pi}{180} = C$$

$$\cos\alpha_{at_3} = \frac{r_{b_3}}{r_{a_3}} = D$$

$$\cos\alpha_{at_1} = \frac{r_{b_1}}{r_{a_1}} = E$$

$$\delta = \left(\frac{z_1}{z_3}\right)\left(inv\alpha_{at_1} - inv\alpha_t\right) + inv\alpha_t$$

$$inv\alpha_{at_1} = \tan\alpha_{at_1} - \frac{\alpha_{at_1} \times \pi}{180} = \tan\alpha_{at_1} - 0.0175\alpha_{at_1} = F$$

$$inv\alpha_t = \tan\alpha_t - 0.0175\alpha_t = 0.02881 = G$$

$$x_1 = inv\alpha_{at_3} = \tan\alpha_{at_3} - 0.0175\alpha_{at_3}$$

$$x_2 = \delta - \theta_{t1r} = \delta - C$$

Condition to avoid tip interference:
If $x_2 > x_1$, then there is no interference.
That is, $x_2 - x_1 \geq 0$

7.5.7 Design Data for Optimization

The input parameters considered for the solar gear design are listed in Table 7.3.

7.5.8 Methodology of Optimization of Wolfrom Second-Stage Solar Gearing

This section describes how the three algorithms, namely, MOPSO, MOSA, and MOACO, are configured for solving the optimization problem of Wolfrom second-stage solar gearing. The results obtained from the three multi-objective optimization techniques are analyzed, and comparisons are made on the performance of these three techniques.

This book utilizes the OPTIMO software package (Gokul Krishnan, 2024), which provides efficient implementations of the algorithms in Python along with the complete problem set corresponding to this case study. Chapter 2 provides detailed instructions on how to obtain and use the software.[6]

7.5.8.1 Variable Bounds

In order to determine the values of design variables, such as module, number of teeth, face width, and helix angle, through the optimization procedure, the upper and lower bounds of the design variables are to be established first.

The following variables are considered, and their lower and upper bounds are fixed. Further, the type of variable – whether discrete (integer) or real-valued (float) – is also specified.

7.5.8.1.1 Data for the Second-Stage Solar Gearing

Number of planet gear teeth, z_1: 20 to 40 (integer)
Number of sun gear teeth, z_3: 70 to 90 (integer)
Profile shift coefficient for planet gear, x_{n1}: 0.0 to 2.0 (float)
Profile shift coefficient for sun gear, x_{n3}: 0.0 to 2.0 (float)
Module, m_n: 0.5 to 2.0 mm (float)
Number of planets, n_p: 2 to 8 (integer)

7.5.8.2 Input Parameters

For this model, the most important data inputs are the gear design parameters of the two mating gears. The decision maker is required to make this decision based on experience and design requirements.

The problem sets provided in this book contain Python code corresponding to the objective functions and design constraints formulated in Sections 7.5.5 and 7.5.6. The upper and lower bounds of design variables are also specified in the same file(s). The seed for the random number generator is fixed at 0.123 for reproducibility of results.

The problem definition corresponding to Chapter 7, second-stage solar gearing, can be found in *problemsets/problem_ch7b.py* file within the software repository. The optimization run is started by entering the command specified in Listing 7.2 from the repository root.

LISTING 7.2 COMMAND TO EXECUTE THE ALGORITHM RUN

```
# macOS / Linux
python3 run.py -a pso -p ch7b
python3 run.py -a sa  -p ch7b
python3 run.py -a aco -p ch7b

# Windows
python.exe run.py -a pso -p ch7b
python.exe run.py -a sa  -p ch7b
python.exe run.py -a aco -p ch7b
```

7.5.9 Interpretation of MOPSO Results

In this section, the results obtained from MOPSO are discussed and compared with the design of existing Wolfrom second-stage solar gearing. For detailed information on the interpretation of the tables and plots in the subsequent sections, refer to Section 3.6.17.

Table 7.11 shows a sample of the MOPSO output. Since constraints are specified in the form of ≥ 0 only, all the columns corresponding to the constraints shown in Table 7.11 indicate results obtained by modifying the constraint equations accordingly. Table 7.11 lists the values of the objective functions, constraints, and design variables from the generated solutions. The output represents a non-dominated Pareto-optimal solution set.

7.5.9.1 Comparison of Results

In Table 7.12, a comparison has been made between the results that are obtained with MOPSO and the design of the existing solar gearing. The solutions correspond to number of teeth of the planet gear, namely, 29 and 33, were selected from the results. The values of design variables and objective functions corresponding to each solution are indicated in Table 7.12. The total weight of solar gearing in kg, the efficiency in percentage, and the center distance in mm were taken for comparison.

The two best results chosen from the solution set of MOPSO output are analyzed for selecting a single best solution. The total weight of the existing solar gearing was 0.1 kg, the center distance was 20.4 mm, and the efficiency was 99.65%.

Referring to Table 7.12, in the case of the first result, by selecting the number of teeth in the planet gear as 29, the weight has reduced to 0.06 kg, the efficiency has reduced to 99.29%, and the center distance has reduced to 15.25 mm. There is a reduction of 40.00% in weight, a reduction of 0.36% in efficiency, and a space reduction of 26.47%. Hence, there is an improvement in weight and center distance, along with marginal efficiency loss, in the solution obtained with MOPSO.

In the case of the second result, by selecting the number of teeth in the planet gear as 33, the weight has reduced to 0.09 kg, the efficiency has reduced to 99.32%, and the center distance has reduced to 17.68 mm. There is a reduction of 10.00% in weight, a reduction of 0.33% in efficiency, and a space reduction of 14.75%. Hence, there is an improvement in weight and center distance, along with marginal efficiency loss, in the solution obtained with MOPSO.

7.5.9.2 Visualization of Results

The formation of the Pareto front from the non-dominated solutions containing nine objective functions is depicted in the subsequent graphs in combinations of three objective functions for every graph. The graphs show the 3D scatter plot of the solution set generated by MOPSO in various combinations of three objective functions.

Referring to Figure 7.16(a), the solutions close to the upper right corner have higher total weight, higher center distance, and higher efficiency, which means they have better efficiency but are heavier and less compact. On the contrary, the solutions near the lower left corner present better compactness and better total

TABLE 7.11
Sample of MOPSO Output for Second-Stage Gearing

			Objectives					
Center Distance	**Weight**	**Efficiency**	**Slide Roll Ratio**	**Specific Sliding**	**Almen Product**	**Flash Temperature**		**Lambda Ratio**
15.69	0.05	99.09	0.59	1.75	1.54	0.60	1.59	0.65
25.52	0.10	99.30	0.60	1.56	2.82	0.35	1.20	1.21
15.25	**0.06**	**99.29**	**0.65**	**1.29**	**2.28**	**0.42**	**0.96**	**1.01**
39.09	0.20	99.38	0.55	1.13	4.94	0.03	0.92	1.69
36.31	0.13	99.27	0.53	1.16	6.82	0.09	1.63	1.30
15.25	0.04	98.75	0.59	2.28	0.11	1.36	2.38	0.58
17.33	0.05	99.29	0.66	1.06	3.45	0.74	1.55	1.17
14.86	0.06	99.12	0.66	2.07	1.54	0.81	1.57	0.95
12.10	0.18	99.64	0.79	0.53	1.13	0.47	0.50	4.35
38.47	0.26	99.37	0.60	1.63	5.30	0.03	0.64	2.06
12.04	0.12	99.61	0.77	1.07	1.00	0.49	0.58	2.62
22.51	0.12	99.23	0.67	1.97	0.57	0.83	1.09	1.65
36.80	0.13	99.20	0.54	1.31	9.90	0.54	1.52	1.34
19.58	0.05	98.99	0.60	1.44	9.30	0.35	2.29	0.85
21.13	0.16	99.60	0.72	1.01	1.70	0.69	0.90	2.80
25.44	0.11	99.34	0.65	1.08	5.13	0.55	1.42	1.81
33.21	0.10	99.10	0.53	1.50	12.50	1.10	1.74	1.09
39.60	0.14	99.24	0.55	1.26	9.93	0.32	1.95	1.48
14.17	0.07	99.25	0.72	1.93	0.22	1.16	1.44	1.49
17.68	**0.09**	**99.32**	**0.73**	**1.34**	**1.27**	**1.36**	**1.47**	**2.33**

(Continued)

TABLE 7.11 (CONTINUED)
Sample of MOPSO Output for Second-Stage Gearing

			Constraints				
Top Land Thickness		Contact Ratio	Bending Stress	Contact Stress	Involute Interference		Tip interference
0.20	0.34	0.55	373.51	1414.28	0.74	0.10	0.01
0.37	0.54	0.60	386.57	1593.15	1.06	0.09	0.01
0.27	**0.34**	**0.72**	**383.59**	**1601.82**	**0.86**	**0.11**	**0.01**
0.50	0.81	0.43	392.21	1705.39	1.21	0.06	0.01
0.45	0.70	0.46	383.28	1557.78	0.96	0.05	0.01
0.16	0.35	0.48	360.04	1255.75	0.98	0.16	0.01
0.30	0.40	0.60	374.94	1544.59	1.16	0.12	0.01
0.26	0.44	0.62	378.06	1486.02	0.97	0.14	0.01
0.24	0.65	0.24	393.42	1879.96	5.04	0.36	0.01
0.62	0.89	0.68	395.51	1751.86	0.86	0.05	0.01
0.35	0.47	0.59	392.05	1808.10	2.71	0.29	0.01
0.29	0.57	0.48	388.72	1669.68	2.63	0.24	0.01
0.51	0.72	0.56	385.44	1560.88	0.59	0.03	0.01
0.33	0.45	0.67	366.52	1343.38	0.39	0.04	0.01
0.42	0.70	0.44	391.18	1774.45	3.26	0.22	0.01
0.47	0.67	0.51	385.46	1643.15	1.49	0.10	0.01
0.46	0.64	0.63	381.63	1476.19	0.33	0.02	0.01
0.55	0.76	0.56	381.52	1519.30	0.84	0.04	0.01
0.27	0.46	0.52	380.74	1608.54	2.23	0.28	0.01
0.15	**0.55**	**0.29**	**381.90**	**1701.60**	**4.71**	**0.40**	**0.01**

Variables					
z_1	z_3	x_{n1}	x_{n3}	m_n	n_p
22	84	0.28	0.21	0.51	7
23	86	0.17	0.18	0.81	5
29	**90**	**0.00**	**0.00**	**0.50**	**8**
20	86	0.00	0.83	1.18	5
20	90	0.00	0.77	1.04	3
22	83	0.42	0.05	0.50	5
28	86	0.00	0.42	0.60	4
24	72	0.24	0.02	0.62	5
40	70	0.36	1.42	0.81	7
22	82	0.06	0.00	1.28	6
36	70	0.21	0.45	0.71	7
27	80	0.51	0.29	0.85	5
20	88	0.00	0.41	1.08	3
22	81	0.00	0.11	0.66	3
29	70	0.25	0.83	1.03	4
24	75	0.00	0.59	1.00	3
20	90	0.02	0.13	0.95	3
21	90	0.00	0.43	1.15	2
29	70	0.46	0.34	0.69	4
33	**76**	**1.00**	**1.28**	**0.82**	**3**

TABLE 7.12
Optimized Design Parameters for Second-Stage Gearing Using MOPSO

	Variables				Objective Functions	
	Module, m (mm)	Number of Teeth (Planet Gear), Z_1	Number of Teeth (Ring Gear), Z_3	Total Weight, W (kg)	Efficiency, η	Center Distance (mm)
Existing second-stage solar gearing design	0.8467	32	81	0.1	99.65	20.74
MOPSO design	0.50	29	90	0.06	99.29	15.25
	0.82	33	76	0.09	99.32	17.68

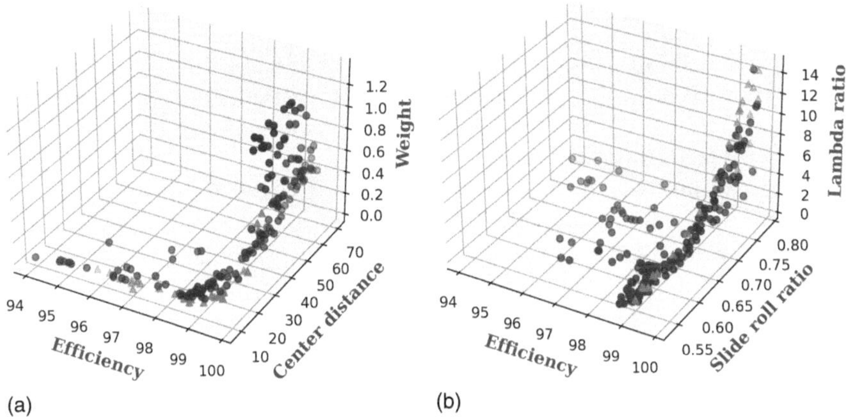

FIGURE 7.16 (a) and (b) 3D scatter plot of MOPSO results for second-stage gearing.

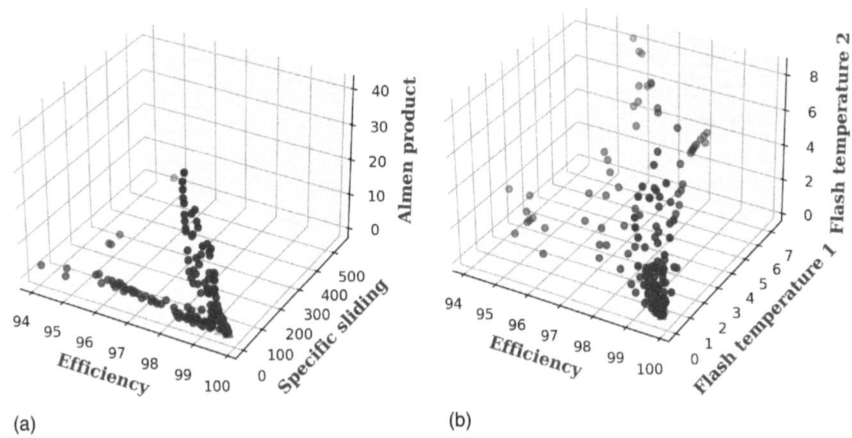

FIGURE 7.17 (a) and (b) 3D scatter plot of MOPSO results for second-stage gearing.

weight but lesser efficiency. The solutions lying in the lower right corner have the best value in terms of higher efficiency, lesser weight, and better compactness.

Referring to Figure 7.16(b), the solutions closer to the upper right corner offer higher efficiency, a lower value of slide roll ratio, and a higher lambda ratio for the planet gear/ring gear engagement.

Referring to Figure 7.17(a), the solutions close to the lower right corner offer higher efficiency, lower value of specific sliding, and lower value of Almen product at SAP and at EAP of the gear tooth profile engagement during running.

Referring to Figure 7.17(b), the solutions close to the lower right corner offer higher efficiency and lower flash temperatures that are balanced at SAP and EAP of the gear tooth profile engagement during running.

7.5.10 Interpretation of MOSA Results

In this section, the results obtained from MOSA are discussed and compared with the design of existing Wolfrom second-stage solar gearing. For detailed information on the interpretation of the tables and plots in the subsequent sections, refer to Section 3.6.17.

Table 7.13 shows a sample of the MOSA output. Since constraints are specified in the form of ≥ 0 only, all the columns corresponding to the constraints shown in Table 7.13 indicate results obtained by modifying the constraint equations accordingly. Table 7.13 lists the values of the objective functions, constraints, and design variables from the generated solutions. The output represents a non-dominated Pareto-optimal solution set.

7.5.10.1 Comparison of Results

In Table 7.14, a comparison has been made between the results that are obtained with MOSA and the design of the existing solar gearing. The solutions correspond to number of teeth of planet gear, namely, 30 and 36, which were selected from the results. The values of design variables and objective functions corresponding to each solution are indicated in Table 7.14. The total weight of solar gearing in kg, the efficiency in percentage, and the center distance in mm were taken for comparison.

The two best results chosen from the solution set of MOSA output are analyzed for selecting a single best solution. The total weight of the existing solar gearing was 0.1 kg, the center distance was 20.4 mm, and the efficiency was 99.65%.

Referring to Table 7.14, in the case of the first result, by selecting the number of teeth in planet gear as 30, the weight has reduced to 0.08 kg, the efficiency has reduced to 99.63%, and the center distance has reduced to 17.11 mm. There is a reduction of 20.00% in weight, a reduction of 0.02% in efficiency, and a space reduction of 17.50%. Hence, there is an improvement in weight and center distance, along with marginal efficiency loss, in the solution obtained with MOSA.

In the case of the second result, by selecting the number of teeth in the planet gear as 36, the weight has reduced to 0.09 kg, the efficiency has increased to 99.70%, and the center distance has reduced to 15.26 mm. There is a reduction of 10.00% in weight, an increase of 0.05% in efficiency, and a space reduction of 26.42%. Hence, there is an improvement in all three objective functions in the solution obtained with MOSA.

7.5.10.2 Visualization of Results

The formation of the Pareto front from the non-dominated solutions containing nine objective functions is depicted in the subsequent graphs in combinations of three objective functions for every graph. The graphs show the 3D scatter plot of the solution set generated by MOSA in various combinations of three objective functions.

TABLE 7.13
Sample of MOSA Output for Second-Stage Gearing

	Objectives							
Center Distance	Weight	Efficiency	Slide Roll Ratio	Specific Sliding	Almen Product	Flash Temperature		Lambda Ratio
17.25	0.05	99.11	0.55	1.11	3.82	0.15	1.57	0.63
15.00	0.05	99.33	0.62	0.85	2.09	0.70	1.31	0.87
24.10	0.11	99.40	0.61	1.42	0.31	0.73	1.14	1.31
14.50	0.06	99.47	0.67	1.15	0.08	0.87	1.09	1.15
17.25	0.04	99.30	0.54	0.87	2.06	1.19	2.55	0.61
17.00	0.04	98.96	0.57	1.97	0.96	0.76	1.90	0.59
13.50	0.07	99.60	0.71	0.93	0.65	0.68	0.85	1.49
24.60	0.19	99.56	0.67	1.26	0.12	0.50	0.63	2.20
17.11	**0.08**	**99.63**	**0.67**	**0.83**	**0.08**	**1.05**	**1.24**	**1.51**
15.25	0.04	98.96	0.59	1.88	0.08	1.06	1.75	0.64
16.50	0.05	98.89	0.59	1.70	0.81	1.13	1.56	0.70
15.50	0.05	99.58	0.64	0.74	0.70	1.31	1.69	1.03
27.30	0.20	99.50	0.65	1.02	2.38	0.26	0.54	2.17
14.14	0.09	99.50	0.69	0.64	1.48	0.80	1.03	1.54
36.65	0.20	99.45	0.58	1.37	0.15	0.70	1.04	1.84
14.84	0.17	99.63	0.76	0.73	1.18	0.40	0.47	3.24
17.46	0.05	99.27	0.61	0.94	3.12	0.73	1.70	0.89
14.42	0.14	99.68	0.76	0.91	0.79	0.47	0.55	2.95
12.50	0.06	99.56	0.70	1.04	0.36	0.89	1.10	1.31
15.26	**0.09**	**99.70**	**0.76**	**0.72**	**1.11**	**1.12**	**1.25**	**2.95**

(Continued)

TABLE 7.13 (CONTINUED)
Sample of MOSA Output for Second-Stage Gearing

		Constraints					
Top Land Thickness		Contact Ratio	Bending Stress	Contact Stress	Involute Interference		Tip interference
0.22	0.34	0.45	370.52	1435.06	0.58	0.06	0.01
0.18	0.38	0.30	375.61	1579.58	1.15	0.13	0.01
0.23	0.53	0.37	387.13	1649.28	2.10	0.17	0.01
0.15	0.33	0.37	380.90	1648.34	2.10	0.25	0.01
0.10	0.45	0.13	347.53	1355.00	1.12	0.11	0.00
0.19	0.35	0.55	368.04	1325.11	0.77	0.11	0.01
0.21	0.33	0.47	384.68	1714.51	2.15	0.25	0.01
0.27	0.57	0.39	393.66	1789.45	3.44	0.24	0.01
0.09	**0.50**	**0.18**	**380.38**	**1692.26**	**2.99**	**0.26**	**0.00**
0.15	0.33	0.42	370.75	1408.08	1.13	0.16	0.01
0.09	0.32	0.32	373.88	1481.89	1.59	0.20	0.01
0.08	0.46	0.12	367.41	1589.25	2.08	0.20	0.00
0.46	0.61	0.57	394.53	1789.29	1.85	0.12	0.01
0.14	0.56	0.13	384.11	1734.94	2.27	0.21	0.00
0.19	0.82	0.25	391.60	1719.38	3.33	0.17	0.01
0.34	0.51	0.48	393.85	1849.08	3.49	0.28	0.01
0.23	0.39	0.40	369.94	1500.60	1.06	0.11	0.01
0.34	0.48	0.53	392.71	1825.86	3.40	0.31	0.01
0.18	0.33	0.41	380.12	1666.43	2.11	0.26	0.01
0.25	**0.61**	**0.29**	**383.14**	**1763.10**	**4.36**	**0.33**	**0.01**

Variables					
z_1	z_3	x_{n1}	x_{n3}	m_n	n_p
21	90	0.00	0.81	0.50	7
24	84	0.00	1.25	0.50	8
23	83	0.50	0.89	0.80	6
30	88	0.61	1.09	0.50	8
20	89	0.00	1.81	0.50	5
22	90	0.31	0.00	0.50	6
35	89	0.33	0.87	0.50	8
29	85	0.60	0.95	0.88	8
30	**87**	**0.61**	**1.87**	**0.60**	**6**
22	83	0.52	0.45	0.50	7
23	89	0.74	0.83	0.50	8
27	89	0.30	2.00	0.50	6
28	88	0.00	0.50	0.91	8
27	73	0.04	1.65	0.61	8
21	83	0.61	1.29	1.18	5
39	79	0.15	0.81	0.74	8
24	87	0.00	0.98	0.55	5
39	79	0.28	0.66	0.72	7
32	82	0.47	1.00	0.50	7
36	**74**	**0.45**	**1.37**	**0.80**	**3**

TABLE 7.14
Optimized Design Parameters for Second-Stage Gearing Using MOSA

	Variables			Objective Functions		
	Module, m (mm)	Number of Teeth (Planet Gear), Z_1	Number of Teeth (Ring Gear), Z_3	Total Weight, W (kg)	Efficiency, η	Center Distance (mm)
Existing second-stage solar gearing design	0.8467	32	81	0.1	99.65	20.74
MOSA design	0.60	30	87	0.08	99.63	17.11
	0.80	36	74	0.09	99.70	15.26

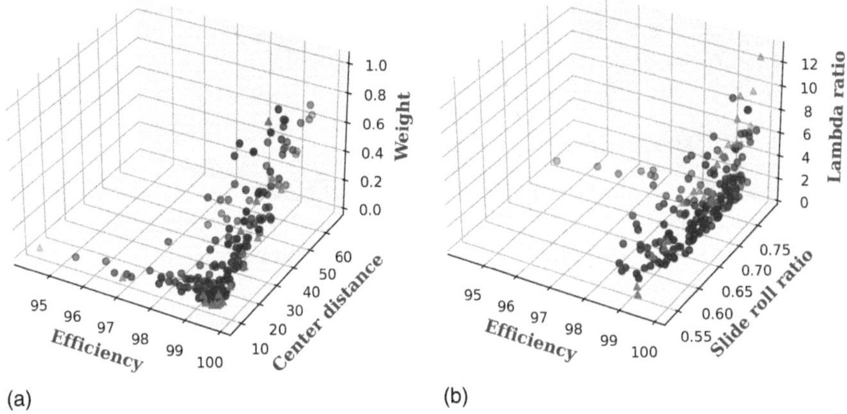

FIGURE 7.18 (a) and (b) 3D scatter plot of MOSA results for second-stage gearing.

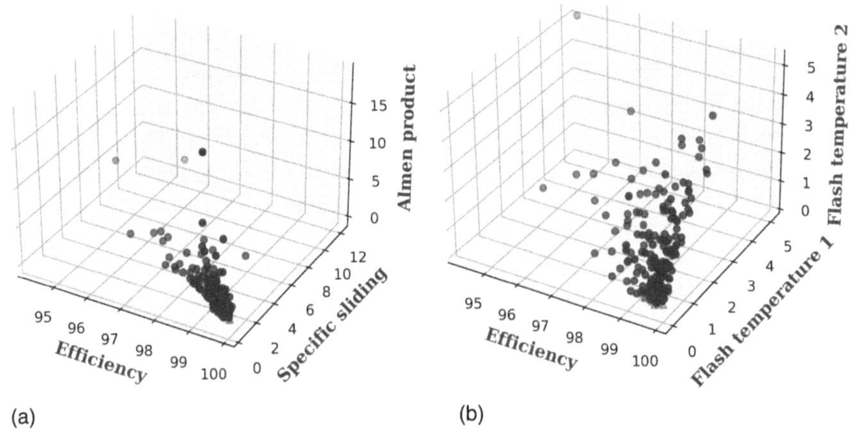

FIGURE 7.19 (a) and (b) 3D scatter plot of MOSA results for second-stage gearing.

Referring to Figure 7.18(a), the solutions close to the upper right corner have higher total weight, higher center distance, and higher efficiency, which means they have better efficiency but are heavier and less compact. On the contrary, the solutions near the lower left corner present better compactness and better total weight but lesser efficiency. The solutions lying in the lower right corner have the best value in terms of higher efficiency, lesser weight, and better compactness.

Referring to Figure 7.18(b), the solutions closer to the upper right corner offer higher efficiency, lower value of slide roll ratio, and higher lambda ratio for the planet gear/ring gear engagement.

Referring to Figure 7.19(a), the solutions close to the lower right corner offer higher efficiency, lower value of specific sliding, and lower value of Almen product at SAP and at EAP of the gear tooth profile engagement during running.

Referring to Figure 7.19(b), the solutions close to the lower right corner offer higher efficiency and lower flash temperatures that are balanced at SAP and EAP of the gear tooth profile engagement during running.

7.5.11 Interpretation of MOACO Results

In this section, the results obtained from MOACO are discussed and compared with the design of existing Wolfrom second-stage solar gearing. For detailed information on the interpretation of the tables and plots in the subsequent sections, refer to Section 3.6.17.

Table 7.15 shows a sample of the MOACO output. Since constraints are specified in the form of ≥ 0 only, all the columns corresponding to the constraints shown in Table 7.15 indicate results obtained by modifying the constraint equations accordingly. Table 7.15 lists the values of the objective functions, constraints, and design variables from the generated solutions. The output represents a non-dominated Pareto-optimal solution set.

7.5.11.1 Comparison of Results

In Table 7.16, a comparison has been made between the results that are obtained with MOACO and the design of the existing solar gearing. The solutions corresponding to number of teeth of the planet gear, namely, 39 and 34, which were selected from the results. The values of design variables and objective functions corresponding to each solution are indicated in Table 7.16. The total weight of solar gearing in kg, the efficiency in percentage, and the center distance in mm were taken for comparison.

The two best results chosen from the solution set of MOACO output are analyzed for selecting a single best solution. The total weight of the existing solar gearing was 0.1 kg, the center distance was 20.4 mm, and the efficiency was 99.65%.

Referring to Table 7.16, in the case of the first result, by selecting the number of teeth in the planet gear as 39, the weight has reduced to 0.07 kg, the efficiency has increased to 99.66%, and the center distance has reduced to 12.10 mm. There is a reduction of 30.00% in weight, an increase of 0.01% in efficiency, and space reduction of 41.66%. Hence, there is an improvement in all three objective functions in the solution obtained with MOACO.

In the case of the second result, by selecting the number of teeth in planet gear as 34, the weight has reduced to 0.09 kg, the efficiency has increased to 99.67%, and the center distance has reduced to 17.00 mm. There is a reduction of 10.00% in weight, an increase of 0.02% in efficiency, and a space reduction of 18.03%. Hence, there is an improvement in all three objective functions in the solution obtained with MOACO.

7.5.11.2 Visualization of Results

The formation of the Pareto front from the non-dominated solutions containing nine objective functions is depicted in the subsequent graphs in combinations of three

TABLE 7.15
Sample of MOACO Output for Second-Stage Gearing

	Objectives							
Center Distance	Weight	Efficiency	Slide Roll Ratio	Specific Sliding	Almen Product	Flash Temperature		Lambda Ratio
35.50	0.27	99.69	0.62	0.89	0.31	0.51	0.67	2.58
13.90	0.14	99.67	0.75	1.07	0.13	0.51	0.57	2.62
13.48	0.15	99.75	0.76	0.74	0.44	0.58	0.63	3.21
24.06	0.36	99.81	0.77	0.67	0.73	0.48	0.51	7.11
16.89	0.24	99.67	0.77	0.45	1.25	0.51	0.55	5.08
27.27	0.31	99.72	0.71	0.63	1.08	0.35	0.42	4.06
31.57	0.21	99.62	0.63	1.08	0.38	0.57	0.79	2.26
20.13	0.42	99.67	0.78	0.68	1.43	0.26	0.30	6.65
29.52	0.32	99.77	0.71	0.78	0.01	0.51	0.57	4.21
41.88	0.27	99.51	0.57	1.34	0.04	0.49	0.77	2.10
18.99	0.34	99.67	0.77	1.03	0.46	0.38	0.39	4.96
33.16	0.65	99.83	0.77	0.81	0.15	0.37	0.39	9.90
24.92	0.19	99.61	0.68	0.75	1.66	0.43	0.61	2.63
12.10	**0.07**	**99.66**	**0.74**	**0.76**	**0.76**	**0.71**	**0.82**	**1.92**
38.65	0.27	99.53	0.59	1.26	0.01	0.55	0.80	2.22
16.74	0.10	99.55	0.71	0.73	1.56	0.57	0.75	2.00
20.90	0.21	99.65	0.72	0.91	0.45	0.60	0.64	3.23
31.16	0.48	99.79	0.76	0.96	0.02	0.40	0.43	7.66
23.92	0.20	99.61	0.72	1.33	0.11	0.46	0.55	3.12
17.00	**0.09**	**99.67**	**0.71**	**0.74**	**0.92**	**0.86**	**1.03**	**2.15**

(Continued)

TABLE 7.15 (CONTINUED)
Sample of MOACO Output for Second-Stage Gearing

Constraints							
Top Land Thickness		Contact Ratio	Bending Stress	Contact Stress	Involute Interference		Tip interference
0.16	0.98	0.15	394.33	1814.33	4.40	0.20	0.00
0.27	0.45	0.45	392.40	1817.29	3.56	0.35	0.01
0.19	0.60	0.25	391.92	1842.72	4.54	0.36	0.01
0.35	1.12	0.24	396.02	1896.41	8.90	0.38	0.01
0.18	0.93	0.10	393.98	1887.34	6.38	0.35	0.01
0.35	0.80	0.29	396.00	1874.88	5.02	0.25	0.01
0.27	0.77	0.28	393.01	1775.79	3.58	0.19	0.01
0.57	0.84	0.47	397.47	1911.61	5.92	0.31	0.01
0.14	1.03	0.16	395.37	1864.11	6.83	0.31	0.00
0.28	0.85	0.30	394.20	1764.84	3.47	0.16	0.01
0.16	0.78	0.25	396.47	1892.32	7.41	0.46	0.01
0.29	1.48	0.22	397.73	1919.49	13.20	0.43	0.01
0.36	0.65	0.37	393.45	1812.64	3.10	0.19	0.01
0.21	**0.36**	**0.42**	**384.83**	**1757.48**	**2.73**	**0.29**	**0.01**
0.18	0.96	0.21	394.18	1776.60	3.84	0.18	0.00
0.28	0.44	0.44	388.52	1764.89	2.45	0.21	0.01
0.04	0.79	0.15	393.54	1841.15	5.82	0.36	0.00
0.52	1.11	0.37	397.22	1898.98	9.70	0.38	0.01
0.41	0.60	0.56	394.48	1816.65	4.16	0.30	0.01
0.22	**0.52**	**0.30**	**385.05**	**1748.50**	**3.31**	**0.26**	**0.01**

Variables					
z_1	z_3	x_{n1}	x_{n3}	m_n	n_p
25	87	0.44	1.88	1.15	7
37	77	0.54	0.86	0.70	8
36	71	0.62	1.52	0.77	7
38	72	0.60	1.56	1.42	4
39	74	0.32	1.84	0.97	6
36	90	0.18	1.46	1.01	8
25	84	0.48	1.36	1.07	6
40	73	0.12	0.78	1.22	7
32	81	0.72	1.94	1.21	6
22	89	0.56	1.18	1.25	6
37	72	1.08	1.60	1.09	8
38	73	0.96	1.74	1.90	4
31	87	0.06	1.14	0.89	7
39	**86**	**0.34**	**1.08**	**0.52**	**7**
21	80	0.58	1.46	1.31	6
35	89	0.06	0.96	0.62	7
32	76	1.00	1.96	0.95	7
38	76	0.70	1.18	1.64	4
37	89	0.42	0.38	0.92	6
34	**84**	**0.36**	**1.42**	**0.68**	**5**

TABLE 7.16
Optimized Design Parameters for Second-Stage Gearing Using MOACO

	Variables			Objective Functions		
	Module, m (mm)	Number of Teeth (Planet Gear), Z_1	Number of Teeth (Ring Gear), Z_3	Total Weight, W (kg)	Efficiency, η	Center Distance (mm)
Existing second-stage solar gearing design	0.8467	32	81	0.1	99.65	20.74
MOACO design	0.52	39	86	0.07	99.66	12.10
	0.68	34	84	0.09	99.67	17.00

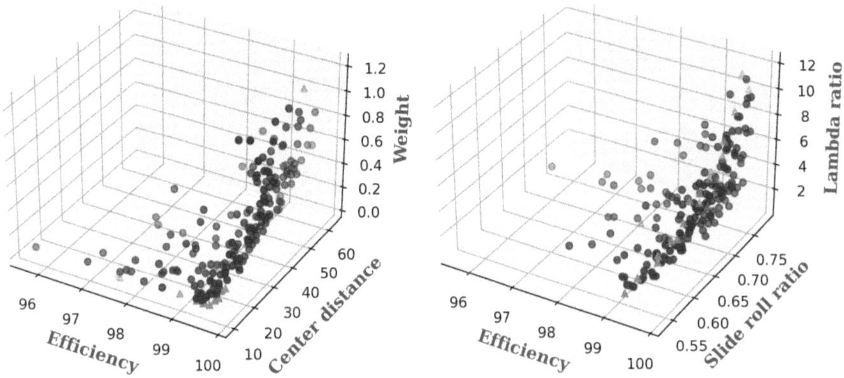

(a) (b)

FIGURE 7.20 (a) and (b) 3D scatter plot of MOACO results for second-stage gearing.

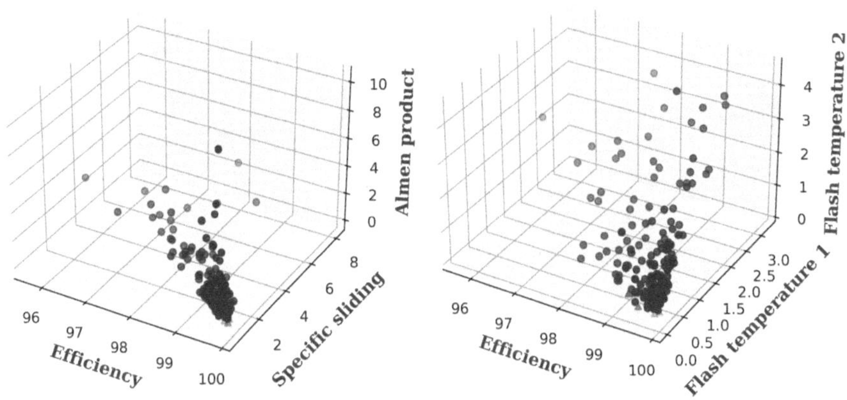

(a) (b)

FIGURE 7.21 (a) and (b) 3D scatter plot of MOACO results for second-stage gearing.

objective functions for every graph. The graphs show the 3D scatter plot of the solution set generated by MOACO, in various combinations of three objective functions.

Referring to Figure 7.20(a), the solutions close to the upper right corner have higher total weight, higher center distance, and higher efficiency, which means they have better efficiency but are heavier and less compact. On the contrary, the solutions near the lower left corner present better compactness and better total weight but lesser efficiency. The solutions lying in the lower right corner have the best value in terms of higher efficiency, lesser weight, and better compactness.

Referring to Figure 7.20(b), the solutions closer to the upper right corner offer higher efficiency, lower value of slide roll ratio, and higher lambda ratio for the planet gear/ring gear engagement.

TABLE 7.17
Comparison of Optimized Design Parameters for Second-Stage Gearing

	Variables				Objective Functions	
	Module, m (mm)	Number of Teeth (Planet Gear), Z_1	Number of Teeth (Ring Gear), Z_3	Total Weight, W (kg)	Efficiency, η	Center Distance (mm)
Existing second-stage solar gearing design	0.8467	32	81	0.1	99.65	20.74
MOPSO design	0.82	33	76	0.09	99.32	17.68
MOSA design	0.60	30	87	0.08	99.63	17.11
MOACO design	0.68	34	84	0.09	99.67	17.00

TABLE 7.18
Acceptable Optimized Design Parameters for Wolfrom Epicyclic Gearing

	Variables				Objective Functions		
	Module, m (mm)	Number of Teeth (Planet Gear), Z_1	Number of Teeth (Sun Gear), Z_2	Number of Teeth (Ring Gear), Z_3	Total Weight, W (kg)	Efficiency, η	Center Distance (mm)
MOPSO design for first-stage planetary gearing	0.50	51	20	82	0.05	99.56	17.75
MOPSO design for second-stage solar gearing	0.82	33	–	76	0.09	99.32	17.68

Referring to Figure 7.21(a), the solutions close to the lower right corner offer higher efficiency, lower value of specific sliding, and lower value of Almen product at SAP and at EAP of the gear tooth profile engagement during running.

Referring to Figure 7.21(b), the solutions close to the lower right corner offer higher efficiency and lower flash temperatures that are balanced at SAP and EAP of the gear tooth profile engagement during running.

7.5.12 Comparison of Wolfrom Second-Stage Solar Gearing Optimization Using MOPSO, MOSA, and MOACO

In this section, a comparison of the results obtained by applying the three optimization techniques, namely, MOPSO, MOSA, and MOACO, for the second-stage gearing is carried out to find out the relative performance of each technique.

From Table 7.17, we can conclude that the three techniques offer a similar range of good quality results, with MOSA and MOACO offering better solutions overall. Also, as already mentioned, since the planet gears of the first stage and second stage are mounted on the same Spider arm, the center distance for both stages must be the same.

7.6 CONCLUSION

Comparing the results obtained for the first stage (Table 7.10) and second stage (Table 7.17) of the Wolfrom epicyclic gearing, since the center distance of the gears in both stages must be the same, the selection of a single best acceptable solution is based on satisfying the above criterion. Hence, the appropriate combination of first-stage and second-stage gearing listed in Table 7.18 is selected as an acceptable solution.

From the above analysis, we can conclude that the solution in Table 7.18 obtained through the design optimization of the Wolfrom epicyclic gearing can be considered the best solution.

NOTES

1 Refer to Table 1.1.
2 Refer to Section 3.6.5.
3 Refer to Section 2.9.
4 Refer to Section 3.6.5.
5 Refer to Section 3.6.8 and onward.
6 Refer to Section 2.9.

REFERENCES

Garcia, P. L., et al., "Customizing Planetary Gear Trains for Human Limb Assistance and Replication," *MATEC Web of Conferences* 287, 01014 (2019), Power Transmissions 2019. https://doi.org/10.1051/matecconf/201928701014. Used under the terms and conditions of the Creative Commons Attribution License 4.0 (http://creativecommons.org/licenses/by/4.0/)

Krishnan, G., OPTIMO software – Algorithms for Multi-Objective optimization. 2024. https://github.com/epicyclic-book/optimo/. Used under the terms and conditions of the MIT license.

Matsuki, H., et al., "Bilateral Drive Gear: A Highly Backdrivable Reduction Gearbox for Robotic Actuators," *IEEE/ASME Transactions on Mechatronics* 24, 6, 2661–2673, December 2019, https://doi.org/10.1109/TMECH.2019.2946403. Used under the terms and conditions of Creative Commons Attribution 4.0 License. https://creativecommons.org/licenses/by/4.0/

8 Optimization Techniques for the Sustainable Manufacture of Epicyclic Gearing

Chandrasekaran G

8.1 INTRODUCTION

In gear manufacturing, optimization techniques are used to make improvements in existing manufacturing methods and also in planning for new product manufacturing. Optimization techniques can be applied in production control during gear manufacturing to achieve quality, cost, volume, and deadline (QCD). Optimization techniques can be applied in gear manufacturing in all the processes starting from the procurement of raw materials and finally up to the packaging and shipping of gearing systems.

Because of worldwide globalization combined with the factor of exponential increase in the world's population and fast-depleting natural resources, there is an urgent need to apply knowledge and exchange information for sustainable gear manufacturing. Because of globalization and a fiercely competitive environment, the manufacturing cost should be reduced. It is highly important to work in optimal conditions and, at the same time, to increase productivity and enhance the manufactured products quality, and numerous studies on the problem of optimizing manufacturing processes have been developed.

Manufacturing industries aim at producing good quality products, at higher productivity in time, at minimum cost. The improvement of the efficiency of manufacturing systems as well as the reduction of manufacturing costs has been important because of ecological and economic reasons for the utilization of resources in manufacturing. The main sources of the world's energy consumption are the fast-depleting fossil fuels. The demand for energy is rising continuously, and with a shortage of energy resources, the energy costs rise, and the burden on the ecosystem increases. The energy, and resource-efficient manufacture of epicyclic gearing is becoming a competitive advantage, and companies are increasingly interested in optimal manufacturing and process operation.

 DOI: 10.1201/9781003144465-8

The traditional objectives of manufacturing were increasing the efficiency of the manufacturing resources (man, material, machine, and energy) and reducing costs. However, in the future, apart from the objectives of reducing the costs and increasing the efficiency of resources, we have to take into account the whole product life cycle and the reduction of the consumption of natural resources and impacts on the environment in the industrial manufacturing as new additional objectives. To achieve all the above objectives during manufacturing, we have to use new technologies and also consider new methods for managing gearing products.

Optimization techniques help in improving performance indicators such as productivity, cost, and energy consumption. In the manufacture of epicyclic gearing, the aim is to improve productivity and product quality and, at the same time, maintain a clean and sustainable environment. Adopting optimization techniques helps in minimizing the number of manufacturing steps by employing advanced and alternative methods, environment-friendly lubricants and lubrication techniques for machining, active waste management, and minimizing energy consumption.

The manufacturing processes for metallic gears can be broadly categorized into three groups, namely,

- i Material removal or machining processes such as hobbing, milling, shaping, broaching, and skiving.
- ii Forming processes such as stamping, forging, cold drawing and extrusion.
- iii Additive processes such as die casting, shell casting and powder metallurgy.

The majority of gears are manufactured by material removal processes, such as hobbing, milling, and shaping. The processing sequence to machine a gear starts with the gear blank preparation in the selected material, followed by the cutting of the teeth. The next step is heat treatment to impart appropriate mechanical properties to the surface and the core. Finally, the finishing processes such as trimming, shaving, grinding, honing, and burnishing, are used for achieving the desired quality (tolerance and surface finish).

In the epicyclic gear machining process, processing parameters such as cutting speed, feed rate, and depth of cut largely determine the quality and production cost. Improper processing parameters may cause defects, such as tooth profile errors, excessive roughness, tooth surface scuffing, tooth size problems, and abnormal hob wear, which will increase costs and energy consumption. In epicyclic gear manufacturing, optimization of the processing parameters of machine tools according to the conditions and requirements is essential to avoid higher manufacturing costs, lower processing efficiency, and difficulty in controlling the gear precision. It is important to reduce the cycle time by increasing the feed rate during machining to achieve lower energy consumption. The optimization techniques are applied to reduce the cutting lubricant and energy costs by shortening the cutting time because of a larger depth of cut and feed rate.

With the rapid improvement of computer hardware performance, optimization techniques are applied to investigate processing parameter optimization methods that are based on data analysis models for the manufacture of epicyclic gearing.

8.2 SUSTAINABLE DESIGN AND MANUFACTURE

Sustainable design and manufacture aim at designing and operating systems in such a way that they utilize energy and assets economically, at a rate that does not destroy the environment, or the capability of future generations to meet their own requirements. The examination and optimization of intra-factory viewpoints related to fabricating plants are essential for designing sustainable manufacturing systems. Economically sound processes must be used for the sustainable design, manufacture and distribution of products and services. A wide range of disciplines including engineering, technology, economics, environmental, health, social desires, government strategies, procedures and policies affect the process of designing environmentally sustainable products. The entire manufacturing enterprise should adopt sustainability objectives at a system-level in decision-making for effective implementation of sustainable manufacturing goals. The sustainable manufacturing aims at rethinking manufacturing based on the three pillars of sustainability: environmental, social and economic.

Optimization techniques for gear manufacturing target overall sustainability and managing natural capital. In order to optimize sustainable decision making process and enhance resource efficiency, optimization techniques have to be developed by identifying potential improvements in the existing systems. Optimization techniques can be effectively applied for green, clean, lean, and sustainable resource utilization during gear manufacturing.

Optimization techniques are needed at the product, process, and system levels to achieve sustainability in manufacturing processes and also in the entire supply chain, including the manufacturing systems across multiple product life cycles. Action is necessary to counteract global warming in the form of sustainable products and processes, sustainable manufacturing systems and enterprises, decision-support tools for sustainability, and eco-intelligent manufacturing. The aim is to promote the use of intelligent decision support optimization techniques for sustainable manufacturing and product life cycle management and the development of advanced information and knowledge-based systems which deliver industrial sustainability. The sense of urgency that now prevails for humanity to take action against climate change has increased manifold, and saving the planet requires nothing short of overhauling civilization's energy infrastructure, which would include a multi-pronged effort to increase energy efficiency by using optimization techniques and advanced renewable technologies (Figure 8.1).

8.3 HEURISTIC ALGORITHMS FOR GEAR MANUFACTURING

Heuristic algorithms are used to optimize production processes and inventory management in gear manufacturing by minimizing waste and maximizing

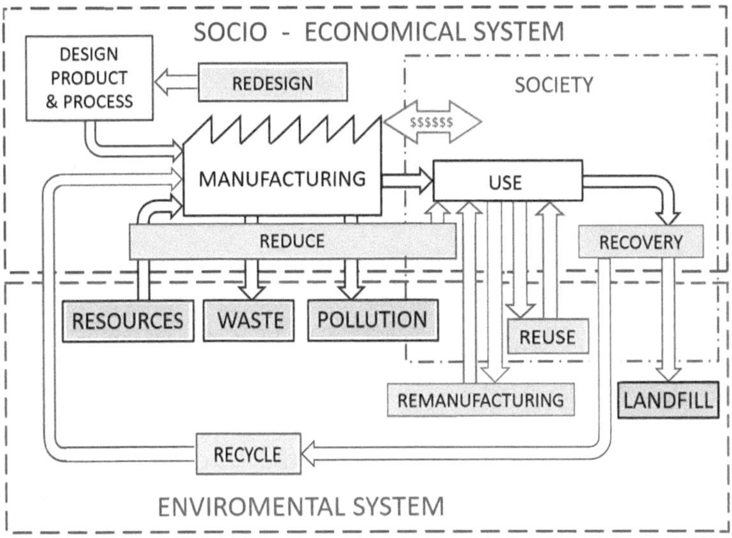

FIGURE 8.1 Sequence of sustainable manufacturing (Sartal, A., et al. 2020).

throughput. Given below are the steps involved in how the heuristic algorithms are used in gear manufacturing to optimize production processes and inventory management:

- *Problem definition*: The first step in applying heuristic algorithms to gear manufacturing is to define the problem to be solved. In the present case, the problem is to optimize gear production processes and inventory management by minimizing waste and maximizing throughput.
- *Data collection*: After defining the problem, the next step is to collect and pre-process data. This may involve gathering information on the gear production processes, equipment utilization, inventory levels, and order volumes. The data may also need to be verified, standardized, and normalized to ensure consistency and accuracy.
- *Selection of heuristic algorithm*: The next step is to select an appropriate heuristic algorithm for gear manufacturing optimization. There are many different types of heuristic algorithms that can be used in gear manufacturing, including particle swarm optimization, tabu search, simulated annealing, and genetic algorithms. The choice of algorithm will depend on factors such as the size and complexity of the gear manufacturing data set, the desired level of accuracy, and the available computational resources.
- *Heuristic algorithm implementation*: Once the heuristic algorithm has been selected, it can be implemented using an algorithm in a proper a programming language or software package. As already explained in previous chapters, the algorithm will typically involve a series of steps, including

initialization, search, evaluation, and iteration, as described in my previous answer.

- *Optimization of production process*: The heuristic algorithm is used to optimize the gear production process by identifying the most efficient sequence of tasks, minimizing waste, and maximizing throughput. The gear manufacturing optimization algorithm may consider factors such as equipment utilization, worker productivity, and product quality in order to improve efficiency and reduce costs.
- *Optimization of inventory management*: The heuristic algorithm can be used to optimize inventory management by minimizing inventory levels while ensuring that customer orders are fulfilled on time. The optimization algorithm may consider factors such as demand variability, lead times, and order sizes in order to minimize inventory holding costs and stock outs.
- *Outcome analysis*: Finally, the results of the heuristic algorithm can be analyzed and evaluated to determine the effectiveness of the production process and inventory management optimization strategies. This may involve comparison of the performance of the optimized processes to previous processes, as well as conducting sensitivity analysis to test the robustness of the algorithm under different conditions.

8.4 CASE STUDY: MULTI-OBJECTIVE OPTIMIZATION OF GEAR CUTTING IN WEDM USING TOPSIS METHOD (K.D. MOHAPATRAA, AND S.K. SAHOO, 2018)

Wire electrical discharge machining (WEDM) is an unconventional machining operation process that uses a spark erosion process for the removal of material from the material piece, where the electrical sparks are developed between the material piece and the tool wire electrode. The wire and the work-piece are flushed in the dielectric fluid in such a way that the work-piece does not touch the wire. The sparks generated evaporate minute amounts of work-piece material, which are then ejected and washed away by the dielectric fluid. Now WEDM is a vast technique in high precision machining of all types of conductive materials like graphite, metals, and metallic alloys. In WEDM, the vibration and movement of the wire electrode is one of the most important phenomena in the machining process. The decrease of the wire deflection and vibration is one of the most important criteria in attaining high shape accuracy and efficient machining. When the spark melts, a small portion of the melted material is removed and washed away. The leftover material re-freezes to form a layer over the surface called the recast layer. The layer contains micro cracks, micro structure, impurities, tensile stress, and other undesirable features when put to use. Different WEDM machining parameters are used while cutting operations, and they can be varied within a certain range. The wire is provided with an upper and lower water nozzle in such a way that the water comes out uniformly between both guides. Many gears are of

TABLE 8.1
Gear Specifications

Gear Specifications	Values
Pressure angle 20°	20°
Pitch circle diameter	20 mm
Base diameter	18.79 mm
Addendum diameter	22.5 mm
Dedendum diameter	16.87 mm
Tooth width	1.96 mm

Source: K.D. Mohapatraa and S.K. Sahoo (2018).

complex shapes and sizes that are difficult to machine by other nonconventional operations and can be easily machined by a wire EDM machine.

In the case study, the experiment is conducted for gear with the specifications listed in **Table 8.1**.

8.4.1 Determination of Material Removal Rate and Kerf width

Material removal rate (MRR) is the volume of the metal detached from the material piece during the cutting process. The volume of the material removed depends upon the cutting speed, thickness of the work-piece, and diameter of the wire. The material is removed by the spark produced from the wire. The MRR (mm^3/min) is given by the following formulae:

$$MRR = V_c \times h \times k$$

where V_c is the cutting speed in mm/min, which depends upon the cutting length and time.

The cutting speed (V_c) is given by the following formula:

$$V_c = 60 \times \frac{l}{t}$$

where l is the cutting length in mm;

t is the machining time in minutes.

The Kerf width is the gap obtained in the work-piece after the machining operation. Generally, the width is more than the wire diameter due to the sparks developed in the wire known as a spark gap. The Kerf width is generally measured in mm. The Kerf obtained is viewed under the optical microscope by using suitable software.

For the optimization process, the MRR should be maximized and Kerf width should be minimized. A multi-objective optimization is used to combine both the

TABLE 8.2
Multi-Objective Optimization Using TOPSIS Method

Run	MRR	Kerf	d_{iw}	d_{ib}	s_{iw}
1	1.476	0.640	0.037	0.0241	0.3890
2	1.416	0.707	0.0379	0.0119	0.2400
3	1.3	0.687	0.0443	0.087	0.1951
4	1.325	0.626	0.0396	0.0221	0.3588
5	1.61	0.747	0.0323	0.0247	0.4333
6	1.67	0.707	0.0236	0.0302	0.5613
7	1.655	0.687	0.0213	0.0302	0.5863
8	1.65	0.687	0.0215	0.0298	0.5805
9	1.605	0.707	0.0266	0.0253	0.4875
8	1.675	0.666	0.0173	0.0331	0.6561
11	1.845	0.707	0.01999	0.0435	0.6852
12	1.81	0.666	0.0125	0.0428	0.7739
13	1.61	0.707	0.0263	0.0257	0.4936
14	1.625	0.631	0.0171	0.0334	0.668
15	1.695	0.606	0.087	0.0405	0.7905
16	1.735	0.602	0.00758	0.0434	0.8513

Source: K.D. Mohapatraa and S.K. Sahoo, 2018.

responses in order to get a single value for the best possible outcome. The need for the optimization is to normalize the results and to get the best combination from the given sets.

The Technique for Order of Preference by Similarity to the Ideal Solution (TOPSIS) is a technique by which the chosen alternatives have the shortest distance from the best solution and the farthest from the negative ideal solution. A TOPSIS solution is the alternative which is closest to the ideal alternative and farthest from the negative ideal.

Table 8.2 shows the multi-objective optimization of the output parameters using the TOPSIS method. From **Table 8.2**, it is seen that the highest alternative is in the 16th run (119-57-4), which is the best alternative among the given set of combinations.

In **Table 8.2**, d_{iw} and d_{ib} are the distances from the target alternative i to the worst and the best conditions, respectively, and s_{iw} is the relative closeness coefficient.

8.4.2 Conclusions of the Case Study

The following conclusions were made based on the results and experimental work carried out. The gear is obtained by minimizing the Kerf width and maximizing the MRR. Pulse on time is found to be the most significant factor in affecting the responses. The MRR and Kerf were measured and optimized using the TOPSIS method. The microstructural analysis was carried out for the material and the wire. It was found that some craters, voids, and pores are formed on the work-piece as well as the wire due to changes in the wire feed rate at high pulse on time.

Addendum and tooth width were also experimented with and analyzed at the optimum settings. The addendum was found to be 2.63 mm. It was also found that the tooth width has an error of 0.852 mm, satisfying the validation at the optimum settings. From the above analysis, it was clear that the optimum setting of the gear obtained can be used to produce high-quality gears and can also be applied for future findings.

8.5 CASE STUDY: OPTIMIZATION OF TRANSMISSION ERROR AND CONTACT PRESSURE (R. HJELM ET AL., 2021)

One of the properties of involute gears is that the nominal gear ratio is constant; that is, a constant angular velocity on the input shaft results in a constant angular velocity on the output shaft. The path of contact in this case becomes a straight line, which tangents the respective base circles. However, this is only true for infinitely stiff gears with perfect involute geometry, whereas physical gears under operation will deviate slightly from the constant gear ratio and straight line of action due to the following reasons:

- *Deformation*: As load is applied to the gear set, the gears will deform. This means that the instant geometry will deviate from the nominal, resulting in a geometry different from the involute, and hence the contact will move away from the nominal line of action.
- *Modifications*: Flank modifications, such as tip relief, render a part of the flank that is not an involute. Thus, when contact occurs on a modified part of the flank, the line of action is not straight, not even without load.
- *Manufacturing errors*: Similar to intentional modifications, manufacturing errors will cause the geometry of the gear to deviate from the nominal involute geometry, thus altering the gear ratio and contact path.

Altogether, the points above add up and cause a non-constant gear ratio. This effect is denoted by transmission error. Typically, static transmission error (STE) should be small for the gears to be silent. Also, to obtain a smooth transmission of motion with little or no jerkiness, sudden changes in the STE derivative should be avoided. This requires careful consideration of manufacturing error tolerances so that deviations from the intended geometry are sufficiently small. However, prescribing tighter tolerances also increases production time and cost. This suggests that with increased knowledge of the relation between STE and manufacturing error tolerances, a better tolerance specification can be found.

For industrial purposes, it would be advantageous to obtain an estimation of the STE of a gear set based on control measurement after it has been produced. On the other hand, simulations can be time-consuming and cumbersome. It is neither realistic to make simulations after a gear has been produced nor to make simulations beforehand of all possible manufacturing outcomes. A solution to this is to make simulations beforehand of a few cases and use a simpler method to estimate STE after production based on those simulations. This paper describes how to simulate STE in gear sets with manufacturing errors, as well as how to estimate STE based on those simulations.

The case study is to investigate the relation between manufacturing tolerances, transmission error, and contact pressure. This is done using the simulation tool developed by Hjelm and Vedmar (2018). This tool includes contact outside the nominal line of action, tip contact, and non-Hertzian contact pressure distribution. Deformation is modeled by a hybrid model, and geometry is based on a parametric description of the reference profile, which includes tooth modifications.

This study will use a spur gear pair, and their properties are summarized in **Table 8.3**.

Different manufacturing errors typically appear together. Figures 8.2–8.4 show cases where the pinion suffers from both pitch error and profile slope error. The combined effect of the manufacturing errors is a shift caused by the pitch error and a twist caused by the profile slope error. As error magnitudes increase, the effect gets more and more pronounced.

TABLE 8.3
Data for the Gear Set Used in the Simulations

Property	Value (Pinion/Gear)
Module, m_n [mm]	4.5
Pressure angle, α_n [∘]	20
Helix angle, β [∘]	0
Number of teeth, z	16/24
Addendum modification, x_n	0.1657/0.1546
Tip radius, R_{tip} [mm]	41.23/59.18
Tip relief start radius, R_{tr} [mm]	40.18/57.96
Face width, b [mm]	14
Center distance, a [mm]	91.5
Normal backlash, δ_0	0.02249
Modulus of elasticity, E [$GPah$]	206
Poisson ratio, v	0.3
Coefficient of friction, μ	0.07
Output torque, M_g [Nm]	200

Source: R. Hjelm et al., 2021, 2018.

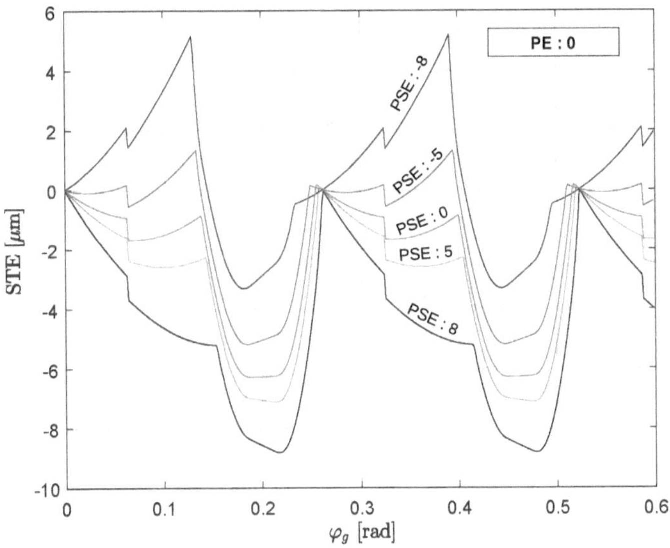

FIGURE 8.2 STE evolution curves for a gear set where the pinion is subjected to positive and negative profile slope errors of classes 5 and 8, together with a nominal case (R. Hjelm et al., 2021)

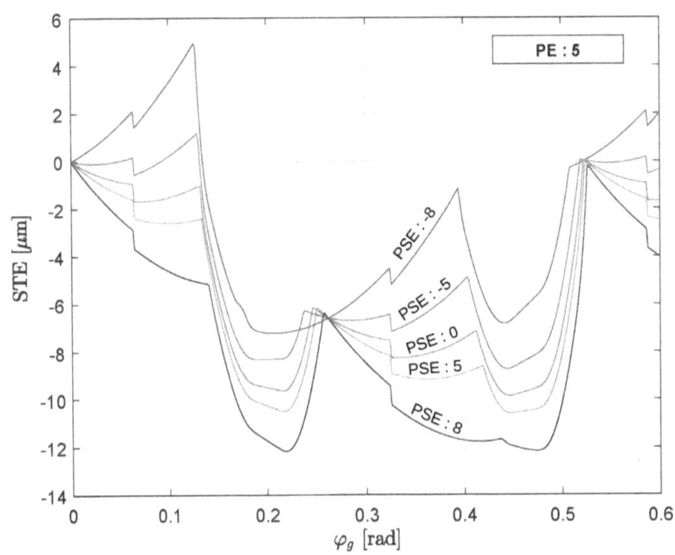

FIGURE 8.3 STE evolution curves for a gear set where the pinion is subjected to pitch error class 5 as well as positive and negative profile slope error of classes 5 and 8 (R. Hjelm et al., 2021)

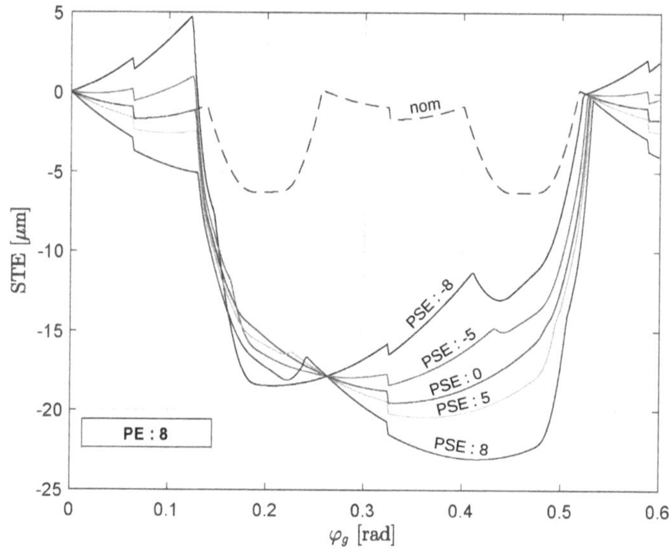

FIGURE 8.4 STE evolution curves for a gear set where the pinion is subjected to pitch error class 8 as well as positive and negative profile slope error of classes 5 and 8. A nominal case (dashed) is shown as a reference (R. Hjelm et al., 2021).

8.5.1 Conclusions of the Case Study

In this study, STE has been simulated and plotted for a number of different manufacturing errors and combinations thereof, with tolerances relevant to typical industrial applications. From this, design curves were made to link PPTE and maximum contact pressure. Interpolations were made based on polynomial fits of simulated values, and finally, a contour plot was made as a design tool for choosing appropriate tolerances to keep the PPTE low.

The following main conclusions can be drawn:

- Pitch error was shown to have a greater impact on PPTE than profile slope error in the numerical example.
- Tip relief magnitude has a low impact on STE.
- Some of the design curves showed clear optima, which suggests an optimal design.
- Interpolations show a very good agreement with simulated values and can thus save a significant amount of time.

REFERENCES

Hjelm, R. and Vedmar, L., "A method for calculating contact pressure and tip contact in spur gear sets with manufacturing errors," *Proceedings of International Gear Conference Lyon*, 1, 2018, 2018, 488–497.

Hjelm, R., et al., "Gear tolerancing for simultaneous optimization of transmission error and contact pressure," *Results in Engineering*, 9, 2021, 100195, https://doi.org/10.1016/j.rineng.2020.100195. Used under the terms and conditions of the Creative Commons Attribution (CC-BY) license (https://creativecommons.org/licenses/by/4.0/)

Mohapatraa, K.D. and Sahoo, S.K., "A multi objective optimization of gear cutting in WEDM of Inconel 718 using TOPSIS method," *Decision Science Letters*, 7, 2018, 157–170, https://doi.org/10.5267/j.dsl.2017.6.002. https://growingscience.com/dsl/dsl.html. Used under the terms and conditions of the Creative Commons Attribution (CC-BY) license (https://creativecommons.org/licenses/by/4.0/)

Sartal, A., et al. "The sustainable manufacturing concept, evolution and opportunities within Industry 4.0: A literature review," *Advances in Mechanical Engineering*, 12(5), 2020, 1–17, https://doi.org/10.1177/1687814020925232. Used under the terms and conditions of the Creative Commons Attribution (CC-BY) license (https://creativecommons.org/licenses/by/4.0/).

Index

Pages in *italics* refer to figures, pages in **bold** refer to tables, and pages followed by "n" refer to notes.

FSC
www.fsc.org
MIX
Papier | Fördert
gute Waldnutzung
FSC® C083411

Kolibri 360 GmbH
Ferdinand-Jühlke-Straße 7
99095 Erfurt, Deutschland
produktsicherheit@kolibri360.de